Honey Bees: Estimating the Environmental Impact of Chemicals

T0187528

Honey Bees: Estimating the Environmental Impact of Chemicals

Edited by

James Devillers

and

Minh-Hà Pham-Delègue

CRC Press

Taylor & Francis Group

Boca Raton London New York

CRC Press is an imprint of the
Taylor & Francis Group, an **Informa** business

A TAYLOR & FRANCIS BOOK

CRC Press
Taylor & Francis Group
6000 Broken Sound Parkway NW, Suite 300
Boca Raton, FL 33487-2742

First issued in paperback 2019

Typeset in Times Ten by Wearset Ltd, Boldon, Tyne and Wear

ISBN-13: 978-0-367-39632-9

Every effort has been made to ensure that the advice and information in this book is true and accurate at the time of going to press. However, neither the publisher nor the authors can accept any legal responsibility or liability for any errors or omissions that may be made. In the case of drug administration, any medical procedure or the use of technical equipment mentioned within this book, you are strongly advised to consult the manufacturer's guidelines.

British Library Cataloguing in Publication Data
A catalogue record for this book is available from the British Library

Library of Congress Cataloging in Publication Data
Honey bees : estimating the environmental impact of chemicals / edited by James Devillers, Minh-Hà Pham-Delègue.
    p. cm.
Includes bibliographical references.
1. Honeybee–Effect of chemicals on. 2. Pollution–Environmental aspects. I. Devillers, J. (James), 1956– II. Pham-Delègue, Minh-Hà, 1958–

QL568.A6 H56 2002
595. 79'9–dc21
                                                2001052295

Cover image reproduced courtesy of Fabrizio Santi, DiSTA – Bologna University, Italy

Visit the Taylor & Francis Web site at
http://www.taylorandfrancis.com

and the CRC Press Web site at
http://www.crcpress.com

Contents

Contributors

G.H. Alnasser, Department of Chemistry, The University of Montana, Missoula, MT 59812–1006, USA.

C. Armengaud, Laboratoire de Neurobiologie de l'Insecte, EA 3037, Université Paul Sabatier Toulouse III, 118 route de Narbonne, 31062 Toulouse Cedex, France.

D. Barišić, Department for Marine and Environmental Research, Rudjer Bošković Institute, Bijenička 54, 10000 Zagreb, Republic of Croatia.

N. Ben Ghouma-Tomasella, CRIIRAD, 471 Avenue Victor Hugo, 26000 Valence, France.

J.J. Bromenshenk, Division of Biological Sciences, The University of Montana, Missoula, MT 59812-1006, USA.

G. Celli, Università degli Studi di Bologna, Dipartimento di Scienze e Tecnologie Agroambientali, via Filippo Re, 6 40126 Bologna, Italy.

S. Cluzeau, ACTA, Association de Coordination Technique Agricole, 149 rue de Bercy, 75595 Paris Cedex 12, France.

A. Decourtye, Laboratoire de Neurobiologie Comparée des Invertébrés, INRA, BP 23, 91440 Bures-sur-Yvette, France.

J. Devillers, CTIS, 3 Chemin de la Gravière, 69140 Rillieux La Pape, France.

J.C. Doré, Laboratoire de Chimie des Substances Naturelles, ESA 8041 CNRS, Muséum National d'Histoire Naturelle, 63 rue de Buffon, 75005 Paris, France.

N. Galand, Laboratoire de Pharmacognosie, Université de Tours, Faculté des Sciences Pharmaceutiques 'Philippe Maupas', 31 Avenue Monge, 37200 Tours, France.

A.M.R. Gatehouse, Department of Agricultural and Environmental Sciences, Ridley building, University of Newcastle, Newcastle upon Tyne NE1 7RU, UK.

E. Gattavecchia, Università degli Studi di Bologna, Istituto di Scienze Chimiche, via S. Donato, 15 40127 Bologna, Italy.

M. Gauthier, Laboratoire de Neurobiologie de l'Insecte, EA 3037, Université Paul Sabatier Toulouse III, 118 Route de Narbonne, 31062 Toulouse Cedex, France.

S. Ghini, Università degli Studi di Bologna, Istituto di Scienze Chimiche, via S. Donato, 15 40127 Bologna, Italy.

S. Girotti, Università degli Studi di Bologna, Istituto di Scienze Chimiche, via S. Donato, 15 40127 Bologna, Italy.

T.K. Haarmann, Los Alamos National Laboratory, University of California, PO Box 1663, MS M887, Los Alamos, NM 87545, USA.

D.C. Jones, Department of Chemistry, The University of Montana, Missoula, MT 59812–1006, USA.

L. Jouanin, Laboratoire de Biologie Cellulaire, INRA, Route de St Cyr, 78026 Versailles Cedex, France.

N. Kezić, Faculty of Agriculture, The University of Zagreb, Svetošimunska 25, 10000 Zagreb, Republic of Croatia.

M. Lambin, Laboratoire de Neurobiologie de l'Insecte, EA 3037, Université Paul Sabatier Toulouse III, 118 route de Narbonne, 31062 Toulouse Cedex, France.

L.A. Malone, The Horticulture and Food Research Institute of New Zealand Ltd, Mt Albert Research Centre, Private Bag 92 169, Auckland, New Zealand.

M. Marenco, Laboratoires Poirier SEROM, La Haute Limougère, 37230 Fondettes, France.

M.H. Pham-Delègue, Laboratoire de Neurobiologie Comparée des Invertébrés, INRA, BP 23, 91440 Bures-sur-Yvette, France.

F. Poirier-Duchêne, Laboratoires Poirier SEROM, La Haute Limougère, 37230 Fondettes, France.

C. Porrini, Università degli Studi di Bologna, Dipartimento di Scienze e Tecnologie Agroambientali, via Filippo Re, 6 40126 Bologna, Italy.

A.G. Sabatini, Istituto Nazionale di Apicoltura, via di Saliceto, 80 40128 Bologna, Italy.

J.C. Sandoz, Laboratoire de Neurobiologie Comparée des Invertébrés, INRA, BP 23, 91440 Bures-sur-Yvette, France.

G.C. Smith, Department of Chemistry, The University of Montana, Missoula, MT 59812–1006, USA.

M. Subirana, CNDA, 149 rue de Bercy, 75595 Paris Cedex 12, France.

J.N. Tasei, Unité de Recherches de Zoologie, INRA, 86600 Lusignan, France.

A. Vertačnik, Department for Marine and Environmental Research, Rudjer Bošković Institute, Bijenička 54, 10000 Zagreb, Republic of Croatia.

C. Viel, Laboratoire de Pharmacognosie, Université de Tours, Faculté des Sciences Pharmaceutiques 'Philippe Maupas', 31 Avenue Monge, 37200 Tours, France.

Preface

Honey bees (*Apis mellifera* L.) are the main pollinating agents for numerous plants and fruit trees and, hence, play a key role in agriculture and more generally in the maintenance of ecological biodiversity. Although these social insects are not the targets of all the different agrochemical treatments used in crop protection, they are widely affected by pesticides. In addition, during their foraging flights, in which they collect nectar, pollen, plant resins, and water, honey bees inadvertently come into contact with a wide array of inorganic and organic pollutants, and these are often taken back to the colony. These xenobiotics can induce lethal and sublethal effects. They can also accumulate in the different members of the colony and in their products such as honey or royal jelly. Because of their foraging activity, honey bees can also be considered as mobile multimedia samplers that average the concentrations of pollutants over time and throughout large spatial areas. Consequently, they can be used as bioindicators of chemical or radioactive contamination.

The 15 chapters of *Honey Bees: The Environmental Impact of Chemicals* discuss these different subjects in depth. Besides presenting various tests and specific methodologies, a huge amount of ecotoxicological data, never previously published, is provided. Similarly, specific problems such as those related to the contamination of non-*Apis* bees, genetically modified plants, or regulatory constraints are addressed.

We hope that the readers will find this book to be a valuable source of reference as well as a source of stimulation for further ecotoxicological studies on terrestrial ecosystems.

We express our sincerest gratitude to the authors for careful preparation of their contributions. We also acknowledge the assistance of the referees for their critical analyses and valuable comments.

James Devillers
Minh-Hà Pham-Delègue

1 The ecological importance of honey bees and their relevance to ecotoxicology

J. Devillers

Summary

In this chapter, the biology, behavior, and ecological role of the honey bee (*Apis mellifera*) are briefly presented. The high degree of social organization of these insects is described. Their pollinating activity, allowing the production of crops and wild plants and the conservation of the biodiversity, is discussed. Finally, the usefulness of the honey bee for detecting xenobiotics in the environment and assessing the adverse effects of agrochemicals and other man-made chemicals is introduced.

Introduction

The contamination of the environment by toxic substances is linked both to industrialization and to intensive agriculture. Xenobiotics reach the aquatic and terrestrial ecosystems from discharges and leaks of industrial products, consumer waste and urban sewage, from farming and forestry runoff, and from accidental spills.

Some of these compounds, because of their high lipophilicity, accumulate in animal tissues, particularly in predators occupying the top of the food chains. In addition, they may be dispersed over great distances by winds and water currents. These chemicals can also affect the ability of living organisms to reproduce, to develop, and to withstand the many other stress factors in their environment, by depressing their nervous, endocrine, and/or immune systems.

Consequently, it is important to estimate the environmental fate and ecotoxicological effects of these different xenobiotics. While the former task can be performed with the use of multimedia models [1] and by measuring the concentrations of these contaminants in the different environmental compartments and the biota, the latter task requires testing the chemicals against representative species in the ecosystems.

Consequently, the aim of this chapter is to show the relevance of the honey bee (*Apis mellifera*) to terrestrial ecotoxicology. After a short presentation of its taxonomical status, the biological cycle of the honey bee colony will be summarized and some of the morphological, physiological, and behavioral characteristics of the different members of the hive will be stressed. Then, the ecological importance of this social insect will be presented and, finally, the value of *A. mellifera* as a bioindicator of environmental contaminations and as a test organism in terrestrial ecotoxicology will be discussed.

Taxonomical position

Bees are insects belonging to the order of Hymenoptera which also includes wasps, ants, ichneumons, chalcids, sawflies, and lesser known types. There are about 25000 described species of bees, divided into 11 families, numerous subfamilies, tribes, and genera, and still more species and subspecies. Most of them are solitary organisms living without social organization. Honey bees belong to the family Apidae, which includes other social bees such as the bumble bees (Bombinae) and the stingless bees (Meliponinae). The subfamily Apinae consists of one tribe, Apini, comprising one genus, *Apis*. There are four species within this genus which are *florea, dorsata, cerana*, and *mellifera*. *Apis florea* (the little honey bee) and *Apis dorsata* (the giant honey bee) are tropical bees building single-comb nests in low bushes and in trees, respectively. The lifestyle of *Apis cerana* (the Eastern honey bee) is similar to that of *Apis mellifera* (the Western honey bee) and hence, *A. cerana* is used in apiculture with modern moveable comb hives. The numerical strength of *A. cerana* colonies is usually much less, and honey yields are smaller. It is therefore being rapidly supplanted by imported *mellifera* races, chiefly *A. m. ligustica*.

Two dozen geographic races of the Western honey bee, *Apis mellifera*, have been recognized, adapted to a range of environments from the cold continental climate of Eastern Europe, through the moist temperate climate of the Atlantic seaboard, the warmth of the Mediterranean, and the heat of the tropics and semi-deserts. Only four of these races need be considered for apiculture in Europe, namely *A. m. mellifera, A. m. ligustica, A. m. carnica*, and *A. m. caucasica* which present different morphological and behavioral characteristics [2, 3].

The biological cycle of the honey bee colony

The common honey bee is a social insect nesting in colonies typically comprising a single queen, drones, and numerous workers (Table 1.1).

The queen and the reproduction process

A queen is easily distinguished from the other members of the colony. Her body is large, especially during the egg-laying period when her abdomen is

Table 1.1 The inhabitants of the hive and their characteristics (adapted from ref. 4)

	Queen	Drone	Worker
Relative size	Large	Large	Small
Number	1	~200 or 0	20 000 to 200 000
Lifespan	2 years depending on number of spermatozoids	21–32 days spring 90 days summer or until mating 0 winter	20–40 days summer 140 days winter
Sex	Female	Male	Sterile female
Functions	Mates with drone Lays 1500 eggs/day (200 000 eggs/year) Secretes pheromone	Mates with young queen	Makes comb Tends larvae, young drones, queen Cleans hive Gathers nectar, pollen, propolis Evaporates nectar Caps cells Defends hive Starves drones Lays drone eggs Moves larvae for making new queen

greatly elongated. Her wings cover only about two-thirds of the abdomen, whereas the wings of both workers and drones nearly reach the tip of the abdomen when folded. A queen's thorax is slightly larger than that of a worker's, and she has neither structures to collect pollen nor functional wax glands. Her sting is longer than that of a worker's, but it presents fewer and shorter barbs.

The main activity of the queen is to lay eggs and to keep the workers uninterested in reproduction through pheromonal control. If the queen stops producing pheromone or laying eggs, one of her most recent eggs will be moved to a specially prepared queen cell to produce a new queen. The queen is constantly attended and fed royal jelly by the workers. A queen generally mates 6 to 10 days following emergence. She may go out on several mating flights, mating with several drones on each flight. Additional mating flights are performed until the spermatheca contains about 5 million to 6 million spermatozoa. The queen controls the release of sperm with her so-called sperm pump. If an egg is fertilized, it will develop into a female bee (queen or worker), but if not fertilized, a male bee will result. Consequently, drones have only one set of chromosomes (haploid) acquired from the queen. Bees are holometabolous. After 3 days, the egg hatches into a voracious worm-like larva growing and molting each day for about 4 days. It then goes into a resting stage, the pupa, which lasts for another few days in a capped cell until the bee emerges as an adult, called imago. This process takes 16–24 days depending on the season and class of bee. The length of the egg stage is the same for all three castes, but the larval and pupal stages are shortest for the queen and longest for the drone.

Egg-laying usually starts within a week after mating, and a queen can continue to lay fertilized eggs throughout most of her life, usually 2 to 5 years. When the sperm supply begins to be depleted, the workers prepare to replace or supersede her. The old queen and her new daughter may both be present in the hive for some time following supersedure [3–5].

The drones

The drones are the largest bees in the colony. The drone's head is much larger than that of either the queen's or the worker's. Drones have no sting, pollen baskets, or wax glands since they are designed for mating only. Drones take their first flights at about 8 days old and are sexually mature at 12 days old. Only a few of them are tolerated in the hive at spring and fall, more in the summer, but none in the winter. In fact, the workers keep them out of the hive to starve to death in the autumn. Their normal lifespan is 8 weeks or less [3–5].

The workers

The workers are the smallest and the most numerous individuals in the colony. They are sexually undeveloped females and under normal hive conditions they do not lay eggs. Worker bees have specialized structures, such as brood food glands, scent glands, wax glands, and pollen baskets, which allow them to perform all the tasks of the hive.

The tasks of the workers depend on their age, their genetic background, and the needs of the colony. Worker bees less than 2 weeks old clean the cells and then feed the larvae from the secretion of their hypopharyngeal glands. The workers also attend the queen. When bees are about 12 to 15 days old, their wax glands become functional and comb building is possible according to a well-defined process.

Older house bees work with honey, pollen, wax, and propolis. Nectar-collecting field bees are met by house bees and trophallactic exchanges occur. The conversion of nectar into honey requires the physical removal of water and the addition to the nectar of the enzyme invertase included in the salivary glands of the bees. This enzyme breaks down sucrose into glucose and fructose. Pollen pellets are deposited in empty cells near the brood nest by the pollen-collecting workers. In the cells, the pollen undergoes a maturing process to what is commonly called bee-bread. Propolis-collecting bees also serve as propolis storage reservoirs [5]. Propolis is not stored in combs or elsewhere, but is removed from the corbiculae of these field bees and used in the hive as needed to fill cracks, to embalm dead animals, and for other purposes.

Workers also use their wings to help to ventilate the hive. While ventilation can be performed by bees of different ages, this activity occurs mainly when workers are about 18 days old. When workers are about 15 to 20 days old, their job is also to defend the hive.

After 2 weeks as house bees, workers take short flights for orientation and defecation. After 3 weeks, the workers become foragers, gathering pollen, nectar, water, or propolis for the colony. Water collectors can comprise 10 percent of all foragers, but this percentage increases considerably during heat stress. Propolis collectors are rarely observed.

The schedule of worker bee activities is highly flexible and depends on physiological, ecological, and behavioral factors [6]. During autumn, a reduction in brood rearing and an increase in pollen consumption result in a population of long-lived "winter" bees having increased fat bodies and protein reserves. The normal 6-week adult life of "summer" bees may be extended to several months in these "winter" bees [3–5].

Sensory organs and communication processes

Reproduction, the search for food, and the social life in the hive require highly developed sensory organs and powerful communication processes.

Eyes and vision

The faceted compound eye of the honey bee is sensitive to ultraviolet radiation. Conversely, it is blind to red light. White flowers, which only partially reflect ultraviolet as a rule, appear colored to a bee. Certain color combinations, while invisible to the human eye, are not only visible but are also very attractive to bees. Some flowers that appear entirely yellow to the human eye reflect ultraviolet from the outer ends of the petals [3]. Only the inner part, then, appears yellow to the bee, directing it to the nectar source. The eye of the honey bee only roughly perceives geometric shapes [3].

Antennae, pheromones, and chemical communication

The honey bees present two types of receptors on their antennae. The first type allows the bees to detect the different odors of flowers. The second type is more specific and is used to detect pheromones produced by the queen and the workers; it allows the recognition of individuals belonging to the same colony since each colony presents a specific odor [7].

The "dance language"

A widely known and main behavioral specificity of honey bees is the communication of information about food sources and the recruitment of foragers by the so-called "dance language" discovered by Karl von Frisch in

1919 [8]. This accurate transmission of information concerning direction and distance of foraging areas leads to efficient exploitation of food sources by the colony.

In brief, after a bee has discovered a new source of food, the insect returns fully loaded to the comb, delivers nectar or pollen, and informs the other bees about the new food source, communicating information about the location and quality of the food source by means of various dance-like movements. Information about the plant species is conveyed by the odor of the flower, which adheres to the bee's body. The other bees detect this information through receptors on their antennae. Information about the quality and quantity of the food source is conveyed by the liveliness and duration of the dance movements of the bee. If the food source is unusually rich and of high quality, certain sounds are also made to convey this information. The location of the food source is indicated by the rhythm of the dance and by the orientation of the axis of the tail with respect to gravity. If the food source is near the hive, a "round" dance is performed. A "tail-wagging" dance indicates that the food source is more than 50–80 m away. This dance transmits precise information about direction as well as distance. The number of dance cycles performed by the bee in a certain length of time is inversely related to the distance of the food source. The sun and gravity are used in conveying directional information. During the flight to the food source the bee determines the angle between the line of flight and the sun. The angle to the vertical at which the bee dances, on the vertical face of the honeycomb, describes the angle between the line of flight to the food source and a line drawn in the direction of the sun. A run 80° to the left indicates that the source is 80° to the left of the sun. Because the position of the sun changes during the day, the dance angle also changes in the course of the day [7]. The tail-wagging dance is also performed when the swarm is searching for a new dwelling [9].

Nutrition

The bee visits flowers in search of nectar and pollen, or visit trees for harvesting resin to make propolis. The propolis is used as glue and caulk to seal cracks in the hive. Bees also collect honeydew secreted by insects of the family Aphididae.

Pollen is stored in broodcomb cells and is the main supply of protein and vitamins for the hive. Pollen is 6 to 28 percent protein by weight and usually contains the 10 amino acids essential for bees [4].

Nectar is from 5 to 80 percent sugar but is less than 0.2 percent in protein, so nectar is the carbohydrate supply for the hive [4]. The conversion of nectar into honey requires the physical removal of water by rapid movements of the wings of the bees and the addition to nectar of the enzyme invertase included in the salivary glands of the bees. When the amount of water remaining in the nectar is less than 18 percent, the mixture is called honey and the bees cap off the cells.

A mixture of honey and pollen is called "bee-bread" and is the food for most larvae and bees. Future queens are fed with large quantities of "royal jelly" which is similar to bee-bread but contains more mandibular gland secretions and more honey (34 percent vs. 12 percent) [4].

Finally, water is also collected by bees and used primarily as a diluent for thick honey, to maintain optimum humidity within the hive, and to maintain appropriate temperatures in the brood area. The amount of water required and collected by a colony is generally correlated with the outside air temperature and relative humidity, strength of colony, and amount of brood rearing in progress [4, 10].

Ecological importance of the honey bee

Wind is the main pollinating agent. In fact, most of the forest trees, almost all grasses and grains, with the exception of some that are completely self-pollinated, and many weeds are wind-pollinated. The flowers of most wind-pollinated plants are either male or female. The male flowers produce an abundance of pollen to be transported by the wind. The female flowers usually have large stigmatic areas to receive the pollen [11].

Nearly 200 000 animal species play roles in pollinating the 250 000 species of wild flowering plants on our planet [12]. Among them, about 1500 species of vertebrates such as birds (e.g. hummingbirds) and mammals (e.g. bats, lemurs) serve as pollinators [12]. However, the main pollinators are insects: they include bees, wasps, moths, butterflies, beetles and so on. Bees are the most efficient and the only dependable pollinators, because they visit flowers methodically to collect nectar and pollen and do not destroy the flower or the plant in the process.

Consequently, bees provide substantial benefits to the maintenance of the biodiversity and the productivity of both natural and agricultural ecosystems [13, 14]. However, with regard to agricultural ecosystems, it is important to stress that only 15 percent of the 100 or so crops that feed the world are serviced by domestic honey bees, while at least 80 percent are pollinated by wild bees and other wildlife [12].

Unfortunately, both wild bees and domestic honey bees are in decline. Thus, for example, the number of commercial US bee colonies plummeted from 5.9 million in the late 1940s to 4.3 million in 1985, and 2.7 million in 1995. The loss of one quarter of all managed honey bee colonies since 1990 signals one of the most severe declines US agriculture has ever experienced in such a short period. There are fewer bee hives in the US today than at any time in the past 50 years [12]. This demise has been brought about by the spread of diseases and parasitic mites, invasion of Africanized honey bees [12], climatic fluctuations, industrialization, and exposure to pesticides and other chemicals. Xenobiotics can either poison the bees or impair their reproduction. These chemicals can also eliminate nectar sources for pollinators and/or deplete nesting materials. Consequently, there is a need to protect the honey bees and the others pollinators because of their ecological importance.

The relevance of the honey bee to ecotoxicology

Sentinels for detecting environmental contamination

Honey bees commonly forage within 1.5 km of their hive and exceptionally as far as 10 to 12 km, depending on their need for food and its availability [3]. During these foraging flights, they randomly sample the environment to gather nectar, pollen, honeydew, resin, water, and so on. They also collect dusts of various origins on their body hair. It is important to note that the honey bees only collect bioavailable contaminants. Consequently, these insects are powerful unbiased samplers which can be used for detecting organic or inorganic chemicals in the environment. In this process, both the bees and their products (i.e. honey, wax, royal jelly) can be used, depending on the physicochemical properties of the pollutant(s) and the goal of the study. However, for hydrophobic chemicals (e.g. polychlorinated biphenyls, PCBs) [15, 16], the use of these other matrices is preferable.

The "Bee Alert" project developed by Bromenshenk *et al.* [17, 18] clearly illustrates the above concept. This system uses domestic honey bees, hives equipped with sensors, chemical analysis, and computer facilities for the early detection of the presence of contaminants in the environment [18].

Estimating the ecotoxicological risk of chemicals in terrestrial ecosystems

In terrestrial ecotoxicology, the two main animals used for assessing the adverse effects of chemicals are the domestic honey bee and birds (e.g. mallard, quail, pheasant). Besides their ecological importance, the honey bees are interesting as test organisms because they are easy to rear and manipulate. Because of its worldwide distribution, the species *A. mellifera* is widely used as a test organism. This point is crucial for regulatory purposes. However, differences of sensibility due to the existence of geographic races must not be underestimated in the comparison of toxicity results. The domestic honey bees are also interesting to use in ecotoxicology because their biological cycle is particularly well known and relatively short. Numerous individuals presenting similar characteristics can be obtained at low cost. With this organism, laboratory tests as well as field experiments can be performed under controlled conditions. Different endpoints can be used (e.g. contact, oral) yielding different ecotoxicological information.

All pesticides used for crop protection have to be tested against the honey bee for estimating their ecotoxicity. In this context, protocols and guidelines have been designed [e.g. 19–21] and hence a huge amount of ecotoxicity data on agrochemicals is available in the literature [22]. However, most of these data only deal with short exposures, i.e. LC_{50} or LD_{50} obtained after 24 or 48 hours of exposure. In fact, there is a lack of information on the sublethal effects of pesticides on honey bees [23, 24]. It is also important to stress that even if pesticides are very important in agriculture, they only represent very few of all the xenobiotics that might contaminate the environment. The number of substances registered with the Chemical Abstracts Service (CAS) was 212000 in 1965, 16×10^6 in 1996, 18×10^6 in 1998, about 23.5×10^6 in April 2000, and more than 29×10^6 in February 2001 [25–27]. For all these man-made chemicals, used for a variety of purposes, there is a total lack of information on their ecotoxicity against the honey bee. Furthermore, while in aquatic toxicology, QSAR (quantitative structure–activity relationship) models [e.g. 28–30] are now widely used to fill the data gaps, in terrestrial ecotoxicology, the number of structure–toxicity models is very scarce, due to a lack of ecotoxicity results for industrial chemicals and other substances not used in crop protection. Thus, an analysis of the QSAR literature reveals that only one QSAR model has been designed for predicting the toxicity of pesticides to the honey bee [31] and it is impossible to find other structure–toxicity models for other classes of chemicals.

Conclusion

The aim of this chapter was to provide some basic information on the biology, behavior, and ecology of the honey bee (*Apis mellifera*). These social insects present a high degree of organization. Because of their pollinating activity, they play a key role in ecology in maintaining the biodiversity and the production of crops and wild plants. Their economic importance is also clearly shown when, for example, we consider the worldwide production of honey [32] (Table 1.2).

Table 1.2 World production of honey (×1000 tons) [32]

Continent	1991	1992	1993	1994	1995	1996	1997	1998
Africa	109	117	129	131	138	142	142	142
North and Central America	222	216	223	195	183	174	178	184
South America	87	87	95	97	105	100	95	95
Asia	334	328	326	354	365	362	386	391
Europe	180	182	181	291	319	278	280	281
Oceania	29	29	30	38	27	35	35	35
Total	961	959	984	1106	1137	1091	1116	1128

Finally, this chapter describes briefly how the honey bees can be used as sentinels for detecting various pollutants or as test organisms for estimating the terrestrial ecotoxicity of xenobiotics. These key roles will be more thoroughly discussed in the other chapters of the book.

References

1 Mackay, D. (1991). *Multimedia Environmental Models. The Fugacity Approach.* Lewis Publishers, Chelsea, p. 257.
2 Milner, A. (1996). An introduction to understanding honeybees, their origins, evolution and diversity. http://www.bibba.com/bibborig.html.
3 Pham-Delègue, M.H. (1998). *Abeilles.* La Martinière, Paris, p. 47.
4 Koning, R.E. (1994). Honeybee biology. Plant physiology website. http://koning.ecsu.ctstateu.edu/plants_human/bees/bees.html.
5 Waller, G.D. Honey bee life history. http://maarec.cas.psu.edu/bkCD/HBBiology/life_history.htm#References.
6 Huang, Z.Y. and Robinson, G.E. (1999). Social control of division of labor in honey bee colonies. In: *Information Processing in Social Insects* (Detrain, C., Deneubourg, J.L. and Pasteels, J.M., Eds). Birkhäuser Verlag, Basel, pp. 165–186.
7 Anonymous (1999). *A Propos de la Santé des Abeilles.* Bayer, p. 109.
8 von Frisch, K. (1967). *The Dance Language and Orientations of Bees.* Harvard University Press, Cambridge, p. 566.
9 Visscher, P.K. and Camazine, S. (1999). The mystery of swarming honeybees: From individual behaviors to collective decisions. In: *Information Processing in Social Insects* (Detrain, C., Deneubourg, J.L. and Pasteels, J.M., Eds). Birkhäuser Verlag, Basel, pp. 355–378.
10 Standifer, L.N. Honey bee nutrition and supplemental feeding. http://maarec.cas.psu.edu/bkCD/HBBiology/nutrition_supplements.htm#requirements.
11 McGregor, S.E. Pollination of crops. http://maarec.cas.psu.edu/bkCD/Pollination/poll_of_crops.html.
12 Ingram, M., Nabhan, G. and Buchmann, S. (1996). Our forgotten pollinators: Protecting the birds and bees. http://www.pmac.net/birdbee.htm.
13 Pimentel, D., Wilson, C., McCullum, C., Huang, R., Dwen, P., Flack, J., Tran, Q., Saltman, T. and Cliff, B. http://www.aibs.org/biosciencelibrary/vol47/dec.97.biodiversity.html.
14 Convention on Biological Diversity, Canada. http://www.biodiv.org.
15 Anderson, J.F. and Wojtas, M.A. (1986). Honey bees (Hymenoptera: Apidae) contaminated with pesticides and polychlorinated biphenyls. *J. Econ. Entomol.* **79**, 1200–1205.
16 Jan, J. and Cerne, K. (1993). Distribution of some organochlorine compounds (PCB, CBz, and DDE) in beeswax and honey. *Bull. Environ. Contam. Toxicol.* **51**, 640–646.
17 Bromenshenk, J.J., Doskocil, J., Olbu, G.J., Degrandi-Hoffman, G. and Roth, S.A. (1991). PC BEEPOP, an ecotoxicological simulation model for honey bee populations. *Environ. Toxicol. Chem.* **10**, 547–558.
18 http://biology.dbs.umt.edu/bees/default.htm.
19 ANPP (1989). *Méthode CEB no. 129. Méthode d'Evaluation, sous Tunnel en*

Plein Air, des Effets à Court Termes des Produits Phytopharmaceutiques sur l'Abeille Domestique Apis mellifera L. First edition January 1989, revised version November 1996, p. 12.

20 ANPP (1996). *Méthode CEB no. 95. Méthode de Laboratoire d'Evaluation de la Toxicité Aiguë Orale et de Contact des Produits Phytopharmaceutiques chez l'Abeille Domestique Apis mellifera L.* First edition April 1982, revised version November 1996, p. 8.

21 EPPO (1992). Guideline on test methods for evaluating the side-effects of plant protection products. No. 170 Honeybee. *OEPP/EPPO Bull.* **22**, 203–216.

22 Devillers, J. and Doré, J.C. (2000). *Etude Bibliographique des Effets Ecotoxicologiques des Xénobiotiques vis-à-vis de l'Abeille.* Contrat: "Programme Communautaire sur l'Apiculture 2000." p. 179.

23 Decourtye, A. (1998). *Etude des Effets Sublétaux de l'Imidachloprid et de l'Endosulfan sur l'Apprentissage Olfactif chez l'Abeille Domestique Apis mellifera L.* DEA "Neurobiologie des Processus de Communication et d'Intégration." Université Montpellier II, p. 18.

24 Pham-Delègue, M.H., Decourtye, A., Kaiser, L. and Devillers, J. (2002). Methods to assess the sublethal effects of pesticides on honey bees. *Apidologie* (in press).

25 Freemanthe, M. (1998). *Chem. Eng. News* **April 27**, 45.

26 Schulz, W. (1998) *Chem. Eng. News* **June 29**, 14.

27 http://www.cas.org/cgi-bin/regreport.pl.

28 Karcher, W. and Devillers, J. (1990). *Practical Applications of Quantitative Structure–Activity Relationships (QSAR) in Environmental Chemistry and Toxicology.* Kluwer Academic Publishers, Dordrecht, The Netherlands, p. 475.

29 Devillers, J. and Karcher, W. (1991). *Applied Multivariate Analysis in SAR and Environmental Studies.* Kluwer Academic Publishers, Dordrecht, The Netherlands, p. 530.

30 Devillers, J. (1998). Environmental chemistry: QSAR. In: *The Encyclopedia of Computational Chemistry* (Schleyer, P.v.P., Allinger, N.L., Clark, T., Gasteiger, J., Kollman, P.A., Schaefer, H.F. and Schreiner, P.R., Eds). John Wiley & Sons, Chichester, Vol. 2, pp. 930–941.

31 Vighi, M., Garlanda, M.M. and Calamari, D. (1991). QSARs for toxicity of organophosphorous pesticides to *Daphnia* and honeybees. *Sci. Total Environ.* **109/110**, 605–622.

32 http://www.apicultura.com/databases/honey-market/world_homey.htm.

2 Volatile and semi-volatile organic compounds in beehive atmospheres

*G.C. Smith, J.J. Bromenshenk,
D.C. Jones, and G.H. Alnasser*

Summary

A colony of honey bees is an effective environmental sampling device for volatile and semi-volatile organic compounds (VOCs and SVOCs) in a complex ecosystem setting. Over the past six years, we have developed a thermal desorption/gas chromatography/mass spectrometry (TD/GC/MS) technique using commercially available carbon molecular sieve tubes to screen beehive atmospheres for the presence of VOCs and SVOCs. Hive air is withdrawn at about $0.100\,dm^3$/min through a small copper tube inserted between frames in the center of the beehive. Besides detecting the compounds normally released by honey bee physiology, hive stores, and hive construction components, we also see a broad range of compounds that are environmental contaminants. These fall into categories of fossil fuel constituents, industrial solvents, pesticides, and explosives. Hives can be deployed over regional landscapes or clustered near known contaminated sites to yield useful guidance on clean-up prioritization. More recent work introduces xenobiotic VOC taggants to feeders as an aid in studying the foraging pattern of bees.

Introduction

Honey bees (*Apis mellifera* L.) are excellent monitors of environmental quality [1–3]. They have been employed as *in situ* monitors of elemental contaminant exposure and associated effects for more than twenty years. Comparative case histories and guidelines for the use of honey bees as sentinel species have been published [2, 4–9]. More recently, the use of bees has been extended to include real-time monitoring of colony condition (i.e. flight activity, temperature regulation in the brood nest) and routine monitoring for volatile and semivolatile organic contaminants in studies for the US Army at Aberdeen Proving Ground, Maryland [10–14].

In the process of monitoring organic contaminants, it has also been necessary to characterize the complex background of organic compounds found naturally inside beehives. Beehives located in uncontaminated environments contain compounds released by the bees themselves (e.g. pheromones, other chemicals released to repel pests and predators, metabolites, etc.), compounds from hive stores (e.g. honey, beeswax, pollen, and propolis), and volatile compounds from the materials out of which hives are constructed (wood, paint, plastic, etc.). We show here that beehive atmospheres also contain compounds from vehicles, farms, industries, and households in the hive vicinity.

This paper summarizes the types of compounds found by our technique while biomonitoring for a variety of volatile and semi-volatile organic contaminant residues. Briefly, hive atmospheres were drawn through multibed sorption traps and subsequently analyzed by thermal desorption/gas chromatography/mass spectrometry (TD/GC/MS).

Methods and materials

Fingerprinting studies

Fingerprinting studies for hive components were conducted at the University of Montana's research apiaries on seven dates during July 1996 (Table 2.1). Ambient air was concurrently sampled so that contaminants present in the urban airshed of our apiary could be identified and accounted for in all other samples.

To fingerprint the active physiology of honey bees by themselves, a stainless steel cage was fabricated to contain about 4000 individuals. The

Table 2.1 Fingerprint studies

Category	Sample dates (1996)
Honey bees	7/5, 7/6, 7/7
Hive stores	
Unoccupied 1995 hive box (no bees or frames)	7/5, 7/6, 7/7
Unoccupied hive 56 (no bees, with frames)	7/11
Propolis A	7/19, 7/20
Propolis B	7/19, 7/20
Hive materials	
Unpainted 1995 wood	7/5, 7/6, 7/7
Unpainted 1996 wood	7/5, 7/6, 7/7
Painted 1996 box	7/5, 7/6, 7/7
Old plastic parts	7/11
New plastic parts	7/11
Vinyl-coated screen	7/19, 7/20
Old condo	7/19, 7/20
New condo	7/19, 7/20
Clock drive assembly	7/13
Aluminum foil	7/13
Ambient air	7/5, 7/6, 7/7, 7/11(2), 7/19, 7/20

top of the cage was outfitted with a syrup bottle to feed the bees during the 8- to 10-hour pumping periods. Pumping was done in the open air, free of any hive enclosure that could contribute extraneous substances.

Hive stores were evaluated by pumping on a previously occupied upper-story box with and without honey frames. These samples had contributions from both hive stores and hive materials. Two samples of propolis from Missoula colonies were placed in loosely capped glass vials for pumping.

Hive components profiled included unpainted wood, painted wood, machined plastic parts, vinyl-coated screen wire, and completely instrumented "condo" units [14]. The effect of aging on the loss of volatile and semi-volatile components from hive boxes was assessed by comparing unpainted wood from 1995 and 1996 lumber inventories. We also compared a condo used during the 1995 field season to a newly completed 1996 model.

Air sampling

Air samples were collected on $11.5\,cm \times 6\,mm$ OD $\times 4\,mm$ ID three-phase Carbotrap 300 thermal desorption tubes (Supelco) or four-phase Carbotrap 400 tubes. These sorbent tubes house a sequence of graphitized carbon and molecular sieves of increasing activity that sorb volatile and semi-volatile organic compounds over a molecular size range from C_1 to C_{30}.

Desorption tubes were connected to constant flow pumps set at rates between 0.080 and $0.150\,dm^3/min$. The distal end of the sorption tube was attached to copper tubing ($2\,mm$ ID $\times 3\,mm$ OD) with a brass compression fitting and a vespel/graphite ferrule. The copper tube was inserted directly into the hive interior between the wooden frames that support the wax combs (Figure 2.1). The outlet end of the sorbent tube was connected to a constant flow pump (SKC, Inc.) with a 1-m section of $5\,mm$ ID $\times 8\,mm$ OD Tygon tubing. Pumping periods ranged from 8 to 12 hours.

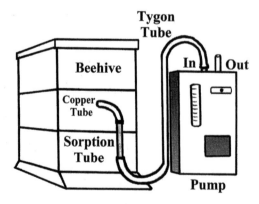

Figure 2.1 Schematic diagram of a hive with air-sampling train.

Sample tubes were sealed in individual vials and stored in a dedicated 4°C sample refrigerator until analyzed.

Thermal desorption analysis

Sample tubes were desorbed in a direction opposite to sampling flow. After a 4-min helium purge to remove incidental moisture, tubes were subjected to a 10-min desorption cycle at 250°C. Each tube was then given a 6-min cooling flush. A helium flow rate of 0.025 dm³/min was used in the desorption tube. Make-up helium flow from other paths on the multi-station desorber (Tekmar LSC2000) yielded a total flow of 0.040 dm³/min going into the focusing trap (10 cm Carbopack B graphitized carbon, 6 cm Carboxen 1000 molecular sieve and 1 cm 1001 molecular sieve). The focusing trap was desorbed and flushed into the gas chromatograph for 1 min (injection port 220°C, septum purge flow of 0.003 dm³/min) and was split 1:20 thereafter.

Chromatographic separations were accomplished on a Hewlett Packard GCD instrument containing a 60 m × 0.32 mm ID Restek RTX-502.2 capillary column (phenylmethyl polysiloxane, 1.8 μm coating). The helium flow was 0.001 dm³/min and the total time for an analysis was 50 min (5 min at initial temperature 40°C, ramp 5°C/min to 220°C, 9 min hold time at 220°C). Mass spectra were collected over a range of 35 to 450 amu.

Computer matches with the National Institute of Science and Technology (NIST) database initially identified compounds. Many, though not all, were subsequently confirmed using commercial mixtures of analytical standards. The concentrations of all compounds were computed on a relative scale (ion abundance/dm³ air sampled) but are not reported here. Compounds of interest to regulatory agencies have been rigorously quantified [11–14].

Results

To place our hive atmosphere findings in perspective, we have compiled lists of specific compounds whose presence in bees and beehives have been documented in the honey bee literature by previous researchers. The data for these tables come from several review articles and selected papers. We have not attempted to conduct a comprehensive review of this large body of work. Honey bees exhibit pheromonal parsimony. The same compound may have different functions in different contexts. Also, many pheromones have not been characterized. As this knowledge base expands, we are changing our interpretation of the function of identified compounds. Queen pheromone may not so much inhibit worker's ovaries as signal the presence of a queen, and brood pheromones may provide the stimulus to prevent workers from laying eggs [15]. Because propolis is highly variable in its composition, we have included compounds reported to be characteristic of different geographic regions [16]. Propolis is a resinous material obtained by bees from woody plants. It is made up of an indeterminate number of substances and has no specific chemical formula [17].

A typical hive atmosphere chromatogram from our TD/GC/MS technique is shown in Figure 2.2. Identified compounds have been systematized into four categories, each with a summary table. Table 2.2 lists compounds reported as honey bee semiochemicals. Semiochemicals are produced in glands that secrete to the exterior of the insect, and include pheromones, which are chemicals used to communication between individuals of the same species. Table 2.3 consists of compounds associated with hive stores. Table 2.4 presents compounds emanating from materials and components from which beehives are assembled. Table 2.5 documents compounds arising from non-bee sources. Within each category, compounds have been listed in formula order. Table 2.6 contains selected levels for hazardous air pollutants that have been collected from hives in our studies in the vicinity of Chesapeake Bay, USA.

Compounds detected with our TD/GC/MS technique are designated with an "X" in the next-to-last column of each table. Compounds that we had analyzed by EPA Methods 8081A (pesticides) and 8082 (PCBs) are designated with a "Y" in the tables. Whenever possible, we also provide CAS numbers for reported compounds. CAS numbers proved difficult to obtain or have not yet been assigned to some of the biologically derived chemicals (i.e. recently discovered semiochemicals and botanicals in propolis).

Discussion

Chromatographic considerations

Because of the general nature of our sampling technique and the subsequent TD/GC/MS analysis, only certain categories of volatile and semivolatile compounds were detectable – nonpolar organics (alkanes, alkenes, alkynes, cycloalkanes, aromatics, terpenes, PAHs, biphenyls), partially oxygenated organics (alcohols, ethers, ketones, aldehydes, acids, esters), organonitrogen and organosulfur compounds (amines, amides, heterocycles), and organochlorine compounds (solvents, pesticides). Highly polar molecules were generally missed with our technique. This is a consequence of choosing sorbents that target nonpolar species and a chromatographic column coated with a substance of intermediate polarity. Use of other sorbents and different column coatings could enhance the ability to find other classes of compounds.

Masses of compounds ranging from 35 amu up to those associated with selected C_{12} organic compounds were detected on the Carbotrap tubes. The molecular weight cut-off was constrained by the maximum tempera-

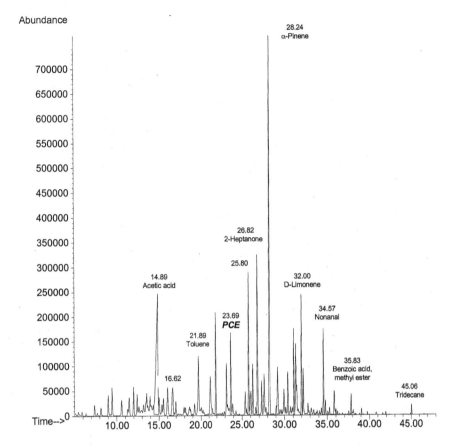

Figure 2.2 Total ion chromatogram of a typical hive atmosphere sample. Selected peaks have been labeled with the identity of the compound and retention time in minutes. Seen here are compounds from bees (nonanal at 34.57 min), from plant resins in propolis or hive boards (α-pinene at 28.24 min), and from non-bee contaminants (toluene at 21.89 min and tetrachloroethene, PCE, at 23.69 min).

ture to which the carbon-based sorbents could be subjected – about 350°C. Higher molecular weights are accessible with silica-based sorbent materials, which can tolerate temperatures up to 600°C. This was demonstrated in side-by-side tests done recently in conjunction with Oak Ridge National Laboratory [111]. Compounds of higher molecular weights, for example many polycyclic aromatic hydrocarbons associated with petroleum and creosote, were seen more readily.

Approximately 25 ng of analyte were needed for detection above background noise in the mass spectrometer. This quantity was usually accumulated during an 8-hour pumping period at a $0.100 \, dm^3/min$ flow velocity. Our current sampling train has added two tubes in front of the Carbotrap

Table 2.2 TD/GC/MS detection of volatile and semi-volatile organic compounds previously reported as honey bee semiochemicals and glandular secretions

Pheromone	Formula	MW	Bees	Function	TD	CAS no.
Mandibular gland						
Hexanoic acid	$C_6H_{12}O_2$	116	nurse	royal jelly antibiotic brood recognition?	x	142-62-1
2-Heptanone	$C_7H_{14}O$	114	guard	alarm, defense, marker	x	110-43-0
Methyl-p-hydroxybenzoate	$C_8H_8O_3$	152	queen	retinue formation		99-76-3
Octanoic acid	$C_8H_{14}O_2$	144	nurse	royal jelly antibiotic brood recognition?	x	124-07-2
4-Hydroxy-3-methoxyphenylethanol	$C_9H_{12}O_3$	168	queen	retinue formation		
9-Oxo-(E)-2-decenoic acid	$C_{10}H_{16}O_3$	184	queen	signals queen presence inhibits queen rearing attracts drones, recognizes queen		334-20-3
S-9-Hydroxy-(E)-2-decenoic acid	$C_{10}H_{18}O_3$	186	queen	retinue formation		
R-9-Hydroxy-(E)-2-decenoic acid	$C_{10}H_{18}O_3$	186	queen	retinue formation		
Menthol	$C_{10}H_{20}O$	156	queen	unknown*	x	89-78-1
10-Hydroxy-(E)-2-decenoic acid	$C_{10}H_{20}O_3$	188	nurse	brood food, antibiotic		334-20-3
Palmityic acid	$C_{16}H_{30}O_2$	256	queen	unknown*		2091-29-4
17-Pentatriacontene	$C_{35}H_{70}$	490	queen	unknown*		6971-40-0
Hydrocarbons	–	–	worker	various	x	–
Nasonov gland						
Geraniol	$C_{10}H_{18}O$	154	worker	orientation	x	106-24-1
(E)-Citral and (Z)-citral	$C_{10}H_{16}O$	152	worker	orientation	x	5392-40-5
Geranic acid	$C_{10}H_{16}O_2$	168	worker	orientation	x	459-80-3
Nerolic acid	$C_{10}H_{16}O_2$	168	worker	orientation	x	
Nerol	$C_{10}H_{18}O$	154	worker	orientation	x	106-25-2
(E,E)-Farnesol	$C_{15}H_{26}O$	222	worker	orientation	x	4602-84-0

Table 2.2 Continued

Koschevnikov gland						
Isopentyl alcohol	$C_5H_{12}O$	88	worker	alarm, defense	x	123-51-3
Butyl acetate	$C_6H_{12}O_2$	116	guard	alarm, defense	x	123-86-4
Benzyl alcohol	C_7H_8O	108	guard	alarm, defense	x	100-51-6
Isopentyl acetate (IPA)	$C_7H_{14}O_2$	130	worker	alarm, defense	x	123-92-2
5,5-Dimethyl-2-hexene	C_8H_{16}	112	young queen	unknown*	x	36382-10-2
1,1,3-Trimethyl cyclopentane	C_8H_{16}	112	young queen	unknown*	x	4516-69-2
3,3-Dimethylhexane	C_8H_{18}	118	young queen	unknown*	x	563-16-6
Octenal	$C_8H_{14}O$	126	young queen	unknown*	x	2548-87-0
Hexyl acetate	$C_8H_{16}O_2$	144	guard	alarm, defense	x	142-92-7
2-Nonanol	$C_9H_{20}O$	144	worker	alarm, defense	x	628-99-9
Benzyl acetate	$C_9H_{10}O_2$	150	guard	alarm, defense	x	140-11-4
Nonanoic acid	$C_9H_{18}O_2$	172	young queen	unknown*	x	112-05-0
p-Menthane-9-ol	$C_{10}H_{20}O$	156	young queen	unknown*	x	89-78-1
2-Propyl-1-heptanol	$C_{10}H_{22}O$	158	young queen	unknown*	x	10042-59-8
Decanoic acid	$C_{10}H_{20}O_2$	172	young queen	unknown*		334-48-5
Octyl acetate	$C_{10}H_{20}O_2$	172	mature worker	attraction*	x	112-14-1
Methyl cyclodecane	$C_{11}H_{22}$	154	young queen	unknown*		143-13-5
2-Nonyl acetate	$C_{11}H_{22}O_2$	186	mature worker	alarm, defense		
4,5-Dimethylnonane	$C_{11}H_{24}$	156	young queen	unknown*	x	5876-87-9
1,11-Dodecadiene	$C_{12}H_{24}$	180	young queen	unknown*		
4,6,8-Trimethyl-1-nonene	$C_{12}H_{24}$	168	young queen	unknown*	x	
2-Decenyl acetate	$C_{12}H_{24}O_2$	200	guard	alarm, defense		67446-07-5
Ethyl decanoate	$C_{12}H_{24}O_2$	200	young queen	unknown*		110-38-3
1,12-Tridecadiene	$C_{13}H_{24}$	180	young queen	unknown*		
2-Methyl-1-dodecanol	$C_{13}H_{28}O$	200	young queen	unknown*		111-82-0
Ethyl dodecanoate	$C_{14}H_{28}O_2$	228	young queen	unknown*		106-33-2
Dodecyl acetate	$C_{14}H_{28}O_2$	228	young queen	unknown*		112-66-3
Hexadecane	$C_{16}H_{34}$	226	young queen	unknown*	x	544-76-3
Hexadecanoic acid	$C_{16}H_{32}O_2$	256	young queen	unknown*		57-10-3

Table 2.2 Continued

Pheromone	Formula	MW	Bees	Function	TD	CAS no.
Ethyl tetradecanoate	$C_{16}H_{32}O_2$	256	young queen	unknown*		124-06-1
6-Cyclohexylundecane	$C_{17}H_{34}$	238	young queen	unknown*		
Heptadecane	$C_{17}H_{36}$	240	young queen	unknown*		629-78-7
9-Octadecen-1-ol	$C_{18}H_{36}O$	268	mature worker	alarm, defense		2774-87-0
Methyl 2-methylhexadecanoate	$C_{18}H_{36}O_2$	284	young queen	unknown*		
2-(Hexadecyloxy)-ethanol	$C_{18}H_{38}O_2$	286	young queen	unknown*		
(Z)-11-Eicosen-1-ol	$C_{20}H_{40}O$	296	worker	alarm, defense		
2,6,10,15-Tetramethylheptadecane	$C_{21}H_{44}$	296	young queen	unknown*		54833-48-6
1-Dotriacontanol	$C_{32}H_{66}O$	466	young queen	unknown*		
17-Pentatriacontene	$C_{35}H_{70}$	490	young queen	unknown*		6971-40-0
3,5,24-Trimethyltetracontane	$C_{43}H_{88}$	604	young queen	unknown*		55162-61-3
Venom sac (Venom oil)						
Histamine	$C_5H_9N_3$	111	worker	defense		51-45-6
Acetylcholine (chloride)	$C_7H_{16}NO_2$	163	worker	defense		60-31-1
Octadecanol	$C_{18}H_{38}O$	270	worker	alarm, defense		112-92-5
(Z)-11-Eicosen-1-ol	$C_{20}H_{40}O$	296	worker	alarm, defense		62442-62-0
Eicosanol	$C_{20}H_{42}O$	299	worker	alarm, defense		629-96-9
Heneicosane	$C_{21}H_{44}$	297	worker	alarm, defense		629-94-7
cis-3-Docosen-1-ol	$C_{22}H_{44}O$	325	worker	alarm, defense		629-98-1
Pentacosane	$C_{25}H_{52}$	353	worker	alarm, defense		629-99-2
Tricosane	$C_{23}H_{48}$	325	worker	alarm, defense		638-67-5
Heptacosane	$C_{27}H_{56}$	381	worker	alarm, defense		593-49-7
Wax gland						
Hydrocarbons	—	—	worker	comb construction		
Monoesters	—	—	worker	comb construction		
Diesters	—	—	worker	comb construction		
Hydroxy polyesters	—	—	worker	comb construction		

Table 2.2 Continued

	Formula	MW	Source	Function		CAS
Tergite gland						
Chemicals mostly unknown	–	–	queen	inhibit worker ovaries / inhibit queen rearing / attract drones, orientation at flowers		
Hexadecanoic acid	$C_{16}H_{32}O_2$	256	young queen	queen recognition		57-10-3
Tarsal (Arnhart's) gland						
12 or more chemicals, unidentified	–	–	queen, worker	swarming, trail marking		
Hexadecanoic acid	$C_{16}H_{32}O_2$	256	young queen	queen recognition?		57-10-3
17-Pentatriacontene	$C_{35}H_{70}$	490	young queen	queen recognition?		6971-40-0
Worker-repellent pheromone						
o-Aminoacetophenone	C_8H_9ON	135	young queen	repel other queens		551-93-9
Brood pheromones						
Dioleoyl-3-palmitoylglycerol	$C_{55}H_{102}O_6$	859	brood	stimulate foraging / brood recognition		
Glyceryl-1,2-dioleate-3-palmitate	–	–	brood	inhibit worker ovaries		
Drone pheromones						
Unknown composition	–	–	drone	mating aggregation		
Beeswax (comb) pheromones						
Oxygenated organics						
Furfural	$C_5H_4O_2$	96		nectar storage?	x	98-01-1
Benzaldehyde	C_7H_6O	106		nectar storage?	x	100-52-7
Octanal	$C_8H_{16}O$	128		nectar storage?	x	124-13-0
Nonanal	$C_9H_{18}O$	142		nectar storage?	x	124-19-6
Decanal	$C_{10}H_{20}O$	156		nectar storage?	x	112-31-2
1-Decanol	$C_{10}H_{22}O$	158		nectar storage?	x	112-30-1

Notes

*Extracted from young queens, does not occur in the alarm pheromone of workers, promotes aggressive behavior of workers towards supernumerary queens.

References: Data by category – general review of bee pheromones [15, 18–25]; mandibular gland [17, 20, 25, 26–50]; Nasanov gland [17, 24, 51–58]; Kuschevnikov gland and venon sac [17, 22, 24, 59–79, 99]; tergite and tarsal glands [25, 79–85]; worker repellent [86]; beeswax pheromones [87–92]; brood and drone pheromones [93–98].

Table 2.3 Volatile and semi-volatile organic compounds found in hive stores

Hive store	Formula	MW	Ref.	TD	CAS no.
Nectar/Honey					
Alcohols			[24]	x	
Alkaloids			[24]		
Ethereal oils			[24]		
Organic acids			[24]	x	
Beeswax					
Alkanes					
n-C$_{23}$ to n-C$_{33}$			[88, 89]		
Alkenes			[88, 89]	x	
Alkadienes			[88, 89]	x	
Diesters			[88, 89]		
Free acids			[88, 89]		
Lipids			[88, 89]		
Monoesters			[88, 89]		
Pollen					
Sterols			[100]		
Propolis					
Hydrocarbons/Acids/Flavonoids					
Propylene glycol	$C_3H_8O_2$	76		x	57-55-6
Isobutyric acid	$C_4H_8O_2$	88		x	79-31-2
N-Methylpyrrole	C_5H_7N	81		x	96-54-8
Tiglic acid	$C_5H_8O_2$	100		x	80-59-1
5-Methyltetrahydrofuran-3-one	$C_5H_8O_2$	100		x	34003-72-0
Benzoic acid	$C_7H_6O_2$	122	[101]	x	65-85-0
Benzaldehyde	C_7H_6O	106		x	100-52-7
Methyl benzoate	$C_8H_8O_2$	136		x	93-58-3
6-Methyl-3,5-heptadien-2-one	$C_8H_{12}O$	124		x	1604-28-0
Phenethyl alcohol	$C_8H_{10}O$	122		x	60-12-8
Acetophenone	C_8H_8O	120		x	98-86-2
4-Methylenecyclohexylmethanol	$C_8H_{14}O$	126		x	1004-24-6
1-Octanol	$C_8H_{18}O$	130		x	111-87-5
Vanillin	$C_8H_8O_3$	152	[101]		121-33-5
Cinnamic acid	$C_9H_8O_2$	148	[101]		621-82-9
Hydrocinnamic acid	$C_9H_{10}O_2$	150	[101]		501-52-0
1-Nonyne	C_9H_{16}	124		x	3452-09-3
3,7-Dimethyl-1,3,6-octatriene	$C_{10}H_{16}$	136		x	29714-87-2
Eucalyptol	$C_{10}H_{18}O$	154		x	470-82-6
β-Myrcene	$C_{10}H_{16}$	136		x	123-35-3
1-Phenyl-2-butanone	$C_{10}H_{12}O$	148		x	1007-32-5
Palmitic acid	$C_{16}H_{32}O_2$	256	[101]		57-10-3
Benzyl cinnamate	$C_{16}H_{14}O_2$	238	[101]		103-41-3
Kaempferid	$C_{16}H_{12}O_6$	300	[101]		
3,4-Dimethoxynaringenin	$C_{17}H_{16}O_6$	316	[101]		
Betuleol	$C_{17}H_{14}O_7$	330	[101]		
1-Nonacosanol	$C_{29}H_{60}O$	424	[101]		25154-56-7
Tetracosyl hexadecanoate	$C_{40}H_{80}O_2$	592	[101]		
Pentacosyl hexadecanoate	$C_{41}H_{82}O_2$	606	[101]		
Heptacosyl hexadecanoate	$C_{43}H_{86}O_2$	634	[101]		
Octacosyl hexadecanoate	$C_{44}H_{88}O_2$	648	[101]		
Nonacosyl hexadecanoate	$C_{45}H_{90}O_2$	662	[101]		
Triacontyl hexadecanoate	$C_{46}H_{92}O_2$	676	[101]		
Dotriacontyl hexadecanoate	$C_{48}H_{96}O_2$	704	[101]		
Tetratriacontyl hexadecanoate	$C_{50}H_{100}O_2$	732	[101]		

Table 2.4 Volatile and semi-volatile organic compounds from hive construction components

Sources	Formula	MW	Ref.	TD	CAS no.
Wood boards					
2,3-Dimethyloxirane	C_4H_8O	72		x	1758-33-4
2,2,3,3-Tetramethylhexane	$C_{10}H_{22}$	142		x	13475-81-5
2-Ethylcyclobutanol	$C_6H_{12}O$	100		x	35301-43-0
Hexanal	$C_6H_{12}O$	100		x	66-25-1
1-(1-Methylethoxy)-2-propanone	$C_6H_{12}O_2$	116		x	42781-12-4
1,2-Diethylcyclobutane	C_8H_{16}	112		x	61141-83-1
α-Pinene	$C_{10}H_{16}$	136		x	80-56-8
β-Pinene	$C_{10}H_{16}$	136		x	127-91-3
3-Carene	$C_{10}H_{16}$	136		x	13466-78-9
4-Carene	$C_{10}H_{16}$	136		x	5208-49-1
D-Limonene	$C_{10}H_{16}$	136		x	5989-27-5
β-Myrcene	$C_{10}H_{16}$	136		x	123-35-3
α-Phellandrene	$C_{10}H_{16}$	136		x	99-83-2
Ocimene	$C_{10}H_{16}$	136		x	29714-87-2
Camphene	$C_{10}H_{16}$	136		x	79-92-5
Vinyl screen					
3-Buten-2-one	C_4H_6O	70		x	78-94-4
1-Methylazetidine	C_4H_9N	71		x	4923-79-9
Methylcyclopentane	C_6H_{12}	84	[102]	x	96-37-7
o-Hexylhydroxylamine	$C_6H_{15}NO$	117		x	4665-68-3
Butylcyclopropane	C_7H_{14}	98		x	930-57-4
Hexyl pentyl ether	$C_{11}H_{24}O$	172		x	32357-83-8
Polyethylene parts					
2-Butoxyethanol	$C_6H_{14}O_2$	118		x	111-76-2
o-Xylene	C_8H_{10}	106		x	95-47-6
Ethylbenzene	C_8H_{10}	106	[102]	x	100-41-4
2-Octene	C_8H_{16}	112		x	111-67-1
1-Ethyl-2-methylbenzene	C_9H_{12}	120		x	611-14-3
1,2,3-Trimethylbenzene	C_9H_{12}	120		x	526-73-8
Propylbenzene	C_9H_{12}	120	[102]	x	103-65-1
1,1,3-Trimethylcyclohexane	C_9H_{18}	126		x	3073-66-3
Tetradecane	$C_{14}H_{30}$	198		x	629-59-4
Painted box					
Isobutyl formate	$C_5H_{10}O_2$	102		x	542-55-2
1-Hexyn-3-ol	$C_6H_{10}O$	98		x	105-31-7
Phenol	C_6H_6O	94		x	108-95-2
2-Methylpyridine	C_6H_7N	93		x	109-06-8
2-Propylfuran	$C_7H_{10}O$	110		x	4229-91-8
Hexyl butanoate	$C_{10}H_{20}O_2$	172		x	2639-63-6
Butyl butanoate	$C_8H_{16}O_2$	144		x	109-21-7
2,2,4-Trimethyl-1,3-pentanediol	$C_8H_{18}O$	146		x	144-19-4
1-(2-Butoxyethoxy)ethanol	$C_8H_{18}O_3$	162		x	54446-78-5
cis-1-Cyclopropyl-2-ethenyl-cyclobutane	C_9H_{14}	122		x	61141-61-5
2,4-Dimethyl-3-heptene	C_9H_{18}	126	[103]	x	2738-18-3
3,4,5-Trimethyl-1-hexene	C_9H_{18}	126		x	56728-10-0
O-Decylhydroxylamine	$C_{10}H_{23}NO$	173		x	29812-79-1
1,4-Dihydro-1,4-methano-naphthalene	$C_{11}H_{10}$	142		x	4453-90-1
1-Methylnaphthalene	$C_{11}H_{10}$	142		x	90-12-0
Clock drive					
Cyclopentanone	C_5H_8O	84		x	120-92-3
cis-Hept-4-enol	$C_7H_{14}O$	114		x	6191-71-5
2-Ethyl-1-decanol	$C_{12}H_{26}O$	186		x	21078-65-9

Table 2.5 Volatile and semi-volatile organic compounds from non-bee sources

Sources	Formula	MW	Refs.	TD	CAS no.
Wood/vegetation combustion sources					
Oxygenates					
2-Butenal	C_4H_6O	70	[104, 105]	x	4170-30-3
2,3-Butanedione	$C_4H_6O_2$	86	[104, 105]	x	431-03-8
Furfural	$C_5H_4O_2$	96	[104, 105]	x	98-01-1
2,3-Pentanedione	$C_5H_8O_2$	100	[104, 105]	x	600-14-6
2,5-Dimethylfuran	C_6H_8O	96	[104]	x	625-86-5
2-Hexanone	$C_6H_{12}O$	100	[104, 105]	x	591-78-6
Isoamyl acetate	$C_7H_{14}O_2$	130		x	123-92-2
2-Methylbenzaldehyde	C_8H_8O	120	[104]	x	529-20-4
Anethole	$C_{10}H_{12}O$	148		x	104-46-1
1-α-Terpineol	$C_{10}H_{18}O$	154		x	10482-56-1
cis-Linalool oxide	$C_{10}H_{18}O_2$	170		x	5989-33-3
Other					
Chloroform	$CHCl_3$	119	[103, 106]	x	67-66-3
1-Butanol	$C_4H_{10}O$	74	[107]	x	71-36-3
Isoprene	C_5H_8	68	[103, 106]	x	78-79-5
Benzene	C_6H_6	78	[103, 106]	x	71-43-2
Toluene	C_7H_8	92	[103, 106]	x	108-88-3
o-Xylene	C_8H_{10}	106	[103, 106]	x	95-47-6
m-Xylene	C_8H_{10}	106	[103, 106]	x	108-38-3
p-Xylene	C_8H_{10}	106	[103, 106]	x	106-42-3
1,3,5-Trimethylbenzene	C_9H_{12}	120	[106]	x	108-67-8
Naphthalene	$C_{10}H_8$	128	[103]	x	91-20-3
1,1'-Biphenyl	$C_{12}H_{10}$	154	[108]	x	92-52-4
Vehicle emissions/creosote					
Alkanes					
n-C_{19} to *n*-C_{33}			[107]		
Dimethyl sulfide	C_2H_6S	62		x	75-18-3
Propane	C_3H_8	44		x	74-98-6
Isobutane	C_4H_{10}	58		x	75-28-5
Pentane	C_5H_{12}	72		x	109-66-0
2-Methylbutane	C_5H_{12}	72		x	78-78-4
3,3-Dimethyloxetane	$C_5H_{10}O$	86	[102]	x	6921-35-3
2-Methylpentane	C_6H_{14}	86		x	107-83-5
n-Hexane	C_6H_{14}	86		x	110-54-3
n-Heptane	C_7H_{16}	100		x	142-82-5
3-Methylhexane	C_7H_{16}	100		x	589-34-4
2,2,4-Trimethylpentane	C_8H_{18}	114		x	540-84-1
n-Octane	C_8H_{18}	114	[102]	x	111-65-9
2-Methylheptane	C_8H_{18}	114		x	592-27-8
3-Ethylhexane	C_8H_{18}	114		x	619-99-8
2,3-Dimethylhexane	C_8H_{18}	114		x	584-94-1
4-Methyloctane	C_9H_{20}	128		x	2216-34-4
2,4-Dimethylheptane	C_9H_{20}	128		x	2213-23-2
n-Nonane	C_9H_{20}	128		x	111-84-2
2-Methylnonane	$C_{10}H_{22}$	142		x	871-83-0
n-Decane	$C_{10}H_{22}$	142		x	124-18-5
n-Dodecane	$C_{12}H_{26}$	170		x	112-40-3
Tridecane	$C_{13}H_{28}$	184		x	629-50-5
4,6-Dimethylundecane	$C_{13}H_{28}$	184		x	17301-23-4
Cycloalkanes					
1-Methyl-2-methylenecyclo-propane	C_5H_8	68		x	18631-84-0

Table 2.5 Continued

Sources	Formula	MW	Refs.	TD	CAS no.
Cyclopentane	C_5H_{10}	70		x	287-92-3
Ethylcyclobutane	C_6H_{12}	84	[102]	x	4806-61-5
Methylcyclohexane	C_7H_{14}	98		x	108-87-2
cis-1,2-Dimethylcyclohexane	C_8H_{16}	112		x	2207-01-4
1,2,4-Trimethylcyclohexane	C_9H_{18}	126		x	2234-75-5
Alkenes					
Divinyl ether	C_4H_6O	70		x	109-93-3
2-Methylpropene	C_4H_8	56		x	115-11-7
1-Pentene	C_5H_{10}	70		x	109-67-1
3,3-Dimethylcyclobutene	C_6H_{10}	82		x	16327-38-1
2-Hexene	C_6H_{12}	84	[102]	x	592-43-8
4-Methyl-1-pentene	C_6H_{12}	84		x	691-37-2
2-Heptene	C_7H_{14}	98		x	592-77-8
4,4-Dimethyl-1-pentene	C_7H_{14}	98		x	762-62-9
1,3,5,7-Cyclooctatetraene	C_8H_8	104		x	629-20-9
3-Methyl-1-heptene	C_8H_{16}	112		x	4810-09-7
2,3-Dihydro-1-methyl-1*H*-indene	$C_{10}H_{12}$	132		x	767-58-8
Alkynes					
5-Methyl-1-hexyne	C_7H_{12}	96		x	2203-80-7
Allenic dienes					
1,2-Butadiene	C_4H_6	54		x	590-19-2
cis,trans-1,3-Pentadiene	C_5H_8	68		x	504-60-9
Cycloheptatriene	C_7H_8	92		x	544-25-2
1,2-Heptadiene	C_7H_{12}	96		x	2384-90-9
1,2-Dihydrobenzocyclobutene	C_8H_8	104		x	694-87-1
3,7-Dimethyl-1,3,6-octatriene	$C_{10}H_{16}$	136		x	29714-87-2
Aromatics					
Benzene	C_6H_6	78	[102]	x	71-43-2
Toluene	C_7H_8	92	[102]	x	108-88-3
Styrene	C_8H_8	104		x	100-42-5
Ethylbenzene	C_8H_{10}	106		x	100-41-4
o-Xylene	C_8H_{10}	106	[103, 106]	x	95-47-6
m-Xylene	C_8H_{10}	106	[103, 106]	x	108-38-3
p-Xylene	C_8H_{10}	106	[103, 106]	x	106-42-3
β-Methoxystyrene	$C_9H_{10}O$	134		x	4747-15-3
Cumene	C_9H_{12}	120		x	98-82-8
Propylbenzene	C_9H_{12}	120		x	103-65-1
Mesitylene	C_9H_{12}	120		x	108-67-8
1,2,4-Trimethylbenzene	C_9H_{12}	120		x	95-63-6
Isopropenyltoluene	$C_{10}H_{12}$	132		x	26444-18-8
tert-Butylbenzene	$C_{10}H_{14}$	134	[102]	x	98-06-6
sec-Butylbenzene	$C_{10}H_{14}$	134		x	135-98-8
n-Butylbenzene	$C_{10}H_{14}$	134		x	104-51-8
p-Cymene	$C_{10}H_{14}$	134		x	99-87-6
1,2,3,5-Tetramethylbenzene	$C_{10}H_{14}$	134		x	527-53-7
1,2-Dimethyl-2-butenylbenzene	$C_{12}H_{16}$	160		x	50871-04-0
Acid and acid derivatives					
Formic acid	CH_2O_2	46		x	64-18-6
Acetic acid	$C_2H_4O_2$	60		x	64-19-7
Isobutyric acid	$C_4H_8O_2$	88		x	79-31-2
3-Furoic acid	$C_5H_4O_3$	112		x	488-93-7
Methoxyacetic acid anhydride	$C_6H_{10}O_5$	162		x	19500-95-9
Hexanoic acid	$C_6H_{12}O_2$	116		x	142-62-1

Table 2.5 Continued

Sources	Formula	MW	Refs.	TD	CAS no.
Benzoic acid	$C_7H_6O_2$	122		x	65-85-0
6-Nonynoic acid	$C_9H_{14}O_2$	154		x	56630-31-0
Amides					
Formamide	CH_3NO	45		x	75-12-7
Acetamide	C_2H_5NO	59		x	60-35-5
Amines					
2,2-Dimethylaziridine	C_4H_9N	71		x	2658-24-4
O-Isobutylhydroxylamine	$C_4H_{11}NO$	89		x	5618-62-2
Benzothiazole	C_7H_5NS	135		x	95-16-9
2-(2-Aminoethyl)pyridine	$C_7H_{10}N_2$	122		x	2706-56-1
Aldehydes					
Acetaldehyde	C_2H_4O	44	[102]	x	75-07-0
Furfural	$C_5H_4O_2$	96		x	98-01-1
4-Pentenal	C_5H_8O	84		x	2100-17-6
Isovaleraldehyde	$C_5H_{10}O$	86		x	590-86-3
5-Methyl-2-furfural	$C_6H_6O_2$	110		x	620-02-0
5-(Hydroxymethyl)-2-furfural	$C_6H_6O_3$	126		x	67-47-0
Benzaldehyde	C_7H_6O	106		x	100-52-7
4-Heptenal	$C_7H_{12}O$	112		x	62238-34-0
2,4-Dimethylpentanal	$C_7H_{14}O$	114		x	27944-79-2
Heptanal	$C_7H_{14}O$	114		x	111-71-7
Terephthalaldehyde	$C_8H_6O_2$	134		x	623-27-8
3-Methoxybenzaldehyde	$C_8H_8O_2$	136		x	591-31-1
2-Ethylhexanal	$C_8H_{16}O$	128		x	123-05-7
Octanal	$C_8H_{16}O$	128		x	124-13-0
Cinnamaldehyde	C_9H_8O	132		x	104-55-2
Dodecanal	$C_{12}H_{24}O$	184		x	112-54-9
Nonanal	$C_9H_{18}O$	142		x	124-19-6
Ketones					
Acetone	C_3H_6O	58		x	67-64-1
1-Hydroxy-2-propanone	$C_3H_6O_2$	74		x	116-09-6
3-Buten-2-one	C_4H_6O	70		x	78-94-4
2-Butanone	C_4H_8O	72		x	78-93-3
3-Hydroxy-2-butanone	$C_4H_8O_2$	88		x	513-86-0
4,4-Dimethyl-2-oxetanone	$C_5H_8O_2$	100		x	1823-52-5
3-Methyl-2-butanone	$C_5H_{10}O$	86		x	563-80-4
2-Pentanone	$C_5H_{10}O$	86		x	107-87-9
2,3-dihydro-3,5-dihydroxy- 6-methyl-4*H*-pyran-4-one	$C_6H_8O_4$	144		x	28564-83-2
2-Hexanone	$C_6H_{12}O$	100		x	591-78-6
5-Methyl-2-hexanone	$C_7H_{14}O$	114		x	110-12-3
2-Heptanone	$C_7H_{14}O$	114		x	110-43-0
Acetophenone	C_8H_8O	120		x	98-86-2
4-Methyl-2-heptanone	$C_8H_{16}O$	128		x	6137-11-7
1-Cyclopropyl-2-(2-pyridinyl)- ethanone	$C_{10}H_{11}NO$	161		x	57276-32-1
Alcohols					
Methanol	CH_4O	32	[102]		67-56-1
Ethanol	C_2H_6O	46	[102]	x	64-17-5
2-Propanol	C_3H_8O	60		x	67-63-0
2-Methyl-1-propanol	$C_4H_{10}O$	74		x	78-83-1
2-Methyl-2-propanol	$C_4H_{10}O$	74		x	75-65-0
1,3-Butanediol	$C_4H_{10}O_2$	90		x	107-88-0

Table 2.5 Continued

Sources	Formula	MW	Refs.	TD	CAS no.
3-Methyl-3-buten-2-ol	$C_5H_{10}O$	86		x	10473-14-0
2-Methyl-1-butanol	$C_5H_{12}O$	88		x	137-32-6
3-Methyl-1-butanol	$C_5H_{12}O$	88		x	123-51-3
2-Pentanol	$C_5H_{12}O$	88		x	6032-29-7
2-Methylcyclopentanol	$C_6H_{12}O$	100		x	24070-77-7
Cyclohexanol	$C_6H_{12}O$	100		x	108-93-0
1-Hexanol	$C_6H_{14}O$	102		x	111-27-3
Benzene methanol	C_7H_8O	108		x	100-51-6
3-Methyl-2-cyclohexen-1-ol	$C_7H_{12}O$	112		x	21378-21-2
trans-2-Hepten-1-ol	$C_7H_{14}O$	114		x	33467-76-4
Benzene ethanol	$C_8H_{10}O$	122		x	60-12-8
2-Ethyl-1-hexanol	$C_8H_{18}O$	130		x	104-76-7
1-Octanol	$C_8H_{18}O$	130		x	111-87-5
Borneol	$C_{10}H_{18}O$	154		x	507-70-0
Hexadecanol	$C_{16}H_{34}O$	242		x	36653-82-4
Ethers					
Heptyl hexyl ether	$C_{13}H_{28}O$	200		x	7289-40-9
Polycyclic aromatic hydrocarbons (PAHs)					
Naphthalene	$C_{10}H_8$	128	[107, 109]	x	91-20-3
Acenaphthene	$C_{12}H_{10}$	154	[109]		83-32-9
1,6-Dimethylnaphthalene	$C_{12}H_{12}$	156		x	575-43-9
2-Ethylnaphthalene	$C_{12}H_{12}$	156		x	939-27-5
1,3-Dimethylnaphthalene	$C_{12}H_{12}$	156		x	575-41-7
1,5-Dimethylnaphthalene	$C_{12}H_{12}$	156		x	571-61-9
1,7-Dimethylnaphthalene	$C_{12}H_{12}$	156		x	575-37-1
2,7-Dimethylnaphthalene	$C_{12}H_{12}$	156		x	582-16-1
1,2,3,4-Tetrahydro-2,6-dimethyl-naphthalene	$C_{12}H_{16}$	160		x	7524-63-2
Biphenyls					
1,1'-Biphenyl	$C_{12}H_{10}$	154	[107]	x	92-52-4
3-Methyl-1,1'-biphenyl	$C_{13}H_{12}$	168	[107]	x	643-93-6
Industrial compounds/solvents					
Halogenated compounds					
Trichlorofluoromethane	CCl_3F	136		x	75-69-4
Tetrachloromethane	CCl_4	154		x	56-23-5
Bromodichloromethane	$CHBrCl_2$	164		x	75-27-4
Dibromochloromethane	$CHBr_2Cl$	208		x	124-48-1
Tribromomethane	$CHBr_3$	253		x	75-25-2
Trichloromethane	$CHCl_3$	118		x	67-66-3
Bromochloromethane	CH_2BrCl	128		x	74-97-5
Dibromomethane	CH_2Br_2	174		x	74-95-3
Dichloromethane	CH_2Cl_2	84		x	75-09-2
Tetrachloroethene	C_2Cl_4	164		x	127-18-4
Hexachloroethane	C_2Cl_6	239		x	67-72-1
Trichloroethene	C_2HCl_3	131		x	79-01-6
1,1,2,2-Tetrafluoroethane	$C_2H_2F_4$	102		x	359-35-3
1,1-Dichloroethene	$C_2H_2Cl_2$	97		x	75-35-4
cis-1,2-Dichloroethene	$C_2H_2Cl_2$	97		x	156-59-2
trans-1,2-Dichloroethene	$C_2H_2Cl_2$	97		x	156-60-5
1,1,1,2-Tetrachloroethane	$C_2H_2Cl_4$	166		x	630-20-6
1,1,2,2-Tetrachloroethane	$C_2H_2Cl_4$	166		x	79-34-5
1-Chloro-1,1-difluoroethane	$C_2H_3Cl F_2$	100		x	75-68-3
1,1,1-Trichloroethane	$C_2H_3Cl_3$	132		x	71-55-6

Table 2.5 Continued

Sources	Formula	MW	Refs.	TD	CAS no.
1,1,2-Trichloroethane	$C_2H_3Cl_3$	132		x	79-00-5
1,2-Dibromoethane	$C_2H_4Br_2$	188		x	106-93-4
1,1-Dichloroethane	$C_2H_4Cl_2$	99		x	75-34-3
1,2-Dichloroethane	$C_2H_4Cl_2$	99		x	107-06-2
1,1-Dichloropropene	$C_3H_4Cl_2$	111		x	563-58-6
cis-1,3-Dichloropropene	$C_3H_4Cl_2$	111		x	542-75-6
trans-1,3-Dichloropropene	$C_3H_4Cl_2$	111		x	10061-02-6
1,2,3-Trichloropropane	$C_3H_5Cl_3$	147		x	96-18-4
1,2-Dichloropropane	$C_3H_6Cl_2$	113		x	78-87-5
1,3-Dichloropropane	$C_3H_6Cl_2$	113		x	142-28-9
2,2-Dichloropropane	$C_3H_6Cl_2$	113		x	594-20-7
1,1,3,4-Tetrachlorobutane	$C_4 H_6Cl_4$	196		x	3405-32-1
1,2,3-Trichlorobenzene	$C_6H_3Cl_3$	181		x	87-61-6
1,2,4-Trichlorobenzene	$C_6H_3Cl_3$	181		x	120-82-1
1,2-Dichlorobenzene	$C_6H_4Cl_2$	147		x	95-50-1
1,3-Dichlorobenzene	$C_6H_4Cl_2$	147		x	541-73-1
1,4-Dichlorobenzene	$C_6H_4Cl_2$	147		x	106-46-7
Bromobenzene	C_6H_5Br	157		x	108-86-1
Chlorobenzene	C_6H_5Cl	113		x	108-90-7
6-Bromo-1-hexene	$C_6H_{11}Br$	162		x	2695-47-8
2-Chlorotoluene	C_7H_7Cl	127		x	95-49-8
4-Chlorotoluene	C_7H_7Cl	127		x	106-43-4
Arochlor-1260	$C_{12}H_3Cl_7$	395	[110]		11096-82-5
Arochlor-1254	$C_{12}H_5Cl_5$	326	[110]		11097-69-1
Arochlor-1248	$C_{12}H_6Cl_4$	292	[110]		12672-29-6
Agrochemicals					
Pesticides					
1,2-Dibromo-3-chloropropane	$C_3H_5Br_2Cl$	236		x	96-12-8
1,4-Dichlorobenzene	$C_6H_4Cl_2$	147		x	106-46-7
Methyl parathion	$C_8H_{10}NO_5PS$	263	[110]		298-00-0
Endosulfan	$C_9H_6Cl_6O_3S$	407	[110]		115-29-7
Heptachlor (bees/pollen)	$C_{10}H_5Cl_7$	373		y	76-44-8
Heptachlor epoxide (bees)	$C_{10}H_5Cl_7O$	389		y	1024-57-3
γ-Chlordane (pollen)	$C_{10}H_6Cl_8$	410		y	57-74-9
Menthol	$C_{10}H_{20}O$	156		x	2216-51-5
Aldrin (bees)	$C_{12}H_8Cl_6$	365		y	309-00-2
Dieldrin (pollen)	$C_{12}H_8Cl_6O$	381		y	60-57-1
Endrin (bees)	$C_{12}H_8Cl_6O$	381		y	72-20-8
Endrin aldehyde (bees)	$C_{12}H_8Cl_6O$	381		y	7421-93-4
Carbaryl	$C_{12}H_{11}NO_2$	201	[110]		63-25-2
4,4'-DDE (bees)	$C_{14}H_8Cl_4$	318		y	72-55-9
4,4'-DDT (bees/pollen)	$C_{14}H_9Cl_5$	355		y	50-29-3
4,4'-DDD (bees)	$C_{14}H_{10}Cl_4$	320		y	72-54-8

Table 2.6 Selected volatile and semi-volatile organic compounds concentrations (ppt by volume) in hive air from colonies located near Chesapeake Bay, USA, during the 1999 summer season

Compound	Maximum level, ppt	Mean level, ppt (n = 17)
Trichloromethane	70	20
Tetrachloromethane	25	4
1,2-Dichloroethane	2	1
1,1,1-Trichloroethane	826	55
1,1,2,2-Tetrachloroethane	4	1
cis-1,2-Dichloroethene	839	149
1,1-Dichloroethene	128	18
Trichloroethene	2	1
Tetrachloroethene	19	7
1,4-Dichlorobenzene	46	6
Bromobenzene	86	6
Benzene	170	78
Toluene	1643	662
Ethylbenzene	146	64
o-Xylene	118	44
Styrene	9239	1594
Naphthalene	16	2
Acetophenone	112	11

400 multibed unit – a drying tube and a Carbotrap 150 "guard column." The drying tube has eliminated samples lost to moisture, a problem on hot days when the bees are using evaporative cooling in the hive for thermoregulation. The Carbotrap 150 tube is a single-bed tube containing graphitized carbon to remove high levels of terpenes and pyrolysis residues of sugar compounds. Without the guard tube, resinous components of propolis and the pine boards in the hive overwhelm the transfer lines between the desorption unit, the focusing trap, and the GC column. Without the Carbotrap 150 tube, transfer lines need to be replaced every 60 samples.

Organic compounds in honey bee semiochemicals

Much communication among honey bees is conducted via the exchange of semiochemicals. Some chemical communication requires direct transfer of fluids from one bee to another; other communication is conducted by the release of volatile compounds inside the hive or externally into the ambient air. Bees have a system of exocrine glands that release mixtures of compounds for specific purposes – e.g. attracting a mate (queen pheromone), suppressing ovaries in workers (queen pheromone), signaling about an intruder (alarm pheromone). Chemical characterization of gland contents has usually been accomplished by physically removing glands and analyzing their concentrated liquid contents.

Of interest to this study is how pervasive semiochemicals are in hive atmospheres. Since some pheromones are effective at very low levels and many are released outside of the hive, they can easily be missed (and were) by our TD/GC/MS methodology. Our hive air sampling, however, picked up those semiochemicals that were produced in large amounts and passed around inside the hive by workers (Table 2.2).

Detectable levels of hexanoic acid, octanoic acid, 2-heptanone, C_4–C_8 acetate esters, C_5–C_9 alcohols, hydrocarbons, and terpenes were probably secreted by the mandibular glands of the hive worker population [26–50]. Many of these compounds are released as alarm and hive defense signals. Constituents of the Koschevnikov gland, also used for alarm and sting purposes (nonanoic acid, octenal, isopentyl acetate, nonanol), were detected. Compounds reported in studies of young queens [71] were also seen – *p*-methane-9-ol, 4,5-dimethylnonane, dimethylhexane, and trimethyl-cyclopentane trimethylnonene. Since there were no significant numbers of isolated, young queens in our hive samples, their presence in hive atmospheres suggests that they are also secreted by the general worker population.

Beeswax pheromones, which give rise to characteristic hive odors, should also be present in substantial quantities in hive air. Among the list of those attributed to comprising hive odor [87–91], we found furfural, benzaldehyde, octanal, nonanal, decanal, and decanol.

Chemicals from the Nasonov glands, used to mark field locations of water and artificial food sources [51–58], were detected by TD/GC/MS. These markers include C_{10} monoterpenoids such as (*Z*)-citral, (*E*)-citral, nerol, geraniol, nerolic acid, and geranic acid.

Organic compounds in hive stores

The TD/GC/MS method described in this paper is designed specifically to target volatile and semi-volatile components of intermediate to low polarity. As such, we did not expect to detect those compounds which partition into aqueous-based phases such nectar and honey. Alcohols and carboxylic acids, which establish significant equilibrium concentrations in both honey and air, were found (Table 2.3). Low molecular weight components of beeswax, e.g. alkenes and alkadienes, were also seen in the hive air.

The most visible bee-foraged material from the standpoint of hive air samples was propolis. A wide variety of hydrocarbons and their partially oxidized breakdown product components were collected in our chemical sorbent beds. These correlated closely with lists of compounds reported in propolis [16, 17, 101] for North America and Europe. Many more compounds appear in propolis from tropical forests [16] that were not examined in this study. Many of these are large waxy esters and would not be seen by our TD/GC/MS method. We expect that specific compounds and their relative amounts will show considerable variation dependent upon the types of vegetation from which saps and resins have been collected. Some analytes reported by us in Table 2.3 that are not reported as propolis components in the bee literature may represent such vegetation-specific compounds. Many of these identifications are tentative. They are based on spectral matches from the NIST library, but have not been confirmed by comparative analysis of standards. Among the proposed compounds are various aliphatic and aromatic acids, ketones, esters, aldehydes, and terpenes.

Organic compounds in instrumented condo components

During the early stages of our bee biomonitoring project, we chemically profiled each component that was used in the construction of our instrumented hive condos (Table 2.4). Unpainted pine boards were rich in terpene peaks. In fact, using artificial neural networks, we were often able to identify from which hive a sample came, based heavily on their individual terpene fingerprints [112, 113]. Vinyl screens gave rise to several ethers. Polyethylene parts released various aromatic and aliphatic derivatives. The application of white paint to the exterior surfaces of hive boxes added some organic acids, alcohols, and additional hydrocarbons to the hive environment.

Organic compounds from non-bee sources

Compounds listed in Table 2.5 are released into the air by well-known, non-honey bee sources. As such, most of them have not previously been considered as part of the hive atmospheres in which colonies live. Hive atmosphere sampling, however, reveals that these compounds are intimately incorporated into the air reservoir used by hive residents. Thus, they should be included in discussions of honey bee ecosystems.

Some contaminants are present because ambient air has suffused into the hive from the outside. Others may be present because honey bee foragers have encountered them during resource collection and brought them back with water, nectar, or pollen [11–14]. Thermoregulation of the hive brood areas, near 35°C for our colonies, is quite effective in volatilizing many organic residues. Studies in our laboratories [112] have demonstrated that water-collecting bee foragers successfully transport an organic film from standing pools of water and moist soil granules. Water is often brought into the hive on warm days and fanned to keep brood areas at the proper temperature. As the cooling water is evaporated, organic film components are vaporized efficiently into the hive atmosphere.

We have categorized non-bee chemicals based on their likely source: (1) compounds arising from wood and biomass combustion [104, 105], (2) petroleum and creosote residues or vehicle emissions [102, 105], (3) industrial compounds [106], and (4) agrochemicals [110]. These lists are by no means comprehensive. Their intent is to provide a sample of representative compounds that fall into these categories.

Compounds assigned to the wood and biomass combustion category include terpenes and oxygenated pyrolysis products such as aldehydes, alcohols, and ketones [104, 105]. Although simple aromatics have been reported in the literature from residential wood burning, our samples were collected during summer when biomass combustion was minimal. It is more likely that detected aromatic products originate from petroleum residuals.

Fossil-fueled vehicles give rise to emissions of unburned fuel and partially oxidized hydrocarbons [102, 106]. Prominent are the BTEX suite of aromatics – benzene, toluene, ethylbenzene, and xylenes. These compounds are ubiquitous in the environment, present in essentially every hive atmosphere we test and often among the most prominent peaks in the chromatogram. To date, it has not been possible to position a bee colony that avoids capture of significant amounts of BTEX. We also detect more biorefractive fuel components in hive air – polycyclic aromatics and biphenyls commonly associated with diesel products [114]. Incompletely burned fuel residuals [102] were also evident as noted in the "Oxygenates" portion of Table 2.5. These comprised aldehydes, ketones, alcohols, and oxides.

A number of halogenated organic compounds found in hive atmospheres by our TD/GC/MS are common industrial solvents. They, too, are fairly ubiquitous in the environment. Chlorination of drinking and wastewater generates some. Many others are components in over-the-counter home products. Engine degreasers used by home mechanics are an especially rich source of these materials. Halogenated contaminants are biorefractive so they persist in the environment after their initial release, probably in a liquid or aerosol form. Tetrachloroethene (PCE), 1,1,2-trichloroethene, and tetrachloromethane have been among the most frequently encountered chlorinated solvents in our environmental biomonitoring work at Aberdeen Proving Ground [11–14]. Contaminant transport experiments [112] have shown that bees readily transport PCE, 1,1,2,2-tetrachloroethane, 1,1,1-trichloroethane, and 1,1,2-trichloroethane from watering stations into the hive.

TD/GC/MS sampling done with carbon-based sorbents has had only limited success at finding pesticides in hive atmospheres. In part this is due to the high molecular weights of some common pest agents such as endosulfan (406.9 g/mol) and chlordane (409.8 g/mol). The pesticides we have seen tend to be the lighter, more volatile agents – *p*-dichlorobenzene (147 g/mol) and menthol (156 g/mol). Silica-based sorbents would extend our range to higher molecular weight values. Sorption and chromatographic considerations decrease sensitivity toward organophosphate and carbamate analytes. Experiments performed in our research apiary during the summer of 1998 demonstrate that methyl parathion doses, sufficient to cause significant bee mortality, are missed by our methodology. Sorption traps that target organophosphate analytes are currently undergoing evaluation.

Xenobiotic VOC taggants

Our ability to successfully detect VOCs and SVOCs in hive atmospheres has led to exploiting this ability in a new direction. Exotic compounds can be added to syrup and pollen feeders. When bee foragers visit these stations and then return to the hive, they carry a chemical signature that we can detect in the hive atmosphere. We employ this technique in studies designed to document foraging patterns or to confirm that honey bees have found "targets" that carry a conditioning scent. This technique is helpful in validating that the bees find unattended targets by means of scent. Bees in our conditioning trials became so accustomed to field personnel being associated with feeders that they would investigate any human in the vicinity for a feeder target.

Two categories of taggants have proved particularly useful – perfluorinated compounds and perdeuterated compounds. We have a sufficient number of taggants that arrays of feeders can be uniquely marked. Taggants can be distributed in feeders on the basis of radial distance from the hive, direction or specific locations. Quantitative analysis of the taggants in the beehive atmosphere allows us to apportion foraging activity among the target categories in the experiment.

Very little taggant is required for any one experiment. We usually add $100\,\mu L$ of taggant to $0.250\,dm^3$ of a $2\,M$ sucrose syrup solution. Figure 2.3 demonstrates hive air contents following deployment of three taggants – perdeuterated benzene, toluene, and heptane. Samples were collected over 30 min intervals. An advantageous property of the perdeuterated taggants is that they clear the hive within about 2 hours. Perfluorinated taggants require about 48 hours to disappear, constraining the number of experiments that can be conducted in a short timeframe.

Conclusion

Sampling by TD/GC/MS has been an effective means of characterizing the mix of volatile and semi-volatile compounds present in hive atmospheres. Hive residents breathe an expected combination of chemicals released by themselves, their forage resources, and the hive walls. What most bee researchers have previously missed is the significant, and often dominant, presence of airborne contaminants from non-bee sources. This becomes apparent when actual hive atmospheres are sampled in place of experiments that characterize the composition of glands or hive stores. High PCE levels in hives at Aberdeen Proving Ground were associated with queen loss in 50 percent of the colonies placed near a hazardous military landfill [14].

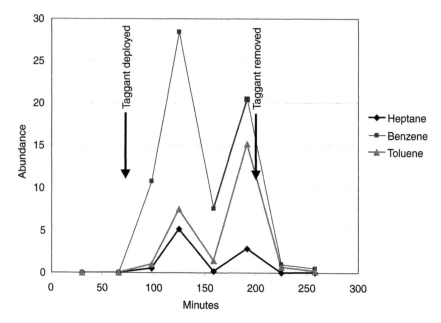

Figure 2.3 Perdeuterated taggants in hive atmosphere. Feeder syrup contained no taggant for points prior to 90 minutes. Taggant was only available for points between 90 and 200 minutes. Non-taggant syrup was restored for points beyond 200 minutes. Note that their presence is seen immediately in the first post-deployment interval and that they disappear within 1 hour of being removed.

Acknowledgments

This work was supported by Contract DAMD 17-95-C-5072, US Army Medical Research and Materials Command, Ft Detrick, MD. Contracting Officers Representative: Tommy R. Shedd, US Army Center for Environmental Health Research (USACEHR), Ft Detrick, MD. Project Oversight: J. Wrobel, C. Powels, D. Green, R. Golding, US Army Directorate of Safety, Health, and Environment, Aberdeen Proving Ground, MD. We thank Hewlett Packard for an institutional gift of the GC/MS portion of the instrumentation to The University of Montana-Missoula. Chris Wrobel engineered the TD/GC/MS facility at the University of Montana. Michelle and Bryon Taylor, Matthew Loeser, Jason Volkmann, and Lennie Hahn helped to manage the honey bee colonies.

References

1 National Research Council (1991). *Animals as Sentinels of Environmental Health Hazards.* National Academy Press, pp. 5, 101.

2 Kapustka, L., LaPoint, T., Fairchild, J., McBee, K. and Bromenshenk, J. (1989). Field assessments: Terrestrial invertebrate surveys. In: *Ecological Assessments of Hazardous Waste Sites* (Warren-Hicks, W., Parkhurst, B.R. and Baker, S.S., Jr., Eds). US EPA 600/3-89/013, pp. 73–88.

3 Winston, G., Di Guilio, R., Bond, J., Van Veld, P., Eldridge, N.E., Glisson, S.R. and Graber, S. (1999). AIBS Peer Review to USACEHR on "The cost and benefit of using sentinel or alternative species for environmental monitoring and human health assessment." *Scientific Peer Advisory and Review Services Report*, Am. Inst. Biol. Sci., Sterling, pp. 1–24.

4 Bromenshenk, J.J., Smith, G.C. and Watson, V.J. (1995). Assessing ecological risks in terrestrial systems with honey bees. *Environ. Sci. Res.* **50**, 9–30.

5 Bromenshenk, J.J. (1992). Site-specific and regional monitoring of pollutants with honey bees. In: *Ecological Indicators* (McKenzie, D.H., Hyatt, D.E. and McDonald, V.J., Eds). Elsevier Applied Science, London, UK, and New York, pp. 689–704.

6 Bromenshenk, J.J. (1988) Regional monitoring of pollutants with honey bees. In: *Progress in Environmental Specimen Banking* (Wise, S., Zeisler, R. and Goldstein, G.M., Eds). NBS Special Publication, 740, pp. 156–170.

7 Johansen, C.A. and Mayer, D.F. (1990). *Pollinator Protection: A Bee and Pesticide Handbook.* Wicwas Press, Cheshire, CT. 212pp.

8 Wallwork-Barber, M.K., Ferenbaugh, R.W. and Gladney, E.S. (1989). The use of honey bees as monitors of environmental pollution. *Am. Bee J.* **12**, 770–772.

9 Celli, G. (1984). L'apa come insetto test della salute di un territorio. *L'apicolt. Mod.* **75**, 133–140.

10 Bromenshenk, J. (2000). Hi tech management. *Bee Culture* **128**, 37–38.

11 Bromenshenk, J.J., Smith, G.C., King, B.E., Seccomb, R.A., Taylor, M., Jones, D.C., Henderson, C.H. and Wrobel, C.L. (2001). *Aberdeen Proving Ground Bee Studies. New and Improved Methods for Monitoring Air Quality and the Terrestrial Environment: Applications at Aberdeen Proving Ground – Edgewood Area.* Annual Report, 1999. US Army Center for Environmental Health Research, Ft Detrick, MD. USA Contract DAMA 17-95-C-5072, 160pp.

12 Bromenshenk, J.J., Smith, G.C., King, B.E., Seccomb, R.A., Jones, D.C., Henderson, C.B., Alnasser, G.H. and Wrobel, C.L. (1999). *New and Improved Methods for Monitoring Air Quality and the Terrestrial Environment: Applications at Aberdeen Proving Ground – Edgewood Area.* Annual Report, 1998. US Army Center for Environmental Health Research, Ft Detrick, MD. USA Contract DAMA 17-95-C-5072, 108pp.

13 Bromenshenk, J.J., Smith, G.C., King, B.E., Seccomb, R.A., Alnasser, G.H., Loeser, M.R., Henderson, C.B. and Wrobel, C.L. (1998*). New and Improved Methods for Monitoring Air Quality and the Terrestrial Environment: Applications at Aberdeen Proving Ground – Edgewood Area.* Annual Report, 1997. US Army Center for Environmental Health Research, Ft Detrick, MD. USA Contract DAMA 17-95-C-5072, 145pp.

14 Bromenshenk, J.J., Smith, G.C., Seccomb, R.A., King, B.E., Alnasser, G.H., Henderson, C.B. and Wrobel, C.L. (1997). *New and Improved Methods for*

Monitoring Air Quality and the Terrestrial Environment. Annual Report, 1996. US Army Center for Environmental Health Research, Ft Detrick, MD. USA Contract DAMA 17-95-C-5072, 173pp.

15 Willis, L.G., Winston, M.L. and Slessor, K.N. (1990). Queen honey bee mandibular pheromone does not affect worker ovary development. *Can. Entomol.* **122**, 1093–1099.

16 Bankova, V.S., De Castro, S.L. and Marcucci, M.C. (2000). Propolis: Recent advances in chemistry and plant origin. *Apidologie* **31**, 3–14.

17 Marcucci, M.C. (1995). Propolis: Chemical composition, biological properties and therapeutical activity. *Apidologie* **26**, 83–99.

18 Blum, M.S. (1992). Honey bee pheromones. In: *The Hive and the Honey Bee* (Graham, J.M., Ed.). Dadant and Sons, Hamilton, pp. 373–398.

19 Blum, M.S. (1982). Pheromonal bases of insect sociality: Communications, conundrums and caveats. In: *Les Médiateurs Chimiques*. Les Colloques de l'INRA, Versailles, pp. 149–162.

20 Blum, M.S. (1974). Deciphering the communicative Rosetta Stone. *Bull. Entomol. Soc. Am.* **20**, 30–35.

21 Gary, N.E. (1982). Chemical mating attractants of the honey bee. *Science* **136**, 773–774.

22 Habermann, E. (1971). Chemistry, pharmacology, and toxicology of bee, wasp, and hornet venoms. In: *Venomous Animals and their Venoms*, Vol. 3 (Bucherl, W. and Buckley, E.E., Eds). Academic Press, New York, pp. 61–63.

23 Hepburn, H.R. and Radloff, S.E. (1966). Morphometric and pheromonal changes of *Apis mellifera* L. along a transect from the Sahara to the Pyrenees. *Apidologie* **7**, 35–45.

24 Morse, R.A. and Hooper, T. (1985). *The Illustrated Encyclopedia of Beekeeping*. E.P. Dutton, Inc., New York, 432pp.

25 Schmidt, J.O. (1998). Mass action in honey bees: Alarm, swarming and the role of releaser pheromones. In: *Pheromone Communication in Social Insects: Ants, Wasps, Bees, and Termites* (Vander Meer, R.K., Breed, M.D., Espelie, K.E. and Winston, M.L., Eds). Westview Press, pp. 258–289.

26 Arnold, G. and Roger, B. (1979). Group effect on the content of 10-hydroxy-dec-2-enoic acid in the head of worker bees. *Apidologie* **10**, 35–42.

27 Barbier, J. and Lederer, E. (1960). Structure chimique de la substance royale de la reine d'abeille (*Apis mellifera* L.) *Ct. R. Acad. Sci. (Paris)* **251**, 1131–1135.

28 Callow, R.K. and Johnson, N.C. (1960). The chemical constitution and synthesis of queen substances of honeybees (*Apis mellifera* L.). *Bee World* **41**, 152–153.

29 Callow, R.K., Chapman, J.R. and Paton, P.N. (1964). Pheromones of the honeybee: Chemical studies of the mandibular gland secretion of the queen. *J. Apic. Res.* **3**, 77–89.

30 Blum, M.S., Novak, A.F. and Taber III, S. (1959). 10-Hydroxy-delta-decenic acid, an antibiotic found in royal jelly. *Science* **130**, 452–453.

31 Boch, R. and Shearer, D.A. (1967). 2-Heptanone and 10-hydroxy-*trans*-dec-2-enoic acid in the mandibular glands of the worker bee. *J. Insect Physiol.* **16**, 17–24.

32 Boch, R., Shearer, D.A. and Shuel, R.W. (1979). Octanoic acid and other volatile acids in the mandibular glands of the honeybee and royal jelly. *J. Apic. Res.* **18**, 250–153.

33 Boch, R., Shearer, D.A. and Petrasovits, A. (1970). Efficacies of two alarm substances of the honey bee. *J. Insect Physiol.* **16**, 17–24.

34 Butler, C.G. (1966). Mandibular gland pheromone of the honey bee. *Nature* **212**, 530.

35 Butler, C.G., Callow, R.K. and Johnston, N.C. (1961). The isolation and synthesis of queen substance, 9-oxodec-*trans*-2-enoic acid, a honeybee pheromone. *Proc. R. Entomol. Soc.* **155**, 417–432.

36 Butler, C.G., Callow, R.K. and Chapman, J.R. (1964). 9-Hydroxydec-*trans*-2-enoic acid, a pheromone stabilizing honey bee swarms. *Nature* **201**, 733.

37 Butler, C.G. and Simpson, J. (1967). Pheromones of the queen honey bee (*Apis mellifera* L.) which enable her workers to follow her when swarming. *Proc. Ry. Entomol. Soc. London (A)* **42**, 149–154.

38 Butenandt, A. and Rembold, H. (1957) Uber den Weiselzellenfuttersaft der Honigbiene, I. Isolierung, Konstitutionsermittlung und Vorkommen der 10-Hydroxy-2-decensaure. *Hoppe-Seyl Z.* **308**, 284–289.

39 Kerr, W.E., Blum, M.S., Pisani, J.F. and Stort, A.C. (1974). Correlation between amounts of 2-heptanone and iso-amyl acetate in honeybees and their aggressive behaviour. *J. Apic. Res.* **13**, 173–176.

40 Moritz, R.F.A. and Crewe, R.M. (1988). Chemical signals of queens in kin recognition of honeybees, *Apis mellifera* L. *J. Comp. Physiol. A.* **164**, 83–89.

41 Morse, R.A. (1972). Honey bee alarm pheromone: another function. *Ann. Entomol. Soc. Am.* **65**, 1430.

42 Nunez, J.A. (1967). Sammelbienen markieren versiegte Futter-guellen durch Duft. *Naturwissenschaften* **54**, 322–323.

43 Pain, J. and Barbier, M. (1960). Mise en evidence d'une substance attractive extraite du corps des ouvières d'abeilles non orphelines (*Apis mellifera* L.). *C. R. Hebd. Seances Acad. Sci.* **250**, 1126–1127.

44 Pain, J. and Roger, B. (1978). Variation annuelle de l'acide hyroxy-10 decene-2-oique dans les testes d'abeilles. *Apidologie* **1**, 29–54.

45 Pain, J., Barbier, M., Bogdanovsky, D. and Lederer, E. (1962). Chemistry and biological activity of the secretions of queen and honeybees (*Apis mellifica* L.) *Comp. Biochem. Physiol.* **6**, 233–241.

46 Shearer, D.A. and Boch, R. (1965). 2-Heptanone in the mandibular gland secretion of the honey-bee. *Nature* **206**, 530.

47 Simpson, J. (1966). Repellency of the mandibular gland scent of worker honey bees. *Nature* **209**, 531–532.

48 Velthuis, H.H.H. (1970). Queen substance from the abdomen of the honey bee queen. *Z. Vergl. Physiol.* **70**, 210–222.

49 Velthuis, H.H.H. (1972). Observations on the transmission of queen substances in the honey bee colony by attendants of the queen. *Behaviour* **41**, 105–129.

50 Yadava, R.P.S. and Smith, M.V. (1971). Aggressive behaviour of *Apis mellifera* L. workers towards introduced queens. I. Behavioural mechanisms involved in the release of worker aggression. *Behaviour* **39**, 212–226.

51 Boch, R. and Shearer, D.A. (1962). Identification of geraniol as the active component in the Nassanoff pheromone of the honey bee. *Nature (London)* **194**, 704–706.

52 Boch, R. and Shearer, D.A. (1965). Attracting honey bees to crops which require pollination. *Am. Bee J.* **105**, 166–167.

53 Butler, C.G. and Calam, D.H. (1968). Pheromones of the honey bee – the secretion of the Nasanoff gland of the worker. *J. Insect Physiol.* **15**, 237–244.

54 Boch, R. and Morse, R.A. (1974). Discrimination of familiar and foreign queens by honey bee swarms. *Ann. Entomol. Soc. Am.* **67**, 709–711.

55 Morse, R.A. and Boch, R. (1971). Pheromone concert in swarming honeybees. *Ann. Entomol. Soc. Am.* **64**, 1414–1417.

56 Pickett, J.A., Williams, I.H., Martin, A.P. and Smith, M.P. (1980). Nasonov pheromone of the honeybee, *Apis mellifera* L. (Hymenoptera:Apidae). Part I. Chemical characterization. *J. Chem. Ecol.* **6**, 426–434.

57 Shearer, D.A. and Boch, R. (1966). Citral in the Nasanoff pheromone of the honey bee. *J. Insect Physiol.* **12**, 1513–1521.

58 Slessor, K.N., Kaminski, L.-A., King, G.G., Borden, J.H. and Winston, M.L. (1988). Semiochemical basis of retinue response to queen honey bees. *Nature* **332**, 354–356.

59 Cole, L.K., Blum, M.S. and Roncadori, R.W. (1975). Antifungal properties of the insect alarm pheromones. *J. Apic. Res.* **20**, 13–18.

60 Collins, A.M. and Blum, M.S. (1982). Bioassay of compounds derived from the honey bee sting. *J. Chem. Ecol.* **8**, 463–470.

61 Collins, A.M. and Blum, M.S. (1983). Alarm responses caused by newly identified compounds derived from the honeybee sting. *J. Chem. Ecol.* **9**, 57–65.

62 Blum, M.S. and Fales, H.M. (1988). Chemical releasers of alarm behavior in the honey bees: informational plethora of the sting apparatus signal. In: *Africanized Honey Bees and Bee Mites* (Needham, G.R., Page, R.E., Delfinado-Baker, M. and Bowman, C.E., Eds). Halsted Press, New York, pp. 141–148.

63 Blum, M.S., Fales, H.M., Tucker, K.W. and Collins, A.M. (1978). Chemistry of the sting apparatus of the worker honeybee. *J. Apic. Res.* **17**, 218–221.

64 Boch, R. and Shearer, D.A. (1965). Alarm in the beehive. *Am. Bee J.* **105**, 206–207.

65 Boch, R. and Shearer, D.A. (1966). Iso-pentyl acetate in stings of honeybees of different ages. *J. Apic. Res.* **5**, 65–70.

66 Boch, R., Shearer, D.A. and Stone, B.C. (1962). Identification of iso-amyl acetate as an active component in the sting pheromone of the honey bee. *Nature* **195**, 1018–1020.

67 Free, J.B. (1961). The stimuli releasing the stinging response of honeybees. *Anim. Behav.* **9**, 193–196.

68 Free, J.B. and Simpson, J. (1968). The alerting pheromones of the honeybee. *Z. Vergl. Physiol.* **61**, 361–365.

69 Ghent, R.L. and Gary, N.E. (1962). A chemical alarm releaser in honey bee stings (*Apis mellifera* L.). *Psyche* **69**, 1–6.

70 Koeniger, N., Weiss, J. and Maschwitz, U. (1979). Alarm pheromones of the sting in the genus *Apis*. *J. Insect Physiol.* **25**, 467–476.

71 Lensky, Y., Cassier P., Rosa, S. and Grandperrin, D. (1991). Induction of balling in worker honeybees (*Apis mellifera* L.) by "stress" pheromone from Koschewnikow glands of queen bees: Behavioral, structural and chemical study. *Comp. Biochem. Physiol.* **110A**, 585–594.

72 Mauchamp, B. and Grandperrin, D. (1982). Chromatographie en phase gazeuse des composés volatils des glandes à phéromones des abeilles: Methodes d'analyse directe. *Apidologie* **13**, 29–37.

73 Morse, R.A., Shearer, D.A., Boch, R. and Benton, A.W. (1967). Observations on the alarm substances in the genus *Apis. J. Apic. Res.* **6**, 113–118.

74 Pickett, J.A., Williams, I.H. and Martin, A.P. (1982). (Z)-11-Eicosen-1-ol, an important new pheromonal component from the sting of the honey bee, *Apis mellifera* L. (*Hymenoptera*, Apidae). *J. Chem. Ecol.* **8**, 163–175.

75 Schmidt, J.O. (1982). Biochemistry of insect venoms. *Annu. Rev. Entomol.* **27**, 339–368.

76 Schmidt, J.O., Slessor, N. and Winston, M. (1993). Roles of Nasanov and queen pheromones in attraction of honeybee swarms. *Naturwissenschaften* **80**, 573–575.

77 Schmidt, J.O., Morgan, E.D., Oldham, N.J., Do Nascimento, R.R. and Dani, F.R. (1997). (Z)-11-Eicosen-1-ol, a major component of *Apis Cerana* venom. *J. Chem. Ecol.* **23**, 1929–1939.

78 Wager, B.R. and Breed, M.D. (2000). Does honey bee sting alarm pheromone give orientation information to defensive bees? *Ann. Entomol. Soc. Am.* **93**, 1330–1332.

79 Lensky, Y. and Slabezki, Y. (1981). The inhibiting effect of the queen bee (*Apis mellifera L.*) foot-print pheromone on the construction of swarming queen cups. *J. Insect Physiol.* **27**, 313–323.

80 Lensky, Y., Cassier, P., Finkel, A., Teeshbee, A., Schlesinger, R., Delorme-Joulie, C. and Levinsohn, M. (1984). Les glandes tarsales de l'abeille mellifique (*Apis mellifera* L.) (Insecta, Hymenoptera, Apidae). *J. Insect Physiol.* **31**, 265–267.

81 Wossler, T.C., Veale, R.B. and Crewe, R. (2000). How queen-like are the tergal glands in workers of *Apis mellifera capensis* and *Apis mellifera scutellata*? *Apidologie* **31**, 55–66.

82 Velthuis, H.H.H. (1970). Ovarian development in *Apis mellifera* worker bees. *Entomol. Exp. Appl.* **13**, 377–394.

83 Vierling, G. and Renner, M. (1977). Die Bedeutung des Sekretes der Tergittaschen-drusen fur die Attraktivitat der Bienenkonigin gegenuber junger Arbeiterinnin. *Behav. Ecol. Sociobiol.* **2**, 185–200.

84 Butler, C.G., Fletcher, D.J.C., Greenway, A.R. and Watler, D. (1970). Hive entrance finding by honeybee (*Apis mellifera*). *Anim. Behav.* **18**, 78–91.

85 Ferguson, A.W. and Free, J.B. (1979). Production of a forage-marking pheromone by the honeybee. *J. Apic. Res.* **18**, 128–135.

86 Page, R.E., Blum, M.S. and Fales, H.M. (1988). *o*-Aminoacetophenone, a pheromone that repels honeybees (*Apis mellifera* L.). *Experientia* **44**, 270–271.

87 Blum, M.S., Jones, T.H., Rinderer, T.E. and Sylverster, H.A. (1988). Oxygenated compounds in beeswax: Identification and possible significance. *Comp. Biochem. Physiol.* **91B**, 237–238.

88 Callow, R.K. (1963). Chemical and biochemical problems of beeswax. *Bee World* **44**, 95–101.

89 Tulloch, A.P. (1980). Beeswax composition and analysis. *Bee World* **61**, 47–62.

90 Rinderer, T.E. (1981). Volatiles from empty comb increase hoarding by the honey bee. *Anim. Behav.* **29**, 1275–1276.

91 Rinderer, T.E. and Baxter, J.R. (1978). Effect of empty comb on hoarding behavior and honey production of the honey bee. *J. Econ. Entomol.* **71**, 435–438.

92 Gerig, L. (1972). Ein weiterer Duftstoff zur Anlockung der Drohnen von *Apis mellifica* (L.). *Z. Angeur. Entomol.* **70**, 286–289.

93 Free, J.B. (1967). Factors determining the collection of pollen by honeybee foragers. *Anim. Behav.* **15**, 134–144.

94 Free, J.B. and Williams, J.H. (1972). Hoarding by honeybees (*Apis mellifera* L.). *Anim. Behav.* **20**, 327–334.

95 Free, J.B. and Winder, M.E. (1983). Brood recognition by honeybee (*Apis mellifera* L.) workers. *Anim. Behav.* **31**, 539–545.

96 Milojevic, B.D. and Filipovic-Moskovljevic, V. (1959). Gruppeneffekt bie Honigbienen. II. Eierstock-entwicklung bei Arbeitsbienen im Kleinvolk. *Bull. Acad. Serbe Sci. Classe Sci. Math. Naturelles.* **25**, 131–138.

97 Jay, S.C. (1970). The effect of various combinations of immature queen and worker bees on the ovary development of worker honeybees in colonies with and without queens. *Can. J. Zool.* **48**, 169–173.

98 Koeniger, N. and Veith, H.J. (1983). Glyceryl-1,2-dioleate-3-palmitate as a brood pheromone of the honey bee (*Apis mellifera* L.). *Experientia* **39**, 1051–1057.

99 Burdock, G.A. (1989). Review of the biological properties and toxicity of bee propolis (propolis). *Food Chem. Toxicol.* **36**, 347–363.

100 Hydak, M.H. (1936). Value of foods other than pollen in nutrition of the honeybee. *J. Econ. Entomol.* **29**, 870–877.

101 Pereira, A.S., Pinto, A.C., Cardoso, J.N. and Aquino Neto, F.R. (1998). Application of high temperature resolution gas chromatography to crude extracts of propolis. *J. High Resol. Chromatogr.* **21**, 396–400.

102 Westerholm, R. and Egeback, K.-E. (1994). Exhaust emissions from light- and heavy-duty vehicles. *Environ. Health Perspect.* **102**, 13–23.

103 Edgerton, S.A., Khalil, M.A. and Rasmussen, R.A. (1986). Source emission characterization of residential wood-burning stoves and fireplaces. *Environ. Sci. Technol.* **20**, 803–807.

104 Sakuma, H., Munakata, S. and Sugawara, S. (1981). Volatile products of cellulose pyrolysis. *Agric. Biol. Chem. J.* **45**, 443–451.

105 Love, S. and Bratzler, L.J. (1966). Tentative identification of carbonyl compounds in wood smoke by gas chromatography. *J. Agric. Food Chem.* **31**, 218–222.

106 Zhang, X.J. and Peterson, F. (1996). Emissions from co-combustion of wood household refuse. *Waste Renew. Energy Comb.* **9**, 997–1000.

107 Rogge, W.F., Hildemann, L.M., Mazurek, M.A. and Cass, G.R. (1997). Source of fine organic aerosol. 8. Boilers burning No. 2 distillate fuel oil. *Environ. Sci. Technol.* **31**, 2731–2737.

108 Russell, J.W. (1975). Analysis of air pollution using sampling tubes and gas chromatography. *Environ. Sci. Technol.* **9**, 1175–1182.

109 Westerholm, R. and Li, H. (1994). A multivariate analysis of fuel-related polycyclic aromatic hydrocarbon emissions from heavy-duty diesel vehicles. *Environ. Sci. Technol.* **28**, 965–972.

110 Anderson, J.F. and Wojtas, M.A. (1986). Honey bees (Hymenpotera: Apidae) contaminated with pesticides and polychlorinated biphenyls. *J. Econ. Entomol.* **79**, 1200–1205.

111 Sigman, M.E., Guerin, M.R., Smith, G.C. and Ilgner, R.H. (1998). Sol-gel sorbents. In: *Proceedings of the First Conference on Engineered Bee Colonies, DARPA, Controlled Biological Systems* (Bromenshenk, J.J., Ed.). The University of Montana, Missoula, MT, pp. 10–17.

112 Alnasser, G.H. (1998). *The Use of Thermal Desorption/Gas Chromatography/Mass Spectrometry (TD/GC/MS), Honey Bees, and Artificial Neural Networks (ANN) in Assessing Ecosystem Contamination.* PhD thesis, The University of Montana-Missoula, Missoula, MT.

113 Smith, G.C. and Wrobel, C.L. (1998). Neural networks in industrial and environmental applications. In: *Soft Computing in Systems and Control Technology* (Tzafestas, S.G., Ed.). World Scientific Press, Hong Kong, pp. 445–466.

114 Lee, L.S., Hagwall, M., Delfino, J.J. and Rao, P.S.C. (1992). Partitioning of polycyclic aromatic hydrocarbons from diesel fuel into water. *Environ. Sci. Technol.* **26**, 2104–2109.

3 Risk assessment of plant protection products on honey bees

Regulatory aspects

S. Cluzeau

Summary

Significant changes have occurred over the past few years in European and French legislation with regard to environmental risk assessment of plant protection products. Regarding honey bees, the tests requested depend on the intrinsic characteristics of the product and its method of use. Whenever the honey bee is likely to be exposed to chemical plant protection products, during or after treatment, laboratory toxicity tests are requested leading to the calculation of the hazard quotient. This value determines the subsequent tests to be carried out: cage tests, tunnel tests, or field tests, according to the test guidelines harmonized at the European level. In addition, specific procedures currently available in the United States are presented.

Introduction

The end of the Second World War was a turning point in the use of plant protection products in agriculture, with the development of synthetic chemicals, especially insecticides, organochlorides, and organophosphates. At this time, the need to produce and therefore to increase yields to ensure national self-sufficiency was the main priority of a rapidly changing agriculture. The chemical industry, also expanding rapidly, developed thousands of new products, providing easy and effective solutions against various crop enemies [1]. In this context the side-effects of the plant protection treatments, such as the risk of resistance phenomena or the impact on beneficial organisms, although inevitable, were not taken into account, or were ignored [2].

In France, in 1943, a law on the registration of plant protection products used in agriculture was adopted. The tests required at this time for plant protection products to become commercialized were limited. They mainly dealt with the biological efficacy of the product and acute toxicity tests on rats. No test was requested on the side-effects of the products on nontargeted insects, such as pollinator insects. Nevertheless, even at this time, the effects of biocides on man and animals, including game, fish and bees, were far from negligible [3].

It was not until 1956, following problems of poisoning in bee colonies observed by beekeepers, mainly on oil seed rape crop, that the public services set up specific regulatory provisions on the protection of honey bees and pollinator insects. The first toxicity tests on honey bees (mainly acute toxicity) were settled during the 1960s. In the 1970s the development of synthetic pyrethroids indicated the occurrence of additional side-effects. Damaging effects on the bees' behavior were suspected, but were difficult to characterize, underlining the need for new types of tests.

Since this period and especially over the past 15 years, legislation on the placing of chemical plant protection products on the market has been continuously developing to cope with European regulations. Directive No. 91/414/CEE of July 15 1991 [4], later transcribed into French law, defines the guidelines of the market approval procedure for chemical plant protection products. The provisions related to the effect on the environment and on beneficial organisms, in particular bees and other pollinators came to special importance.

After briefly outlining the different effects that chemical plant protection products can have on bees, we present the regulatory provisions that are currently applied in France to limit this impact. We then describe the European regulation to assess plant protection product risks on bees and the setting up of new procedures for market approval of chemical preparations in Europe. Finally, we mention the regulatory provisions applied in the United States.

Bees and plant protection products: a difficult interaction

The economic value of honey bees relies on the products derived from its activity (honey, pollen, royal jelly, etc.). Moreover, the honey bee plays a major role as a pollination agent. Crops of major importance, in fields or in greenhouses, benefit significantly from the activity of bees: fruit trees, oil crops (sunflower, oilseed rape), vegetable crops, seed production, etc. Although the benefits brought through pollination are difficult to put into figures, in 1982, Borneck and Bricout [5] estimated that entomophilous crops represented about 27 billion French francs, with 12 percent (i.e. 3 billion French francs) being attributed to pollinator insects. In addition, the honey bee significantly contributes to the maintenance and development of the biodiversity of ecosystems, notably in wild flora.

Bees can be contaminated by chemical plant protection products or other xenobiotics, either directly or indirectly, immediately or with delayed effects [6]. They can be poisoned directly during the spraying of these products in the blooming period through contact with the spray. They can also encounter product residues when foraging on the plants or by eating polluted nectar, honeydew, or pollen. In addition, the bee can contaminate the hive by bringing back polluted food, which will be stored and poison hive bees. In the case of growth regulator insecticides this effect will be particularly harmful to the larvae. Not only can the domestic honey bees be affected, but also bumble bees and other insects, such as the alfalfa leafcutting bee, which is important in the United States.

Over the past 15 years, an increasing number of studies have shown so-called sublethal effects in the laboratory and under controlled conditions: delayed mortality, "disappearing disease" [7], decreasing laying rates [8], effects on the viability of larvae, disturbances of flight, orientation, and communication [9], and so on. These effects are also strongly suspected by beekeepers under field conditions, without any formal proof since no objective means of measurement are yet available.

Although insecticides are the most often involved in damage caused to the pollinator insects, other plant protection products might also have negative effects. Thus, fungicides, notably when applied in combination with insecticides – the most well-known case being the mixture of deltamethrin and prochloraz, which in the 1980s was responsible for significant bee mortality [10] – can be responsible for poisoning. With regard to herbicides, they contribute to a decrease in pollinator insects' food sources and sometimes can have an insecticide action [6].

Plant protection products and regulatory provisions on honey bees and pollinator insects in France

As mentioned in the introduction, the first regulatory provisions which took into account the risk assessment of chemical plant protection products on bees started in 1956. They were progressively modified, leading to the publication of the July 5 1985 Decree [11], which modified the February 25 1975 Decree [12], and which is still in force.

The July 5 1985 Decree

This text specifies that the use of insecticides and acaricides attacks on all crops and forestry plantations is forbidden during the flowering period and the production of honeydew due to aphid attacks. Moreover, products known to be dangerous to bees should be labelled as being "dangerous to bees and other pollinator insects" (Table 3.1).

During the flowering period and the production of honeydew, only products with a special exemption can be used. This exemption is granted by the Ministry of Agriculture on the advice of the *Comité d'homologation* for a given utilization and treatment dose level. The *Comité d'homologation* bases its opinion on the recommendations of the "bee" section of the *Commission d'étude de la toxicité des produits antiparasitaires à usage agricole*. This advice is established after the examination of test results provided by agrochemical companies (the list of tests is given below). The exemption

Table 3.1 Extract from February 25 1975 Decree relating to the application of plant protection products for agricultural use [11]

Part II – Provisions relative to the protection of bees and other pollinating insects (July 5 1985 Decree)

Art. 8 § 1. With a view to protecting bees and other pollinating insects, treatments involving insecticides and acaricides are forbidden, whatever the product and application apparatus used, on all crops and forestry plantations visited by these insects during flowering and during the honeydew production period following attacks from aphids.

§ 2. As an exemption to this provision, only those insecticides and acaricides with the following wording: "Use authorized during flowering and honeydew secretion periods following aphid attacks, provided the doses, means of application and precautions set out in the sales authorization are respected" may be used during these periods. These particular wordings must appear on the packaging.

§ 3. Amongst other things, all insecticides and acaricides recognized as being dangerous to bees and other pollinating insects must carry the wording: "Product dangerous to bees and other pollinating insects."

§ 4. When melliferous plants are in flower under trees or in the middle of crops to be treated, they should be cut or uprooted before treatment.

appears on the label as "use authorized during the flowering and honeydew production periods following aphid attacks."

Tests required to obtain exemption

Companies requesting authorization to use their product during the flowering period submit to the relevant authorities a toxicological file for each preparation and for each requested use, allowing the toxicity of the product for bees to be assessed.

The main tests required are the following:

- Laboratory assessment of LD_{50} (at 24 and 48 hours) through contact or ingestion of the active ingredient and the formulated product; the official guideline being the CEB test No. 95 method of the *Commission des essais biologiques* (CEB) [13].
- Assessment of the effects of application of the product outdoors under a wire meshed tunnel; the official guideline in France being CEB test No. 129 [14].

Based on the results of these tests, insecticides and acaricides can be classified into three groups (Table 3.2). The preparations considered as dangerous (Group 1) and as presumed dangerous (Group 2) are forbidden during flowering and honeydew production periods. The Group 1 preparations are recognized as toxic and should be labeled "dangerous to bees." Only the Group 3 preparations can benefit from exemptions for their use during

Table 3.2 Bee toxicity groups used in France to classify insecticides and acaricides [27]

Group	Hazard group	Conditions of use	Wording on label
1	Product recognized as dangerous to bees	Not to be used on plants visited by bees or pollinating insects	Dangerous to bees and other pollinators
2	Product presumed dangerous to bees	Not to be used on plants visited by bees or pollinating insects	–
3	Formulations having received at least an exemption or a classification for a specific use	Authorized for use in situations benefiting from the classification	Utilization authorized during flowering or honeydew secretion periods following aphid attacks for a specific use provided the doses, means of application and precautions set out in the sales authorization are respected

flowering and honeydew production periods. However, Group 2 preparations, which are presumed dangerous and consequently are forbidden during flowering, have no indication on their labeling, which leads to a certain ambiguity.

Limitations in current regulations and problems encountered

Several problems have been encountered which indicate the limitations in the test methods being used and the difficulty of studying the effects of plant protection products on bees.

- In the laboratory, the LD_{50} of certain plant protection products can vary by a factor of 10 to 100, and even 1000, depending on the biological material used [15].
- Field tests hardly allow the assessment of the effects of certain molecules, shown to be toxic in the laboratory, in particular the synthetic pyrethroids. With this kind of product, used at very low doses per hectare (a few grams), mortality is almost never observed in or near the hive.
- Tunnel tests allow the estimation of the risk to bees with a product applied under conditions similar to agricultural practice. However, these tests are not often repeated because of the complexity of their set-up and their high cost. Replicates can be conducted over time but they are subjected to climate variations.
- The problems recently encountered in France with imidacloprid, a sys-

temic insecticide treatment applied on sunflower seeds which is suspected to induce sublethal effects on bees' behavior, have stressed the limitations of current test methods.

- The 1985 Decree sets out provisions relating to the use of products and only concerns insecticides and acaricides. Fungicides and herbicides, which represent a significant part of plant protection product sales and which may have also adverse effects on bees, are usually not tested.

European regulation to assess risks of plant protection products on bees

The European Directive No. 91/414/CEE of July 15 1991 concerning the registration of plant protection products sets out the conditions which must be fulfilled to obtain market authorization (Table 3.3). This approval is carried out in two steps. Approval for the active substance is given at the level of the European Commission. Authorizations for preparations remain at the national level. Simplified procedures are planned for the future to allow authorization to be extended from one country to another.

In France, the European Directive has been transferred into national law by different decrees and laws. The decree of September 6 1994 [16]

Table 3.3 Main headings of European Directive 91/414/EEC [4]

Annex I	Active substances authorized for incorporation in plant protection products
Annex II	Requirements for the file to be submitted by applicants for the inclusion of an active substance in Annex I
....	
	8. Ecotoxicological studies [28]
	8.3 Effects on bees
	8.3.1 Acute toxicity
	8.3.2 Bee brood feeding test
....	
Annex III	Requirements for the file to be submitted by applicants for the authorization of a plant protection product
....	
	10. Ecotoxicological studies
	10.4 Effects on bees
	10.4.1 Acute oral and contact toxicity
	10.4.2 Residue test
	10.4.3 Cage tests
	10.4.4 Field tests
	10.4.5 Tunnel tests
....	
Annex IV	Risk phrases
Annex V	Safety phrases
Annex VI	Uniform principles for the evaluation of plant protection products

defines the conditions to be fulfilled in constituting an approval request file for plant protection preparations. The Decree annexes list the tests required to register an active substance on a positive list and to ask for approval for the preparation. The tests required for bees are detailed in Part 8.3.1 of Annex I for the active substance and in Part 10.4 of Annex II for the preparation.

Different national and international institutions have established methods to assess the environmental risk of plant protection products. The EPPO (European and Mediterranean Plant Protection Organization) and the Council of Europe have worked to harmonize test methods for the environmental side-effects evaluation especially on bees. This work has led to the publication of a decision-making scheme for the environmental risk assessment of plant protection products [17] (Figure 3.1). This scheme is common to all the member countries in Europe, but the methods used are in fact determined by each country.

European regulation is based on the concept of risk which has replaced that of innocuousness, toxicity being an inherent property of any plant protection product [18]. Thus, the assessment of the risks involved with a plant protection product can only be done after laboratory and field tests, which lead to classifying the product into one of the following categories: high-, medium-, and low-risk products. Each stage of the procedure leads to either an assessment of the risk (and therefore classification in one of the risk categories previously mentioned) or to further tests. This assessment takes into account several factors such as the doses of the product used, its method of application, and the crop being treated.

Laboratory tests

Acute toxicity of the active substance

The LD_{50} values for oral or contact acute toxicity with the active substance are indicators of the potential impact on bees. Their measurement is compulsory for all plant protection products once bees are susceptible to being exposed to them. These tests are therefore not requested for products with particular uses not involving contact with bees (foodstorage in closed buildings, treatment of seeds or transplanted plants and bulbs, wound sealing and healing treatment, rodenticidal baits, use in greenhouses with no pollinating insects).

Acute toxicity of the commercial formulation

Determining LD_{50} values (after 48 hours) for a commercial formulation administered by contact or ingestion routes can be requested when the formulation contains several active substances or when its toxicity cannot be predicted from comparisons with those of similar formulations.

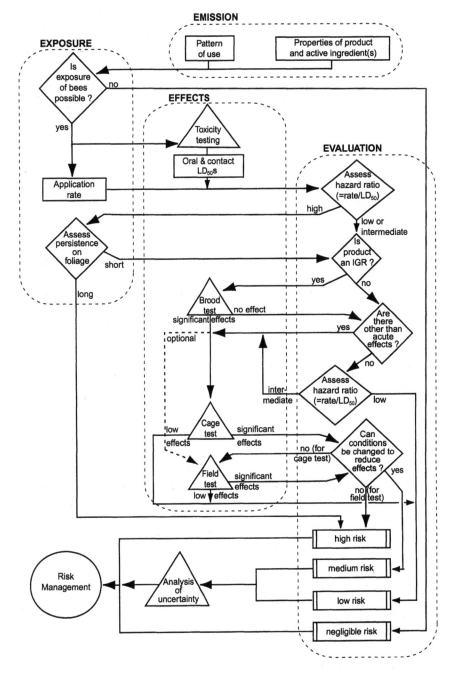

Figure 3.1 Decision-making scheme for the environmental risk assessment of plant protection products [17].

Determining hazard quotients

Establishing LD_{50} values allows us to determine the Hazard Quotients following oral or contact exposure (Q_{HO} and Q_{HC}) which are calculated from the LD_{50} values and the doses recommended for field treatment.

Calculation of the Hazard Quotient

Preparation containing only one active substance (a.s.)

Q_{HO} (µg a.s./bee) = Dose (g a.s./ha)/oral LD_{50} (µg a.s./bee)

Q_{HC} (µg a.s./bee) = Dose (g a.s./ha)/contact LD_{50} (µg a.s./bee)

Preparation containing several active substances

Q_{HO} (µg preparation/bee) = Dose (g preparation/ha)/oral LD_{50} (µg preparation/bee)

Q_{HC} (µg preparation/bee) = Dose (g preparation/ha)/contact LD_{50} (µg preparation/bee)

Dose: maximum application dose for which the authorization is requested, expressed as grams of active substance per hectare.

The value of the Hazard Quotient is critical for the next step of the assessment procedure. Different types of tests are requested when the value obtained for the Hazard Quotient is above a threshold value of 50.

The case of insect growth regulators

For these types of substances a test to evaluate the risk for bee brood is required, unless it has been shown that exposure is highly unlikely (the official method used is the ICPBR (International Commission for Plant–Bee Relationships) 1992 method [19]). On the basis of the results observed, additional cage or tunnel tests may be requested.

Test of residues on foliage

In addition to the other tests, this test can be requested when the risk is considered as being high (Q_{HO} or $Q_{HC} > 50$). It aims to provide information to assess the potential toxicity of product residues on crops for foraging bees. This test should allow us to establish the 50 percent Lethal Time (LT_{50}) expressed in hours. This value gives us an idea of the time period during which the residues remain toxic to bees.

Tests under controlled conditions and field tests

According to the EPPO, the most reliable risk assessment is based on data gathered under conditions which are most similar to agricultural practice by setting up field tests. However, owing to the difficulty and cost involved in setting up such tests, as far as possible they are replaced by laboratory or cage tests, with the results of the latter taken as being decisive.

Cage tests

These tests are carried out in accordance with the EPPO 170 Directive [20]. They should allow us to obtain enough information to assess the potential risks of the plant protection product on the survival and behavior of bees. They are requested when the risk is considered as high (Q_{HO} and $Q_{HC} > 50$, or in the case of significant effects on the brood or of suspected indirect effects: delayed effects, behavior modification, etc.).

Tunnel tests

These tests aim to assess the impact of contaminated flowers or honeydew on foraging bees. They can be requested when particular effects (toxicity for larvae, long-term delayed effects, disorientation of bees) have been observed during field tests. In France, the CEB 129 method is used.

Regulatory situation in the USA

Authorization for plant protection products is governed in general by the EPA (Environmental Protection Agency) and more precisely by the OPP (Office of Pesticide Programs). The market approval procedure is initially at the federal level and then at the level of each state. Thus, the company requesting approval first submits a request file at the federal level, and once federal approval has been granted, the company files in each state in which it intends to commercialize the product. Certain states may request additional tests, not previously requested by the federal authority.

The reference text which regulates approval in the United States is the FIFRA (Federal Insecticide, Fungicide and Rodenticide Act), which is equivalent to European Directive No. 91/414. This text, which presents the outlines for the market approval of plant protection products, was considerably amended in 1986 by the Food Quality Protection Act. To compile their approval registration file, companies must carry out ecotoxicity tests to assess effects on the environment and in particular on nontargeted organisms such as bees.

Tests on bees are requested for all products, whether they are insecticides, fungicides, herbicides, rodenticides, or disinfectants, etc., once they are to be used outdoors and bees may come into contact with them. Three official test methods, harmonized with those of the OECD [21, 22] and certified by the OPP, are available (OPPTS Nos 850.3020 [23], 850.3030 [24] and 850.3040 [25]).

Acute toxicity tests

If bees can come into contact with the plant protection products, the contact LD_{50} must be measured. Similarly, in the case of possible ingestion of the product by bees, the oral LD_{50} must be established. The method used is defined by OPPTS Guideline No. 850.3020.

The case of growth regulators

If the product has a growth regulating action, brood food tests are required.

On the basis of the values obtained for the LD_{50}, additional tests may be requested. Toxicity is measured on the Atkins scale [15] If the LD_{50} value is equal to or above $11\,\mu g$/bee, the product is considered as being almost nontoxic. For LD_{50} values between 2 and $11\,\mu g$/bee, the product is considered as moderately toxic. Below $2\,\mu g$/bee, the product is highly toxic.

Tests of residues on foliage

When the LD_{50} is below $11\,\mu g$/bee, additional tests to assess the toxicity of residues on foliage may be required. The recommended method is OPPTS No. 850.3030.

Field tests

It sometimes happens that field tests are requested, especially when the acute toxicity of the product is high, when the residues are highly toxic or when the toxicity for other nontargeted species is high or strongly suspected. These tests may be carried out on several economically important pollinating species: honey bees (*Apis mellifera*), alfalfa leafcutting bees (*Megachile rotundata*), and alkali bees (*Nomia melanderi*).

Based on the different test results, the EPA decides on the classification of the product (Table 3.4) and whether the label should carry an indication that the product must not be used on crops or weed flowers visited or foraged by bees.

Currently, the EPA is revising the wordings on labels concerning the protection of bees. According to the proposals under review, the label should, in addition, mention how long (in days or hours) the product residues are dangerous for bees.

Conclusion

Apart from the fact that the honey bee is a source of income, it is being considered increasingly as a bioindicator of the quality of the environment into which it is introduced [26].

Table 3.4 Honey bee toxicity groups and cautions in the US (EPA)

Toxicity group	Precautionary statement if extended residual toxicity is displayed	Precautionary statement if extended residual toxicity is not displayed
I Product contains any active ingredient with acute LD_{50} of less than $2\,\mu g$/bee.	This product is highly toxic to bees exposed to direct treatment or residues on blooming crops or weeds. Do not apply this product or allow it to drift to blooming crops or weeds if bees are visiting the treatment area.	This product is highly toxic to bees exposed to direct treatment on blooming crops or weeds. Do not apply this product or allow it to drift to blooming crops or weeds while bees are actively visiting the treatment area.
II Product contains any active ingredient(s) with acute LD_{50} of $2\,\mu g$/bee but less than $11\,\mu g$/bee.	This product is toxic to bees exposed to direct treatment or residues on blooming crops or weeds. Do not apply this product if bees are visiting the treatment area.	This product is toxic to bees exposed to direct treatment. Do not apply this product while bees are actively visiting the treatment area.
III All others	No bee caution required.	No bee caution required.

Over the past few years, regulations concerning the protection of bees have changed greatly, tending towards a reinforcement in protection. In Europe, the regulations, based on the fact that the potential toxicity is an inherent characteristic of the product, have introduced the notions of risk and nonintentional effects on the environment.

A strict procedure has been put into place. However, questions remain, particularly in relation to the reliability of the tests requested and their feasibility. The assessment of the effects of plant protection products is based on the calculation of the Hazard Quotient, which in turn is based on the LD_{50}. As the latter can vary to a significant degree, the validity of the Hazard Quotient may be questionable.

Concerning the test methods, some development is required. Questions encountered recently in France with imidacloprid, a systemic seed treatment applied to sunflowers, have shown the limitations of the existing methods. In fact, the suspected sublethal effects (changes in bee behavior) involve extremely low doses and are difficult to demonstrate with currently available methods.

Similarly, the problem of possible synergic effects, due to product mixtures, is not taken into account. Finally, another point which must be addressed concerns the cumulative effect of successive treatments.

54   *S. Cluzeau*

References

1 Bain, C., Bernard, J.L. and Fougeroux, A. (1995). *Protection des Cultures et Travail des Hommes.* Éd. Le Carrousel, p. 264.

2 Leufeuvre, J.C. (1985). Intensification de l'agriculture et usage des produits insecticides: évaluation des risques pour les équilibres biologiques, la faune et la santé humaine. In: *Coll. Mode d'Action et Utilisation des Insecticides, Angers.* Éd. ACTA, 19–22 nov. 1985, pp. 9–70.

3 Borneck, R. (1987). Rôle et protection des abeilles – Organisation de la profession apicole – Efforts internationaux. *La Défense des Végétaux,* **243–244**, 3–7.

4 EEC (1991). Council directive 91/414/EEC of 15 July 1991 concerning the placing of plant protection products on the market. *Off. J. Eur. Comm.* **230**, 1–32.

5 Borneck, R. and Bricout, J.P. (1984). Évaluation de l'incidence économique de l'entomofaune pollinisatrice. *Bull. Tech. Apic.* **47**, 117–124.

6 Taséi, J.N. (1996). Impact des pesticides sur les abeilles et les autres insectes pollinisateurs. *Courrier de l'environnement de l'INRA* **29**, 9–18.

7 Bendahou, N., Fléché, C. and Bounias, M. (1999). Biological and biochemical effects of chronic exposure to very low levels of dietary cypermethrin (Cymbush) on honeybee colonies (*Hymenoptera: Apidae*). *Ecotoxicol. Environ. Saf.* **44**, 147–153.

8 Lensing, W. (1986). Veränderungen bei den Arbeiterinnen der Honigbiene nach Fütterung von Dimethoat in subletalen Dosen. *Apidology* **17**, 339–342.

9 Vandame, R., Meled, M., Colin, M.E. and Belzunces, L.P. (1995). Alteration of the homing-flight in the bee *Apis mellifera* L. exposed to sublethal dose of deltamethrin. *Environ. Toxicol. Chem.* **14**, 855–860.

10 Colin, M.E. and Belzunces, L.P. (1992). Evidence of synergy between prochloraz and deltamethrin: A convenient biological approach. *Pestic. Sci.* **36**, 115–116.

11 Journal officiel de la République Française (1985). Arrêté du 5 juillet 1985 relatif à la délivrance d'autorisation d'emploi de produits antiparasitaires à usage agricole. *JO du 12 juil. 1985*, p. 7867.

12 Journal officiel de la République Française (1975). Arrêté du 25 février 1975 fixant les dispositions relatives à l'application des produits antiparasitaires à usage agricole. *JO du 7 mar. 1975*, p. 2563.

13 ANPP (1982). Méthode CEB no. 95 – Méthode de laboratoire d'évaluation de la toxicité aiguë orale et de contact des produits phytopharmaceutiques chez l'abeille domestique *Apis mellifera* L. Première édition, avril 1982, révision nov. 1996, p. 7.

14 ANPP (1989). Méthode CEB no. 129 – Méthode d'évaluation, sous tunnel en plein air, des effets à court terme des produits phytopharmaceutiques sur l'abeille domestique *Apis mellifera* L. Première édition, janvier 1989, révision nov. 1996, p. 12.

15 Atkins, E.L., Kellum, D. and Atkins, K.W. (1981). Reducing pesticide hazards to honey bees: mortality prediction techniques and integrated management strategies. University of California, Division of Agricultural Sciences. Leaflet, 883, p. 22.

16 Journal officiel de la République Française (1994). Arrêté du 6 septembre 1994 portant application du décret no. 94–359 du 5 mai 1994 relatif au contrôle des produits phytopharmaceutiques, *JO du 23 déc. 1994*.

17 EPPO (1999). Decision-making scheme for the environmental risk assessment of plant protection products. *Bull. OEPP/EPPO* **23**, 59–65.

18 Sauphanor, B., Belzunces, L. Brun, J., Chabert, A. and Colin, M.E. (1992). Produits phytosanitaires et organismes utiles, une approche européenne. *Phytoma-la Défense des Végétaux*, **445**, 13–22.

19 Oomen, P.A., De Ruijter, A. and Van Der Steen, J. (1992). Method for honeybee brood feeding tests with insect growth-regulating insecticides. *Bull. OEPP/EPPO* **22**, 613–616.

20 OEPP (1992). Guideline for the efficacy evaluation of plant protection products – Side-effects on honeybees – directive PP 1/170 (2). *Bull. OEPP/EPPO* **22**, 203–216.

21 OECD (1998). 213 Guidelines for the testing of chemicals – Honeybees, acute oral toxicity test. Adopted 21 September 1998, p. 8.

22 OECD (1998). 214 Guidelines for the testing of chemicals – Honeybees, acute contact toxicity test. Adopted 21 September 1998, p. 8.

23 US EPA (1996). Honey bee acute contact toxicity test (OPPTS 850.3020). Ecological effects test guidelines. EPA 712-C-96-147. Washington DC, USA.

24 US EPA (1996). Honey bee toxicity of residues on foliage test (OPPTS 850.3030). Ecological effects test guidelines. EPA 712-C-96-148. Washington DC, USA.

25 US EPA (1996). Field testing for pollinators (OPPTS 850.3040) Ecological effects test guidelines. EPA 712-C-96-150. Washington DC, USA.

26 Rivière, J.L. (1993). Les animaux sentinelles. *Courrier de l'environnement de l'INRA* **20**, 59–67.

27 ACTA (1990). Réglementation française des produits phytosanitaires, 4e édition, p. 448.

28 EEC (1996). Commission directive 96/12/EC of 8 March 1996 amending Council directive 91/414/EEC of 15 July 1991 concerning the placing of plant protection products on the market. Off. *J. Eur. Comm.* **65**, 20–37.

4 Acute toxicity of pesticides to honey bees

J. Devillers

Summary

Honey bees are beneficial arthropods playing a key role in pollinating wild and crop plants. Unfortunately, during their foraging activity they can be exposed to pesticides. The members of the colony can also be poisoned indirectly by contaminated food brought back to the hive by the foragers. The aim of this chapter is to discuss some aspects of the acute toxicity of pesticides to *Apis mellifera*.

Introduction

In many parts of the world, but especially in North America and western Europe, drastic changes in farming practice occurred in the 1950s, shortly after the Second World War, when tractors replaced horses, chemical fertilizers replaced organic manure, and aerial application of pesticides became commonplace. Many farmers were encouraged to stop crop rotation and to devote large acreages of their farms to the cultivation of a single crop which undoubtedly required the use of increasing quantities of synthetic fertilizers and pesticides [1, 2].

Broadly speaking, pesticides include many different chemical structures and are used to control pest plants and animals. They are generally classified according to their target as follows: insects (insecticides), nematodes (nematicides), mollusks (molluscicides), weeds (herbicides), bacteria (bactericides), fungi (fungicides), and so on.

Unfortunately, the successes of intensive agriculture have been bought at a price. The massive use of pesticides has induced resistances in target organisms, contamination of the ecosystems, and adverse effects on non-target species. Among this category of organisms, the honey bees occupy a place of choice due to their economic and ecological importance.

Pesticides act in two ways to reduce bee populations. First, many pesticides necessary in crop production are highly toxic to honey bees. In this category, we principally find insecticides. Second, the use of herbicides reduces the acreages of attractive plants for the bees to forage on. Pesticide damage to bee colonies takes many forms. Bees can be poisoned directly when they feed on nectar or pollen contaminated with certain pesticides. They can also be poisoned when they fly through a cloud of pesticide dust or spray, or walk on treated parts of a plant [3]. The other members of the colony can also be contaminated directly or indirectly by pesticides. Thus, if the pesticide is brought back to the hive by the foragers, the nurse bees are killed when they feed on contaminated honey or pollen and the brood will exhibit symptoms of neglect or poisoning. The queen can also be contaminated and the brood too.

It is obvious that the nature of the toxicological effects, the persistence, and more generally the environmental behavior of the pesticides depend on their physicochemical properties (e.g. water solubility, 1-octanol/water partition coefficient ($\log P$), vapor pressure) [4–9].

The aim of this chapter is to discuss some basic aspects of the acute toxicity of pesticides to the honey bee (*Apis mellifera*).

Experimental determination

The classical way for estimating the acute toxicity of chemicals is to determine their LD_{50} (i.e. the dose inducing 50 percent of mortality in the tested animals) or LC_{50} (i.e. the concentration inducing 50 percent of mortality in the tested animals) in laboratory tests. Official guidelines have been designed for estimating the above endpoints on the honey bee (see Chapter 3 and, e.g. [10]). In these tests, pesticides are administered by oral and contact routes in order to represent the different types of exposure under field conditions and the mortality is recorded after 24 or 48 hours of exposure. Indeed, absorption through the bee's integument is the basis for contact toxicity. The physicochemical properties of a pesticide and especially its formulation are largely responsible for the relative hazard from this mode of entry into the bee. Ingestion of contaminated pollen and nectar offers yet another route of entry. The alimentary tract may become altered or paralyzed, making feeding impossible, or the bee's gut may cease to function.

In these laboratory tests, dimethoate is generally used as reference compound due to the high acute toxicity of this organophosphorus insecticide against the honey bee. Thus, the contact LD_{50} values after 24 and 48 hours of exposure of technical dimethoate to workers are 0.162 and 0.152 µg active ingredient (a.i.)/bee, respectively [11]. An oral contamination yields LD_{50} values of 0.177 and 0.166 µg a.i./bee after 24 and 48 hours of exposure, respectively [11].

Laboratory tests offer the most convenient way for rapidly estimating the toxicity of pesticides to honey bees but they do not reflect the reality observed in the fields. Consequently, different methodologies have been developed to estimate the acute toxicity of pesticides to honey bees under more realistic environmental conditions (Chapter 3). Thus, in France, the tests performed under tunnels [12] and the use of free-flying bees being allowed to forage treated or untreated plants allow the contamination processes to be simulated in a simplified natural environment. The bee survival rate, the foraging activity before and after contamination, and the effects of various parameters on the colony development (e.g. population size, brood surface) can be recorded. It is important to note that the tunnel test is based only on a comparison of the results obtained with the treated and untreated assays. Thus, no toxic reference substance is used. In addition, because of the cost and complexity of this test, no simultaneous repetitions are made. Nevertheless, this test provides interesting results.

Effects on the different members of the colony

Honey bees are social insects, hence generally all the members of the colony will be affected directly or indirectly by pesticide contamination. Foraging bees are affected primarily by pesticides. Depending on the degree of toxicity, the bees can die in the field, on the flight back to the hive, or later at the hive entrance. The number of bees dying away from the hive is difficult to estimate, yielding false conclusions on the state of the hive. The useful aspect of bees dying away from the hive is that poisoned nectar and/or pollen are not brought into the hive, where they might be fed to the immature bees (brood) or stored in the combs. In all cases, it is noteworthy that if foraging bees are killed, then young bees are forced into the field earlier than normal, disrupting and thus disorienting the colony [13].

Hive bees are usually poisoned by contaminated nectar and pollen which are collected in the field, brought back and stored in the hive. When the hive bees die, the brood will show signs of neglect or poisoning and many, or all, immature bees still in the cells may die, especially the larvae which are very sensitive to pesticides. If a colony loses most of its foraging bees, nectar and pollen collection will be drastically reduced, but the population could recover in a few weeks. Conversely, if the foragers and hive bees are lost, the colony may never recover and will perish during the winter. As hive bees are killed, there are fewer bees to tend the brood and a further decline in population results [13].

The queen in the hive can also be contaminated by the pesticides and reduce egg laying, and hence will be killed by the workers.

More generally, it is important to note that honey bees can be subjected to seasonal variations in their susceptibility to pesticides due to changes in their physiology and ethology [14].

Type of pesticide and formulation

The toxicity of a specific pesticide is a composite of its physicochemical properties, the method of formulation, and the inherent ability of the honey bee to deal with the material internally.

Organophosphorus insecticides such as dimethoate, diazinon, malathion, fenitrothion, and parathion are the most toxic to bees. They inhibit the enzyme cholinesterase, which mediates the transmission of nerve signals. These chemicals induce problems of regurgitation (bees are wet), distended abdomen, disorientation, lethargy, paralysis, and so on [15]. Pyrethroids (e.g. permethrin, esfenvalerate) are also generally very toxic to honey bees through a neurotoxic action [1]. In the honey bee, they induce erratic movements, inability to fly, and stupefaction, followed very soon by paralysis, moribundity, and death [15]. The organochloride compounds (e.g. aldrin, DDT, chlordane, dieldrin, heptachlor, lindane, toxaphene) present different levels of toxicity against *A. mellifera*. They act as neuroactive agents on the transmission of nerve impulses, inducing erratic movements, abnormal activities, and trembling [15]. The carbamates (e.g. carbaryl, aminocarb, methomyl), like the organophosphorus compounds, also act on the cholinesterasic system. However, whereas the organophosphorus compounds work by phosphorylating the enzyme, the carbamates often seem to compete with acetylcholine for the enzyme surface [1]. The carbamates present different degrees of toxicity to honey bees. They are responsible for aggressive behaviors, erratic movements, inability to fly, paralysis, morbidity, and death [15].

Insecticides which are generally toxic to bees are not used as technical material but are used with different carriers. More generally, for optimizing their action, pesticides are used through different formulations. These formulations often vary considerably in their toxicity to bees. Thus, granular insecticides generally are not hazardous to honey bees. Dust formulations are typically more hazardous than emulsifiable concentrates because they adhere to the bee's body hairs and are carried back to the hive. Emulsifiable concentrates are less hazardous than wettable powders. Ultra-low-volume (ULV) formulations are usually more hazardous than other liquid formulations [13, 16]. Microencapsulated insecticides are much more toxic to honey bees than the other formulations. Because of their size, these capsules are carried back to the colony and there can remain toxic for long periods. Penncap-M®, containing methyl parathion, illustrates this problem perfectly [17].

Most herbicides and fungicides are not directly toxic to honey bees. In the same way, sex lures, attractants, and other hormones usually cause no acute toxicity problems for adult bees. Occasionally, a few honey bees and bumble bees have been found in traps containing Japanese beetle lures [18].

Relative acute toxicity of pesticides to honey bees based on laboratory and field tests

Most pesticides have been tested for their toxicity to honey bees (e.g. [19–21]). However, laboratory and field results generally do not coincide, due to peculiarities of bee behavior, abiotic factors, variations in the behavior of the pesticide studied, effects of different formulations, and so on. Atkins and collaborators [22], accounting for these factors, have classified the most commonly used pesticides according to their acute toxicity to honey bees estimated from both laboratory and field tests. Three different groups of toxicity were defined (Table 4.1). Group I deals with highly toxic pesticides ($LD_{50} < 2 \mu g/bee$). In this case, severe losses may be expected if they are used when bees are present at treatment time or within a day thereafter. This group is mainly characterized by the organophosphorus insecticides. Group II includes moderately toxic pesticides ($LD_{50} = 2$ to $11 \mu g/bee$). They can be used around bees if dosage, timing, and method of application are correct, but should not be applied directly on bees in the field or on the hives. Endrin, DDT, and chlordane are, for example, included in this group. Group III deals with relatively nontoxic pesticides ($LD_{50} > 11 \mu g/bee$) which can be used around bees with minimum injury. Most of the herbicides and fungicides belong to this last class.

Other factors affecting the acute toxicity of pesticides

The residual activity of an insecticide is an important factor in determining its safety to bees. An insecticide which degrades within a few hours can generally be applied with minimal risk when bees are not actively foraging. Conversely, applying insecticides with extended residual activity when bees are not foraging will not prevent bee injury if bees visit the crop during the period of residual activity [16]. The residual activity of a pesticide depends on its physicochemical properties and abiotic factors such as the temperature. If temperatures after a treatment are unusually low, insecticide residues can remain toxic to bees for much longer than if normal temperatures prevail. Conversely, it is well known that the toxicity of chemicals increases with temperature. When the target crop is not blooming or is unattractive to bees, insecticide drift can cause significant bee poisoning if it reaches adjacent flowering crops or weeds visited by bees [16]. This drift is dependent on the wind. Thus, it is obvious that abiotic factors widely affect the acute toxicity of pesticides to honey bees.

Similarly, different pesticides can be applied to a crop on which bees may be foraging. In addition, the environment is contaminated by a huge number of xenobiotics, most of them being undoubtedly toxic to bees. Consequently, the field honey bees are generally exposed to mixtures of chemicals instead of simple ones as in a classical laboratory test. Unfortunately, this can change the toxicity of the compound [23]. Thus, for

Table 4.1 Relative toxicity of pesticides to honey bees determined by laboratory and field tests (adapted from ref. 22)

Group I (Highly toxic: $LD_{50} < 2 \mu g$/bee)
Pesticides
Aldrin*
Ambush®* (permethrin**)
Arsenicals*
Avermectin®
Azodrin®* (monocrotophos)
Baygon®* (propoxur)
Baytex®* (fenthion)
Bidrin®* (dicrotophos)
Bux® (bufencarb)
Cygon®* (dimethoate)
Cythion®* (malathion)
Dasanit® (fensulfothion)
DDVP®* (dichlorvos)
Decis®* (decamethrin)
De-Fend®* (dimethoate)
Dibrom®* (naled)
Dieldrin*
Dimecron®* (phosphamidon)
Dursban®* (chlorpyrifos)
Ekamet® (etrimfos)
EPN*
Ethyl Guthion® (azinphos-ethyl)
Famophos® (famphur)
Ficam® (bendiocarb)
FMC-35001 (carbosulfan)*
Folithion® (fenitrothion)
Furadan®* (carbofuran)
Gardona®* (stirofos)
Guthion®* (azinphos-methyl)
Heptachlor*
Imidan®* (phosmet)
Lannate®* (methomyl)
Lorsban® (chlorpyrifos)
Malathion*
Matacil® (aminocarb)
Mesurol® (methiocarb)
Monitor®* (methamidophos)
Nemacur® (fenamiphos)
Nudrin®* (methomyl)
Orthene®* (acephate)
Parathion*
Pay-Off® (flucythrinate)
PennCap®* (methyl parathion)
Phosdrin®* (mevinphos)
Pounce®* (permethrin)
Pydrin®* (fenvalerate)
Sevin®* (carbaryl)
Spectracide®* (diazinon)
Sumithion® (fenitrothion)

Sumithrin® (*d*-phenothrin)
Supracide®* (methidathion)
Synthrin® (resmethrin)
Tamaron®* (methamidophos)
Temik®* (aldicarb)
Tepp*
Vapona®* (dichlorvos)

Group II (Moderately toxic: $LD_{50} = 2$ to 11 μg/bee)
Insecticides
Abate®* (temephos)
Agritox® (trichloronate)
Bolstar® (sulprofos)
Carzol®* (formetanate hydrochloride)
Chlordane*
Ciodrin® (crotoxyphos)
Counter® (terbufos)
Croneton® (ethiofencarb)
Curacron® (profenofos)
DDT*
Di-Syston®* (disulfoton)
Dyfonate® (fonofos)
Endrin*
Korlan® (ronnel)
Larvin®* (thiodicarb)
Metasystox-R®* (oxydemeton-methyl)
Mocap® (ethoprop)
Perthane® (ethylan)
Pyramat®
Sevimol®* (carbaryl)
Sevin® 4-oil* (carbaryl)
Systox®* (demeton)
Thimet®* (phorate)
Thiodan®* (endosulfan)
Trithion®* (carbophenothion)
Vydate®* (oxamyl)
Zolone® (phosalone)

Group III (Relatively nontoxic: $LD_{50} > 11 \mu g$/bee)
Insecticides and acaracides
Acaraben® (chlorobenzilate)
Altosid® (methoprene, insect growth regulator)
Baam® (amitraz)
Birlane® (chlorfenvinphos)
Comite® (propargite)
Delnav® (dioxathion)
Dessin® (dinobuton)
Dimilin® (diflubenzuron, insect growth regulator)

Table 4.1 Continued

Dylox®* (trichlorfon)
Ethion
Fundal® (chlordimeform)
Galecron® (chlordimeform)
Kelthane® (dicofol)
Kryocide® (cryolite)*
Marlate® (methoxychlor)*
Mavrik®* (fluvalinate)
Mitac® (amitraz)
Morestan® (oxythioquinox)
Morocide® (binapacryl)
Murvesco® (fenson)
Nicotine*
Omite® (propargite)
Pentac® (dienochlor)
Pirimor®* (pirimicarb)
Plictran®* (cyhexatin)
Pynamin® (allethrin)
Pyrethrum (natural)
Rotenone*
Sabadilla*
Sayfos® (menazon)
Sevin® SL* (carbaryl)
Sevin® XLR* (carbaryl)
Smite® (sodium azide)
Tedion® (tetradifon)
Tetram®
Tokuthion® (prothiofos)
Torak® (dialifor)
Toxaphene*
Zardex® (cycloprate)

Fungicides
Afugan®* (pyrazophos)
Arasan® (thiram)
Bayleton® (triadimefon)
Benlate® (benomyl)
Bordeaux mixture*
Bravo® (chlorothalonil)
Captan
Copper oxychloride sulfate
Copper 8-quinolinolate
Copper sulfate
Cuprex® (dodine)
Cupric oxide
Delan® (dithianon)
Dessin® (dinobuton)
Difolatan® (captafol)
Dithane® D-14 (nabam)
Dithane® M-22 (maneb)
Dithane® M-45 (manzeb)
Dithane® Z-78 (zineb)

Du-Ter® (fentin hydroxide)
Dyrene® (anilazine)
Ferbam
Glyodin
Hinosan® (edifenphos)
Indar® (butrizol)
Karathane® (dinocap)
Kocide® (cupric hydroxide)
Lesan® (fenaminosulf)
Morestan® (oxythioquinox)
Morocide® (binapacryl)
Mylone® (dazomet)
Phaltan® (folpet)
Plantvax® (oxycarboxin)
Polyram® (metiram)
Ridomil®
Sisthane® (fenapanil)
Smite® (sodium azide)
Sulfur*
Thyfural
Thylate® (thiram)
Vitavax® (carboxin)
Zerlate® (ziram)

Herbicides, defoliants, and desiccants
Aatrex® (atrazine)
Accelerate® (endothall, sodium salt)
Alachlor
Alanap® (naptalam)
Alopex® (clofop-isobutyl)
Amex® 820 (butralin)
Amiben® (chloramben)
Amitrole
Ammate® (AMS)
Aquathol K® (endothall, dipotassium)
Avenge® (difenzoquat)
Balan® (benefin)
Banvel® (dicamba)
Basagran® (bentazon)
Basalin® (fluchloralin)
Betanal® (phenmedipham)
Betanex® (desmedipham)
Bladex® (cyanazine)
Blazer® (acifluorfen)
Butachlor
Butam
Cacodylic acid
Cambilene® (TBA)
Caparol® (prometryn)
Casoron® (dichlobenil)
Chloro IPC® (chlorpropham)
Cotoran® (fluometuron)

Table 4.1 Continued

2,4-D*
DEF®
Desiccant L-10® (arsenic acid)
Devrinol® (napropamide)
Dichlorprop (2,4-DP)
Dinoseb (dinitrobutylphenol)
Diquat
Dual® (metolachlor)
Eptam® (EPTC)
Eradicane® (EPTC + safener)
Evik® (ametryn)
Evital® (norflurazon)
Folex® (merphos)
Garlon® (triclopyr)
Goal® (oxyfluorfen)
Hoelon® (diclofop-methyl)
Hydrothol 191® (endothall
   monopotassium salt)
Hyvar® (bromacil)
Igran® (terbutryn)
IPC® (propham)
Karmex® (diuron)
Kerb® (pronamide)
Lasso® (alachlor)
Lorox® (linuron)
Maloran® (chlorbromuron)
MCPA
Methar® (DSMA)
Milogard® (propazine)
Modown® (bifenox)
MSMA
Mylone® (dazomet)
Nortron® (ethofumesate)
Paarlan® (isopropalin)
Paraquat
Planavin® (nitralin)
Pramitol® (prometon)
Preforan® (fluorodifen)

Princep® (simazine)
Probe® (methazole)
Prowl® (pendimethalin)
Ramrod® (propachlor)
Randox® (CDAA)
Ronstar® (oxadiazon)
Roundup® (glyphosate)
Sancap® (dipropetryn)
Sencor® (metribuzin)
Silvex (2,4,5-TP)
Sinbar® (terbacil)
Smite® (sodium azide)
Surflan® (oryzalin)
Sutan® + (butylate)
2,4,5-T*
Telvar® (monuron)
Tenoran® (chloroxuron)
TOK® (nitrofen)
Tolban® (profluralin)
Tordon® (picloram)
Treflan® (trifluralin)
Turf Herbicide® (endothall, disodium)
Vegadex® (CDEC)
Zorial® (norflurazon)

**Nematicides and miscellaneous
   (e.g. plant growth regulators,
   nitrification inhibitors)**
Endothall
Exhalt® 800
Gibberellic acid
Mocap® (ethoprop)
Mylone® (dazomet)
N-Serve® (nitrapyrin)
Polaris® (glyphosine)
Smite® (sodium azide)
Sustar®

Notes
*Laboratory and field tested mainly on alfalfa, citrus, cotton, ladino clover, milo, and sweet corn; all other chemicals were laboratory tested only.
**Common name.

example, a synergistic toxicity has been found between deltamethrin, a type II pyrethroid insecticide, and prochloraz which is an imidazole fungicide [24]. In this case, summer bees seem to be eight times more susceptible than winter bees to the synergistic action of these two chemicals [14].

Concluding remarks

The estimation of the acute toxicity of pesticides against the honey bees is an important task due to the economic and ecological importance of these social insects. In this context, laboratory, semi-field, and/or field standardized tests can be used successfully (Chapter 3). The results obtained can be integrated into methodological schemes to better assess the impact of these chemicals on the terrestrial ecosystems (e.g. [25]). While these different acute toxicity tests provide interesting information, they suffer from limitations. Indeed, the results obtained can present a high degree of variability due to a lack of repeatability in the experiments. They do not satisfactorily represent natural field conditions of exposure. Thus, for example, it is obvious that in the environment, temperature, humidity, wind, and rainfall influence the behavior of the honey bee and interact with the contamination process by pesticides. Finally, while mortality is a crucial criterion for estimating the acute toxicity of pesticides to honey bees, it is not sufficient for estimating with accuracy the hazards of pesticides to these social insects. In fact, it has been shown that sublethal concentrations of pesticides can significantly alter the behaviors of the honey bees [26–31]. Consequently, these sublethal effects also have to be integrated in the methodological schemes for the assessment of the short-term effects of pesticides on *A. mellifera*.

References

1 Hassall, K.A. (1990). *The Biochemistry and Uses of Pesticides*. Second Edition, VCH, Weinheim, p. 536.
2 Wilson, W.T., Sonnet, P.E. and Stoner, A. Pesticides and honey bee mortality. http://maarec.cas.psu.edu/bkCD/Pollination/Pesticides_Mortality.html.
3 http://maarec.cas.psu.edu/bkCD/Pollination/Factors_to_kills.html.
4 Domine, D., Devillers, J., Chastrette, M. and Karcher, W. (1992). Multivariate structure–property relationships (MSPR) of pesticides. *Pestic. Sci.* **35**, 73–82.
5 Domine, D., Devillers, J., Chastrette, M. and Karcher, W. (1993). Estimating pesticide field half-lives from a backpropagation neural network. *SAR QSAR Environ. Res.* **1**, 211–219.
6 Bintein, S. and Devillers, J. (1996). Evaluating the environmental fate of atrazine in France. *Chemosphere* **32**, 2441–2456.
7 Bintein, S. and Devillers, J. (1996). Evaluating the environmental fate of lindane in France. *Chemosphere* **32**, 2427–2440.
8 Devillers, J. (2000). Prediction of toxicity of organophosphorus insecticides

against the midge, *Chironomus riparius*, via a QSAR neural network model integrating environmental variables. *Toxicol. Methods* **10**, 69–79.

9 Devillers, J. and Flatin, J. (2000). A general QSAR model for predicting the acute toxicity of pesticides to *Oncorhynchus mykiss*. *SAR QSAR Environ. Res.* **11**, 25–43.

10 ANPP (1996). *Méthode CEB no. 95. Méthode de Laboratoire d'Evaluation de la Toxicité Aiguë Orale et de Contact des Produits Phytopharmaceutiques chez l'Abeille Domestique Apis mellifera L.* First edition April 1982, revised version November 1996, p. 8.

11 Gough, H.J., McIndoe, E.C. and Lewis, G.B. (1994). The use of dimethoate as a reference compound in laboratory acute toxicity tests on honey bees (*Apis mellifera* L.) 1981–1992. *J. Apic. Res.* **33**, 119–125.

12 ANPP (1989) Méthode CEB no. 129. Méthode d'évaluation, sous tunnel en plein air, des effets à court terme des produits phytopharmaceutiques sur l'abeille domestique *Apis mellifera* L. Première édition janvier 1989, révision novembre 1996, p. 12.

13 Sanford, M.T. (1993). Protecting honey bees from pesticides. Circular 534, Florida Cooperative Extension Service, Institute of Food and Agricultural Sciences, University of Florida. April 25, 1993. http://edis.ifas.ufl.edu/scripts/htmlgen.exe?DOCUMENT_AA145.

14 Meled, M., Thrasyvoulou, A. and Belzunces, L.P. (1998). Seasonal variations in susceptibility of *Apis mellifera* to the synergistic action of prochloraz and deltamethrin. *Environ. Toxicol. Chem.* **17**, 2517–2520.

15 http://maarec.cas.psu.edu/bkCD/Pollination/Pesticide_Poisoning.html.

16 Ellis, M.D., Baxendale, F.P. and Keith, D.L. Protecting bees when using insecticides. http://ianrwww.unl.edu/pubs/insects/g1347.htm#t5.

17 http://maarec.cas.psu.edu/Penncap_M.html.

18 Tew, J.E. (1996) Protecting honey bees from pesticides. http://www.aces.edu/department/ipm/beesnpesticides.html.

19 Tomlin, C. (1995). *The Pesticide Manual. Incorporating the Agrochemicals Handbook.* Tenth Edition, Crop Protection Publications, BCPC/RSC, p. 1341.

20 Devillers, J. and Doré, J.C. (2000). *Etude Bibliographique des Effets Ecotoxicologiques des Xénobiotiques vis-à-vis de l'Abeille.* Contrat: "Programme Communautaire sur l'Apiculture 2000." p. 179.

21 http://ace.orst.edu/info/extonet/.

22 Atkins, E.L., Kellum, D. and Atkins, K.W. (1981). *Reducing Pesticide Hazards to Honey Bees: Mortality Prediction Techniques and Integrated Management Strategies.* Division of Agricultural Sciences, University of California, Leaflet 2883, p. 22.

23 Yang, R.S.Y. (1994). *Toxicology of Chemical Mixtures. Case Studies, Mechanisms, and Novel Approaches.* Academic Press, San Diego, p. 720.

24 Chalvet-Monfray, K., Auger, P., Belzunces, L.P., Fleché, C. and Sabatier, P. (1996). Synergy between deltamethrin and prochloraz in bees: Different mechanisms of action tested by modeling. *SAR QSAR Environ. Res.* **5**, 185–211.

25 Kovach, J., Petzoldt, C., Degni, J. and Tette, J. (2000). A method to measure the environmental impact of pesticides. http://www.nysaes.cornell.edu/ipmmet/ny/program_news/EIQ.html.

26 Schricker, B. and Stephen, W.P. (1970). The effect of sublethal doses of

parathion on honeybee behaviour: I. Oral administration and the communication dance. *J. Apic. Res.* **9**, 155–164.

27 MacKenzie, K.E. and Winston, M.L. (1989). Effects of sublethal exposure to diazinon and temporal division of labor in the honeybee. *J. Econ. Entomol.* **82**, 75–82.

28 MacKenzie, K.E. and Winston, M.L. (1989). The effects of sublethal exposure to diazinon, carbaryl and resmethrin on longevity and foraging in *Apis mellifera* L. *Apidology* **20**, 29–40.

29 Vandame, R., Meled, M., Colin, M.E. and Belzunces, L.P. (1995). Alteration of the homing-flight in the honey bee *Apis mellifera* L. exposed to sublethal dose of deltamethrin. *Environ. Toxicol. Chem.* **14**, 855–860.

30 Decourtye, A. (1998). *Etude des Effets Sublétaux de l'Imidachlopride et de l'Endosulfan sur l'Apprentissage Olfactif chez l'Abeille Domestique Apis mellifera L.* DEA "Neurobiologie des Processus de Communication et d'Intégration." Université Montpellier II, p. 18.

31 Pham-Delègue, M.H., Decourtye, A., Kaiser, L. and Devillers, J. (2002). Methods to assess the sublethal effects of pesticides on honey bees. *Apidologie* (in press).

5 The proboscis extension response

Assessing the sublethal effects of pesticides on the honey bee

A. Decourtye and
M.H. Pham-Delègue

Summary

The risk assessment of chemical pesticides on honey bees relies mainly on acute toxicity tests. Besides mortality, various aspects of the behavior of honey bees may be affected by sublethal doses of pesticides. Among the bees of a colony, foragers are the most likely to be exposed to chemicals. The foraging behavior is known to be based on a conditioning process, floral cues being associated with the food, memorized, and used for flower recognition during the following trips. The conditioning process occurring on the flower can be reproduced under laboratory conditions by using the olfactory conditioning of the proboscis extension response on restrained individuals. This bioassay has been adapted to screen the effects of various chemicals at sublethal concentrations. It allows threshold concentrations to be established above which a significant decrease in the olfactory learning abilities is observed. This method appears to be very promising for screening out pesticides, using a standard laboratory procedure. However, a wider range of compounds should be tested and the reliability of the assay still needs to be validated under more natural conditions before it can be proposed as a new method for regulatory guidelines.

Introduction

Among conventional pesticides, many neurotoxic compounds are used for crop protection against pest insects. These compounds target the nervous system and therefore affect insect behavior [1]. Whereas numerous studies have been conducted on the efficiency of such molecules on target pest insects, fewer studies have considered the potential effects on non-target organisms. Pollinating insects such as the honey bee (*Apis mellifera*) are especially exposed to chemicals when visiting melliferous plants. Special attention must be paid to their protection not only for their ecological importance by contributing to the maintenance of wild plant biodiversity but also for their economic value as honey producers and crop-pollinating agents [2]. Therefore, their potential exposure to pesticides in the field may adversely affect their effectiveness as pollinators by reducing their survival or modifying their behavior. Current methods for assessing the toxicity of pesticides to bees mainly involve the determination of mortality in acute toxicity tests, as described in the method CEB No. 95 [3]. The acute lethal concentration estimate (median lethal concentration, LC_{50}, i.e. the concentration that induces 50 percent death at short term) is the most common endpoint for measuring toxicity in the honey bee. However, the LC_{50} estimate is an incomplete measure of the negative effects because of the limited number of parameters examined (mortality) and the short duration of these tests (1 to 3 days in most cases). Such an estimate would only account for a situation where foragers are exposed to high-dose/short-term treatment. Nevertheless, hive worker bees may also be exposed to the chemicals since foragers collect potentially contaminated food to be stored inside the hive. As stored food originates from different plants, a dilution of toxic compounds occurs; however, they can be present in the hive at lower concentrations but for longer periods than on plants. Therefore, it is important to examine the effect of ecologically relevant sublethal exposure on various aspects of honey bees' behavior in order to develop robust assays mimicking realistic conditions. Such assays could be standardized and proposed for pesticide risk-assessment procedures. We discuss here the possibility of using a bioassay based on the conditioned proboscis extension response in restrained individuals for assessing the sublethal behavioral effects of insecticides on the honey bee.

Classical methods of assessing sublethal toxicity in the honey bee

Under natural conditions, the foraging behavior of bees relies on the learning of floral cues such as odor and color while visiting the flower [4], and on a communication process within the hive between foragers and newly recruited bees, by which distance, direction, and relative profitability of the food source are transmitted [5]. Studying the impact of sublethal doses of insecticides on the foragers is especially relevant since the foragers are directly exposed to pesticide applications in the field but may not die from the treatment, and may become the agents by which the whole colony can be contaminated when feeding on stored food. Furthermore, the foraging behavior involves a high functionality of sensory and integrative systems which can be the target of neurotoxic compounds in particular. The deleterious impact of pesticide spraying on the foraging activity and on the behavior of bees on the crop and around the hive, as well as on the brood rearing, is in fact, already taken into account, these being subject to official guidelines [6, 7]. These bioassays are developed under semi-field and field conditions (cage and tunnel tests, field trials) and mainly evaluate the repellent reaction after pesticide spraying on flowering crops, since it is expected that bees would avoid toxic substances. Although the approach is global, it provides information on potential specific abnormal behaviors. However, the identification of precise effects requires additional investigations using specific methods to make appropriate evaluations of the hazards. Thus, a method for evaluating the side-effects of plant protection products on a honey bee brood may be recommended, especially when products with insect growth-regulating properties are concerned [6]. Based on such methods, the long-term consumption of diflubenzuron or carbofuran was shown to have negative effects on brood rearing [8–10]. Also, Barker and Waller [11] found that methyl-parathion and parathion in water and sugar syrup produced deleterious sublethal effects on the brood production. Assays based on recording the longevity of the bees were also proposed to assess the sublethal effect of insecticides such as malathion and diazinon [12]. Together with pesticide treatment, honey bees' age (newly emerged *versus* older workers) and rearing conditions (small cage or hive) significantly affected workers' longevity. Thus, in newly emerged workers, carbaryl and resmethrin at sublethal doses can affect both longevity and the age at which the workers start to forage [13]. Sublethal effects can also be found on behavioral traits, such as a decrease in the foraging activity, a disruption in the communication process, or an alteration in the spatial orientation. An orally administered sublethal dose of parathion disrupted the communication of the food source direction by the foragers to the potentially recruited worker bees within the hive [14]. Under normal conditions, directional information on the food source is communicated to other bees by the angle at which the wagtail dance is performed relative to the vertical comb. After returning from a feeding station, the treated bees carried out a wagtail dance indicating the position of the source at a wrong angle. In fact, parathion prevented the foragers from making a translation from photomenotaxis (directed movement at an angle relative to light) to geomenotaxis (directed movement at an angle relative to gravity) [15]. A sublethal dose of parathion also disrupted the time sense and the wagtail dance rhythm of the foragers [14, 16, 17]. Honey bee foragers treated topically with a sublethal dose of permethrin exhibited a significantly higher percentage of time spent in self-cleaning and the trembling dance, and a lower percentage of time spent in walking, trophallaxy, and foraging, compared to untreated bees [18]. Moreover, most of the foraging bees that were treated with a sublethal dose of permethrin became so disoriented that they could not return to the hive. Another pyrethroid, deltamethrin, altered the homing flight in treated bees at sublethal doses [19]: in an insect-proof tunnel, the percentage of flights back to the hive decreased in treated foragers, the deltamethrin-treated bees flying in the direction of the sun, without using the local landmarks. The authors assumed that the disorientation was due to incorrect acquisition or integration of the visual patterns. This work indicates that toxic agents can have deleterious effects on sensory and integrative systems involved in the social communication and the spatial orientation of honey bees.

The conditioning proboscis extension assay

Principle

In the course of foraging a learning process occurs during which floral parameters such as location, shape, color, and smell of flowers are associated with a reward [4]. These floral cues are memorized by the forager and used for flower recognition during the following trips. Consequently, individual associative learning processes are important for the effective accomplishment of foraging activities. The associative learning of workers may therefore be regarded as having a high ecological significance because it is a prerequisite to the foraging success of the whole colony. Under laboratory conditions, learning and memory can be analyzed using a bioassay based on the olfactory conditioning of the proboscis extension (CPE) response on restrained individuals. This assay tentatively reproduces what happens in the honey bee–plant interaction: when landing on the flower, the forager extends its proboscis as a reflex when the gustatory receptors set on the tarsae, antennae, or mouthparts are stimulated with nectar. This reflex leads to the uptake of nectar and induces the memorization of the floral odors diffusing concomitantly. This response has been reproduced successfully under artificial conditions [20, 21], and has become a valuable tool for studying various aspects of the neurobiology of bees, including memory mechanisms and duration [22–25], neural bases of learning [26, 27], genetic variations in learning performances [28], and complex mixture recognition [29, 30]. Furthermore, the CPE procedure has given results well correlated with the responses of free-flying foragers under more natural conditions [30, 31]. This suggests that responses gained under controlled conditions may be transferred to more realistic situations.

These different considerations have led us to assume that this method would be useful to investigate the behavioral effects of toxicants in preference to more natural approaches such as studies in field or semi-field conditions because it allows better control of treatment and conditioning parameters. Indeed, precise quantification of behavior is essential for determining whether a specific non-environmental variable affects the normal behavior. The sublethal effects of chemical pesticides have already been studied using restrained workers in the CPE assay [32–35]. It remains to establish whether the use of the CPE response as a measure of the sublethal effects of chemicals on honey bees can be a reliable indicator of the hazards associated with the exposure to sublethal doses of toxic compounds, and consequently can be included in standard screening procedures of chemical pesticides. Furthermore, basic knowledge on the neural mechanisms of learning can be gained by using the CPE assay and analyzing the impairment of memory consecutive with the exposure to toxic compounds [26, 27].

The classical odor conditioning of the proboscis extension reflex, as described for example, by Bitterman *et al.* [22] and Sandoz *et al.* [25], is based on the temporal paired association of a Conditioned Stimulus (CS) and an Unconditioned Stimulus (US). During conditioning, the proboscis extension reflex is elicited by contacting the gustatory receptors of the antennae with a sucrose solution (US), an odor (CS) being delivered simultaneously (Figure 5.1). The proboscis extension is immediately rewarded (Reward R) by the uptake of the sucrose solution. Bees can develop the proboscis extension response as a Conditioned Response (CR) to the odor alone after even a single pairing of the odor with a sucrose reward.

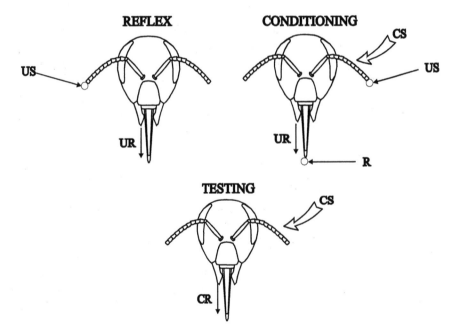

Figure 5.1 Conditioning proboscis extension (CPE) assay. The proboscis extension reflex (Unconditioned Response, UR) is elicited by contacting the antennae with a sugar solution (Unconditioned Stimulus, US). For the conditioning trials, this reflex is elicited during the delivery of odor stimulation (Conditioned Stimulus, CS). The honey bee is immediately rewarded by the uptake of sugar solution (Reward, R). During the testing trials, if the bee is properly conditioned, the delivery of the CS alone induces a conditioned proboscis extension response (Conditioned Response, CR).

Application to pesticide evaluation

Tested organisms

As in all tests involving behavioral responses, the CPE assay requires control treatments with rigorous uniformity of the testing environment. The influence of non-experimental variables should be taken into consideration in the development of the CPE assay to reduce variation and increase precision of measurement. In most studies using the CPE assay for pesticide toxicity assessment, the authors tested worker bees of unknown age [32–35]. However, experiments have proved the variability of olfactory learning performances in the CPE assay according to the age of the bees. Pham-Delègue *et al.* [36] have shown that bees between 12 and 18 days of age exhibited higher levels of conditioned responses than younger and older groups. Ray and Ferneyhough [37] found that younger workers until 10 days have lower performances than adult foragers. More recently, Laloi *et al.* [38] found that the performances of the youngest bees (2 days and 4 days old) significantly differed from those of older individuals. However, few studies have explored the variability of pesticide sensitivity according to the age of the bees. Only Delabie *et al.* [39] demonstrated that the sensitivity of the bees to cypermethrin increased with their age (LD_{50} of 2–6-day-old bees was 1.8 times that of 12–18-day-old bees). These studies indicate that it is necessary to standardize the age of the bees tested for both behavioral and toxicological reasons. Thus, we recommend the use of emerging worker bees collected on a comb of a sealed brood from a healthy, varroacide-untreated and queen–right colony. The bees should be maintained in groups (30–60 individuals) of homogeneous age and kept in an incubator (temperature: 33°C, relative humidity: 55 percent, in the dark) until an age of 14–15 days old. At this age worker bees generally become foragers under natural conditions [40] and give the most consistent performances in the CPE assay [36]. Bees are provided with sucrose solution and with fresh pollen during the first 8 days. Special attention must be paid to the origin of the food and its preservation. Wahl and Ulm [41] have shown that the degree of sensitivity of the worker bee to pesticides depends on its pollen diet in the first days of life, and a pollen feed varying in nutrient quality leads to the highest pesticide sensitivity. During bee rearing under laboratory conditions, the olfactory environment of the individuals must be strictly controlled in order to limit the early olfactory experience which can influence later learning performances in the CPE assay [42]. Also the subspecies of bees and the season of collection must be controlled, since the learning performances and the sensitivity to pesticides can be influenced by genetic and seasonal factors [24, 37, 41, 43, 44]. Consistently, using a CPE procedure, the No Observed Effect Concentration (NOEC) of imidacloprid on the learning performances was lower in summer bees than in winter bees, although these latter bees originated from hives maintained in a heated apiary (A. Decourtye, unpublished data). This study suggests that bees subjected to the CPE assay, following a subchronic treatment with imidacloprid at sublethal doses (1 to 48 ppb), have a higher sensitivity to the toxic material during summer than during winter. The physiological mechanisms underlying these variations in sensitivity are not yet known, but the use of worker bees collected preferentially in spring or summer is recommended.

Chemical treatment

The toxicant exposure can be carried out before, during, or after the CPE procedure. The pre-conditioning treatment leads to the determination of whether an insecticide exposure applied prior to a learning task may influence components of the learning process such as the acquisition and/or the recall of the learned response. In an ecological context this type of exposure corresponds to the case of bees newly involved in foraging duties based on their learning ability, after being fed contaminated food within the hive. Most studies have evaluated the impact of acute pre-conditioning exposure by using an instantaneous administration [34, 35] or 16 to 24 h exposure [32, 33]. Other authors [45, 46] have tested the effect of longer-term exposure to toxicants (11 to 12 days) in order to induce chronic intoxication. This is an attempt to simulate what young hive bees would experience when feeding on contaminated stored food, before becoming foragers, since it is commonly known that bees become foragers at an age of 15 days on average [40]. Long-term exposure to sublethal doses of chemicals may affect different physiological functions. When neurotoxic compounds are involved, the nervous system can be disrupted, the later foraging behavior therefore being affected. To elucidate the mechanisms underlying possible negative effects on learning, investigations have been conducted on the mode of chemical action and the targeted receptors of the nervous system [26, 27].

The toxic substance can also be delivered in the sucrose solution used as the reward during the CPE procedure [35]. These studies hypothesized that the contamination would occur while foragers collect the nectar and investigated the acute effects on the olfactory learning involved in the foraging activity. It assumes that foragers would react, on the one hand, to an antifeedant effect of the chemical associated with the food. The value of the reward being decreased, the paired CS/US–R association would be less efficient, leading to low learning performances. On the other hand, the chemical might be toxic enough to induce rapid disruption of nervous mechanisms, resulting in a rapid change in the learning abilities. The CPE assay would then be sensitive enough to detect such effects.

Complementarily, the products can be associated with the scent used as the CS to determine whether the insecticides have a repellent effect [35].

The results indicated that none of the insecticides tested (Endosulfan, Decis®, Baythroid®, Sevin®) was repellent when associated with the CS; that is the olfactory conditioning efficiency was not affected by the pure chemicals or by other volatile compounds potentially emitted by the insecticides. It is interesting to discuss this point since the potential repellent effect of chemicals may be useful to control the behavior of pollinating insects, by avoiding their visits during crop treatment when toxicity to pollinators is suspected. However, at least in a laboratory CPE test, it is unlikely that bees would be disturbed by changes in the olfactory quality of the CS, as long as it is associated with a satisfactory food reward. Only chemicals with high volatility and potential adverse effects on the peripheral olfactory receptors would produce a detectable effect in this assay.

Post-conditioning treatment to permethrin has been conducted, before subjecting the bees to the test trials, in order to study the recovery period needed for treated bees to resume normal learning ability [33]. This aimed to examine how chemical treatment can interfere with the memory process, which gives an indication of the way foragers will be able to come back to a crop where they have been exposed to the toxic material while they were collecting food and memorizing the floral cues.

The CPE assay also enables comparative studies of the responses to different chemical treatments to be carried out. Thus, Taylor et al. [32] have used the CPE assay to evaluate the learning performances of honey bees previously exposed to a range of six pyrethroid insecticides (fluvalinate, fenvalerate, permethrin, cypermethrin, cyfluthrin, flucythrinate). The treatment consisted of a 24-hour exposure in a Petri dish containing an insecticide-treated piece of filter paper at the LC_{50}. Pyrethroid-treated bees learned at a slower rate than untreated bees during the CPE assay. The conditioned responses were least affected by fluvalinate and most seriously affected by flucythrinate and cyfluthrin; permethrin, fenvalerate, and cypermethrin had intermediate effects. However, misinterpretation might arise from the use of concentrations derived from lethal concentration estimates to study sublethal effects. Thus, the exposure to fairly high concentrations of a toxic substance can result in a selection of worker bees staying alive because they are less sensitive to the pesticide tested. Such resistant bees may give responses in the CPE assay not representative of these of normal bees. Moreover, the use of LC_{50} seems to be not very realistic compared to concentrations potentially met in natural conditions. The use of sublethal concentrations can provide a better approximation of potential intoxication in the field. In addition, in most work using the CPE assay, the authors have tested only one concentration of insecticide. Thus, concentration–response relationships and the determination of threshold concentrations to specific chemicals are not established systematically. We consider this information as crucial to relate laboratory data and exposure under field conditions. Such an evaluation has been conducted by Decourtye et al. [46] who showed that honey bees surviving a subchronic treatment of endosulfan (tarsal contact exposure for 11 days in cages of 50–60 individuals) had reduced olfactory learning performances at 25 ppm treatment concentration and not at 5 ppm. After 11 days of oral treatment with imidacloprid or hydroxy-imidacloprid, one of the main imidacloprid metabolites [47], the NOEC for the conditioned responses in the CPE assay were established at 24 and 60 ppb, respectively [48]. However, the CPE responses may not be directly related to contaminant concentrations. For example, Decourtye *et al.* [49] observed reduced learning performances among bees exposed to deltamethrin at $LC_{50}/120$ dosage, while a higher concentration ($LC_{50}/24$) did not significantly reduce the learning performance. Nevertheless, these studies indicated that the CPE assay can enable the discrimination of different sublethal concentrations of chemicals inducing more or less graduate effects on the learning performances. Thus the establishment of threshold concentrations is important to evaluate the sensitivity of the bioassay and to define the no-effect concentrations in this assay. Although sublethal and more realistic concentrations have been used, the experiments mentioned previously referred to contact or ingestion treatment administered under artificial conditions where bees were forced to encounter the chemicals. These conditions can be considered as worst-case conditions, which do not reflect the natural conditions. Therefore, we were concerned about testing the CPE responses after more realistic exposure conditions in a standard crop protection agronomic system. Therefore, we designed an experiment under tunnels following the CEB No. 129 [50]: in one tunnel ($20 \times 8 \times 3.5$ m), four parcels of oilseed rape were treated with mix Decis® Micro-Sportak® 45 CE and in another tunnel the crop received only water treatment. Bees foraging on the crop were collected in both tunnels before the treatment, 1 hour after the treatment, and 1 day after. All bees were caged and subjected to the CPE assay. We found differences between the bees collected in treated and control tunnels, but further replicates are needed to confirm these data. These preliminary results (unpublished) indicate the possibility of subjecting the bees to the CPE assay after an exposure to chemical pesticides under agronomic conditions. This may be a means to validate this laboratory assay by establishing the responses of the bees in the CPE assay after an exposure under realistic conditions and comparing these responses to those obtained in the worst-case conditions. Also the range of concentrations tested in the laboratory would be compared to the doses used for crop treatment as well as to residue analysis. The value of this assay conducted under laboratory conditions to predict the effects of crop treatment would be better assessed, and experiments are in progress to provide data in this respect.

Behavioral measurements

The conditioned proboscis extension response involves gustatory, olfactory, and motor functions, as well as integrative processes underlying memory acquisition and recall of learned information. Therefore, depending on the physiological action of the xenobiotic, different behavioral parameters should be considered. In the standard CPE procedure [25] the responses are recorded during two successive phases: the acquisition phase where paired US–CS are presented, and the extinction phase where only the CS is delivered (Figure 5.2). Each phase comprises several trials lasting 6 s each, with about 15 min intertrial duration. During the acquisition period, the bees that did not initially respond to the CS (first trial C1), rapidly exhibit the conditioned response (CR), so that up to 80–100 percent of the tested individuals respond after one to five conditioning trials. No more trials are needed since after standard starving conditions (2–4 hours prior to testing), the motivation of the bees to get food would not overpass the fifth trial, the level of the CR then starting to decrease. Most often the level of CR reaches a maximum by the third trial. This acquisition phase relies on the memorization process, the learned information passing from the short-term memory to the long-term memory [51]. Then the conditioning trials are followed by testing trials during which the level of the CR slowly decreases down to the initial level of spontaneous response to the CS. This extinction process expresses the fact that bees

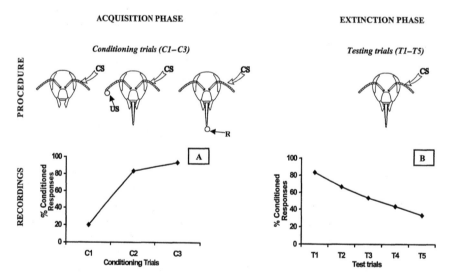

Figure 5.2 A model of the learning curve built into the CPE assay. During the acquisition phase, the level of the CR increased up to a maximum value at the third conditioning trial (A). This value is an indicator of the bee's ability to get conditioned properly, and can be compared according to the treatments. During the extinction phase, the level of the CR slowly decreased, back to the initial level of spontaneous response (B). This expresses the resistance of the bee's response to successive presentations of the unrewarded CS. Values in T1 and T5 are commonly used to compare responses of bees subjected to different treatments.

stop responding to the unrewarded odor stimulus, which has lost its predictive value of the occurrence of food delivery. However, this extinction of the CR does not necessarily mean that bees have forgotten the CS, since later presentation of the learned odor would again induce a high level of response [22]. Based on this general kinetic of responses, even slight modulations following chemical treatment are indicated.

The most commonly measured parameter is the level of conditioned responses during the acquisition phase of the CPE assay. Statistical comparisons of treated and untreated groups at the maximum value of the CR during the acquisition phase reveal sublethal effects of chemicals on the memorization of the CS. Honey bees exposed to pyrethroids at the LC_{50} exhibited maximum CR levels of 30–50 percent, while bees exposed only to acetone-treated filter paper (control) showed 90 percent responses [32]. With permethrin, a decrease in the CR level in bees surviving to one-fourth of the LD_{50} has been reported by Mamood and Waller [33]. After one acquisition trial 69 percent of the control bees gave a CR and 100 percent responded during the last conditioning trial, while 34 percent of the permethrin-contaminated bees gave a CR after the first conditioning trial and the responses slowly increased up to 81 percent CR at the last conditioning trial. Also honey bees surviving the dosage suggested on the manufacturer's label of dicofol had reduced CR in the CPE assay [34].

To evaluate the value of CPE responses as a routine measure for toxicity assessment, it is necessary to compare these responses to standard measures of toxicity such as mortality data, but few works have documented this point. Learning performances after contact treatment with endosulfan were decreased at 25 ppm, in contrast to the survival recordings which were not affected at the same concentration [46]. The NOEC of hydroxy-imidacloprid on the mortality was established at 120 ppb ($LC_{50}/120$) whereas the NOEC on the conditioned responses was established at 60 ppb ($LC_{50}/240$) [48]. On average, the differences between LC_{50} values and NOEC values on the conditioned responses was of a factor of 120–240 for endosulfan, imidacloprid, hydroxy-imidacloprid, and prochloraz [46, 48, 49]. From these studies it was found that differences between acute LC_{50} and NOEC for CPE responses were variable. Nevertheless, it is more often found that the NOEC values on the CPE responses are significantly lower than LC_{50} values determined by standard toxicity tests.

The CPE assay can involve associative and non-associative phenomena. The associative nature of proboscis extension reflex conditioning can be established by demonstrating that only forward pairing of CS–US sequences are effective to establish proper conditioning, compared to various control procedures, such as unpaired CS–US presentation [52]. The effects of an imidacloprid exposure can be shown not only on the bees' performances in an associative learning task [53] but also in a non-associative learning procedure such as habituation: imidacloprid at sublethal doses alters the number of trials needed to habituate the bees (i.e. extinguish the response) to repeated sucrose stimulation [54]. In the assessment of dicofol effects, parallel to a classical conditioning procedure, an unpaired conditioning procedure was conducted to ensure that any increase in the rate of proboscis extension responses was the result of associative processes and not of a non-associative process such as sensitization [34]. The unpaired conditioning procedure showed a high probability of obtaining proboscis extension responses after dicofol treatment, which indicated that the high learning response level in the classical conditioning procedure may be due to sensitization. Furthermore, a differential conditioning paradigm was used to evaluate whether the animals treated with dicofol can discriminate between two explicit conditioned stimuli (one odor associated with a reward and one odor not associated with a reward). In contrast to a classical conditioning procedure, the differential conditioning did not demonstrate differences between control and treated groups. It was suggested that the neurotoxic action of dicofol increased the value of the experimental design background signals that might serve as potential conditioned stimuli. Thus, in treated bees the need to "extract" the significant signal from the background stimuli would make the learning of a single conditioned stimuli more difficult than the discrimination between two CSs. These results clearly indicate task-dependent behavioral effects of sublethal concentrations of insecticides.

The extinction process, when the CS is delivered alone, can also be used to indicate potential effects of toxic compounds. The acquisition phase shows the ability of treated animals to learn the temporal relation between the US and the CS, whereas the extinction phase indicates their resistance to extinguish the response to a CS no longer associated with a reward. Ingestion of dicofol [34], endosulfan [35, 46], imidacloprid, and hydroxy-imidacloprid [48] significantly reduced the level of conditioned responses in both acquisition and extinction phases. By contrast, the response level was not reduced in bees conditioned prior to an exposure to permethrin [33]. Therefore, permethrin did not affect bees' ability to recall information previously learned. However, prior ingestion of prochloraz and deltamethrin–prochloraz in combination did not affect the CR level in the acquisition process but the decrease of response level in the extinction phase occurred more rapidly compared to the control group [49]. These studies show that acquisition and extinction are two independent processes that can be differentially affected by toxic exposure. This may rely on the fact that different steps of the memorization process are involved, the acquisition covering the information storage in the short-term memory, while long-term memory is already established when the extinction phase occurs, if we refer to the model of memory temporal schedule in the honey bee as described by Menzel and Greggers [51]. Some chemicals would affect the first step of information storage, others interfering with the memory already consolidated.

Another means to evaluate the effects of pesticides on bees' behavior is to measure their impact on the gustatory and motor functions of the proboscis extension reflex, prior to conditioning. This can be investigated by comparing the number of proboscis extension responses obtained when the antennae are contacted with a sucrose solution (unconditioned responses or reflex responses), in treated and control bees. Some works have documented the potential effects of chemicals on sensory-motor activity underlying the proboscis extension reflex [26]. Prior administration of permethrin induced deleterious effects on the conditioned responses but not on the reflex responses [33]. In contrast to conditioned responses, the reflex responses of bees were not affected by chronic exposure to imidacloprid with concentrations of 48 and 24 ppb [48]. This suggests that the exposure to the insecticides tested disrupted only the bee's ability to learn the odor–sucrose reward association and not the peripheral nervous system controlling the proboscis extension reflex.

Furthermore, the impairment of olfactory learning performances can result from the disruption of olfactory functions by a toxic substance, which can be shown using electroantennogram recordings (corresponding to the pooled responses of all the antennae neuroreceptors detecting the odor stimulus) [55]. Thus, the contact treatment with endosulfan at $LD_{50}/14$ has impaired the olfactory learning performances in a CPE assay and electroantennogram responses were decreased as well in the treated bees [46]. Considering the concomitant modifications in the learning capacity and in the olfactory sensitivity, it may be assumed that the decrease in antennae sensitivity after endosulfan treatment may be involved in the decrease of learning performances, although the neural processes have not yet been identified.

Conclusion

Measurements of behavioral endpoints in honey bees should provide an effective assessment of hazards caused by crop protection chemicals especially when applied to melliferous plants. Under laboratory conditions, the conditioned proboscis extension (CPE) assay provides detectable sublethal effects due to pesticides, and also to gene products potentially used in plant genetic engineering (see other chapters of this book). Impairment in olfactory learning abilities have been shown for chemical concentrations at which no additional mortality occurred. Thus, the use of the CPE assay as a method to evaluate the potential effect on the honey bees' foraging behavior can help to assess the toxicity of chemicals in a more comprehensive way than by considering the mortality endpoint alone. The CPE procedure can be used to compare responses to different chemicals (Table 5.1) and to different concentrations of the same chemical, and to determine the no-effect concentrations. However, the CPE assay does not always show clear dose-related responses. In summary, CPE responses seem to be valid indicators of sublethal toxicity in honey bee. This assay can also be used to carry out investigations on the nervous circuitry underlying the olfactory learning processes, when neurotoxic molecules that affect peripheral or central nervous system are used. The CPE recordings are applicable to various races (*Apis mellifera ligustica*, *Apis mellifera capensis*, Africanized honey bees) of honey bees [25, 28, 35], and even to bumble bees [56]. Moreover, this method is simple to carry out, easily standardized, and needs low-cost stimulation and recording devices. As with other ecotoxicological endpoints, the extrapolation of behavioral responses gained in the CPE assay to colony and field conditions remains questionable. However, preliminary studies indicate that the decrease in learning performances induced by imidacloprid observed at the individual level in the CPE assay was confirmed at the colony level in an olfactory discrimination task [53]. Moreover, the sublethal effects of imidacloprid on the CPE responses can be related to a reduction in the foraging activity and to changes in the dancing behavior, when sucrose solution containing imidacloprid at a concentration higher than 20 ppb was fed to forager bees [57]. Thus, the CPE assay can also predict effects that might occur in the field. But further work is needed to establish a better correlation between the behavioral responses observed under laboratory conditions and those

Table 5.1 Pesticides tested in the CPE assay as cited in the text

Pesticide	Chemical class	Major target sites	Ref.
Cyfluthrin Flucythrinate Permethrin Fenvalerate Cypermethrin Fluvalinate	Pyrethroid[1]	Voltage-gated sodium channel	[32]
Permethrin	Pyrethroid[1]	Voltage-gated sodium channel	[33]
Dicofol	Chlorinated hydrocarbon[2]	Octopamine	[34]
Endosulfan	Organochlorine[1]	GABA receptor	
Carbaryl (Sevin®)	Carbamate[1]	Acetylcholinesterase	[35]
Deltamethrin (Decis®)	Pyrethroid[1]	Voltage-gated sodium channel	
Cyfluthrin (Baythroid®)	Pyrethroid[1]	Voltage-gated sodium channel	
Deltamethrin	Pyrethroid[1]	Voltage-gated sodium channel	
Prochloraz	Imidazole[3]	Cytochrome P-450	[49]
Mix deltamethrin-prochloraz	Pyrethroid[1] and imidazole[3]	Voltage-gated sodium channel and cytochrome P-450	
Imidacloprid	Chloronicotinyl[1]	Nicotinic acetylcholine receptor	[48]
Endosulfan	Organochlorine[1]	GABA receptor	
Imidacloprid	Chloronicotinyl[1]	Nicotinic acetylcholine receptor	[46]
OH-imidacloprid Olefin	Metabolites of imidacloprid		

Notes
1 Insecticide.
2 Insecticide–acaricide.
3 Fungicide.

observed in field studies. Nevertheless, the CPE assay can be considered as a quantifiable and reliable method to assess sublethal toxicity, and could be easily incorporated into test protocols to expand the range of existing toxicity tests.

References

1 Haynes, K.F. (1988). Sublethal effects of neurotoxic insecticides on insect behavior. *Annu. Rev. Entomol.* **33**, 149–168.

2 Williams, R.R. (1970). Factors affecting pollination in fruit trees. In: *Physiology of Tree Crops* (Luckwill, L.C. and Cutting, C.V., Eds). Academic Press, London, pp. 193–207.

3 CEB (1996). Méthode de laboratoire d'évaluation de la toxicité aiguë orale et de contact des produits phytopharmaceutiques chez l'abeille domestique *Apis mellifera* L. Méthode no. 95. ANPP, Paris, p. 8.

4 Menzel, R. and Müller, U. (1996). Learning and memory in honeybees: From behavior to neural substrates. *Annu. Rev. Neurosci.* **19**, 379–404.

5 von Frisch, K. (1967). *The Dance Language and Orientation of Bees*. Harvard University Press, Cambridge, p. 566.

6 OEPP/EPPO (1993). Guideline on test methods for evaluating the side effects of plant protection products on honeybees. *Bull. OEPP/EPPO* **22**, 203–215.

7 ICPBR (2000). Guideline for the efficacity evaluation of plant protection products. In: *Hazards of Pesticides to Bees* (Pélissier, C. and Belzunces L.P., Eds). IOBC wprs Bulletin, Avignon, pp. 51–55.

8 Johansen, C.A. (1977). Pesticides and pollinators. *Annu. Rev. Entomol.* **22**, 177–192.

9 Stoner, A. and Wilson, W.T. (1982). Diflubenzuron (dimilin): Effect of long-term feeding of low doses of sugar-cake or sucrose syrup on honey bees in standard-size field colonies. *Am. Bee J.* **122**, 579–582.

10 Stoner, A., Wilson, W.T. and Rhodes, H.A. (1982). Carbofuran: Effect of long-term feeding of low doses of sucrose syrup on honey bees in standard-size field colonies. *Environ. Entomol.* **11**, 53–59.

11 Barker, R.J. and Waller, G.D. (1978). Sublethal effects of parathion, methyl parathion, or formulated methoprene fed to colonies of honey bees. *Environ. Entomol.* **7**, 569–571.

12 Smirle, M.J., Winston, M.L. and Woodward, K.L. (1984). Development of a sensitive bioassay for evaluating sublethal pesticides effects on the honey bee (Hymenoptera: Apoidea). *J. Econ. Entomol.* **77**, 63–67.

13 MacKenzie, K.E. and Winston, M.L. (1989). The effects of sublethal exposure to diazinon, carbaryl and resmethrin on longevity and foraging in *Apis mellifera* L. *Apidologie* **20**, 29–40.

14 Schricker, B. and Stephen, W.P. (1970). The effects of sublethal doses of parathion on honeybee behaviour. I. Oral administration and the communication dance. *J. Apic. Res.* **9**, 141–153.

15 Stephen, W.P. and Schricker, B. (1970). The effect of sublethal doses of parathion. II. Site of parathion activity, and signal integration. *J. Apic. Res.* **9**, 155–164.

16 Schricker, B. (1974). Der Einfluss Subletaler Dosen von Parathion (E 605) auf das Zeitgedächtnis der Honigbiene. *Apidologie* **5**, 385–398.

17 Schricker, B. (1974). Der Einfluss Subletaler Dosen von Parathion (E 605) auf die Entfernungsweisung bei der Honigbiene. *Apidologie* **5**, 149–175.

18 Cox, R. and Wilson, W.T. (1987). The behavior of insecticide-exposed honey bees. *Am. Bee J.* 118–119.

19 Vandame, R., Meled, M., Colin, M.E. and Belzunces, L.P. (1995). Alteration of the homing-flight in the honey bee *Apis mellifera* L. exposed to sublethal dose of deltamethrin. *Environ. Toxicol. Chem.* **14**, 855–860.

20 Frings, H. (1944). The loci of olfactory end-organs in the honey-bee, *Apis mellifera* Linn. *J. Exp. Zool.* **97**, 123–134.

21 Takeda, K. (1961). Classical conditioned response in the honey bee. *J. Insect Physiol.* **6**, 168–179.

22 Bitterman, M.E., Menzel, R., Fietz, A. and Schäfer, S. (1983). Classical conditioning of proboscis extension in honeybees (*Apis mellifera*). *J. Comp. Psychol.* **97**, 107–119.

23 Menzel, R., Greggers, U. and Hammer, M. (1993). Functional organization of appetitive learning and memory in a generalist pollinator, the honey bee. In: *Insect Learning* (Papaj, E.D.R. and Lewis, A.C., Eds). Chapman Hall, New York, pp. 79–125.

24 Bhagavan, S., Benatar, S., Cobey, S. and Smith, B.H. (1994). Effect of genotype but not of age or caste on olfactory learning performance in the honey bee, *Apis mellifera*. *Anim. Behav.* **48**, 1357–1369.

25 Sandoz, J., Roger, B. and Pham-Delègue, M.H. (1995). Olfactory learning and memory in the honeybee: Comparison of different classical conditioning procedures of the proboscis extension response. *C. R. Acad. Sci. Paris, Sci. Vie* **318**, 749–755.

26 Cano-Lozano, V., Bonnard, E., Gauthier, M. and Richard, D. (1996). Mecamylamine-induced impairment of acquisition and retrieval of olfactory conditioning in the honeybee. *Behav. Brain Res.* **81**, 215–222.

27 Cano-Lozano, V. and Gauthier, M. (1998). Effects of muscarinic antagonists atropine and pirenzepine on olfactory conditioning in the honeybee. *Pharmacol. Biochem. Behav.* **59**, 903–907.

28 Brandes, C. (1988). Estimation of heritability of learning behavior in honey bees (*Apis mellifera capensis*). *Behav. Genet.* **18**, 119–132.

29 Le Metayer, M., Marion-Poll, F., Sandoz, J.C., Pham-Delègue, M.H., Blight, M.M. Wadhams, L.J., Masson, C. and Woodcock, C.M. (1997). Effect of conditioning on discrimination of oilseed rape volatiles by the honey bee: use of a combined gas chromatography-proboscis extension behavioural assay. *Chem. Senses* **22**, 391–398.

30 Laloi, D., Bailez, O., Blight, M.M., Roger, B., Pham-Delègue, M.H. and Wadhams, L. (2000). Recognition of complex odors by restrained and free-flying honey bees, *Apis mellifera*. *J. Chem. Ecol.* **26**, 2307–2319.

31 Mauelshagen, J. and Greggers, U. (1993). Experimental access to associative learning in honeybees. *Apidologie* **24**, 249–266.

32 Taylor, K.S., Waller, G.D. and Crowder, L.A. (1987). Impairment of classical conditioned response of the honey bee (*Apis mellifera* L.) by sublethal doses of synthetic pyrethroid insecticides. *Apidologie* **18**, 243–252.

33 Mamood, A.N. and Waller, G.D. (1990). Recovery of learning responses by

honeybees follows a sublethal exposure to permethrin. *Physiol. Entomol.* **15**, 55–60.

34 Stone, J.C., Abramson, C.I. and Price, J.M. (1997). Task-dependent effects of dicofol (kelthane) on learning in the honey bee (*Apis mellifera*). *Bull. Environ. Contam. Toxicol.* **58**, 177–183.

35 Abramson, C.I., Aquino, I.S., Ramalho, F.S. and Price, J.M. (1999). The effect of insecticides on learning in the Africanized honey bee (*Apis mellifera* L.). *Arch. Environ. Contam. Toxicol.* **37**, 529–535.

36 Pham-Delègue, M.H., De Jong, R. and Masson, C. (1990). Effet de l'âge sur la réponse conditionnée d'extension du proboscis chez l'abeille domestique. *C. R. Acad. Sci. Paris* **310**, Série III, 527–532.

37 Ray, S. and Ferneyhough, B. (1997). The effects of age on olfactory learning and memory in the honey bee *Apis mellifera*. *NeuroReport* **8**, 789–793.

38 Laloi, D., Gallois, M. and Pham-Delègue, M.H. (1999). Etude comparée des performances d'apprentissage olfactif chez des ouvrières, des reines et des mâles d'abeille. *Actes Colloq. Insectes Soc.*, Tours, France, pp. 95–104.

39 Delabie, J., Bos, C., Fonta, C. and Masson, C. (1985). Toxic and repellent effects of cypermethrin on the honeybee: Laboratory, glasshouse and field experiments. *Pestic. Sci.* **16**, 409–415.

40 Seeley, T.D. (1982). Adaptative significance of the age polyethism schedule in honeybee colonies. *Behav. Ecol. Sociobiol.* **11**, 287–293.

41 Wahl, O. and Ulm, K. (1983). Influence of pollen feeding and physiological condition on pesticide sensitivity of the honey bee *Apis mellifera carnica*. *Oecologia* **59**, 106–128.

42 Sandoz, J.C., Laloi, D., Odoux, J.F. and Pham-Delègue, M.H. (2000). Olfactory information transfer in the honeybee: Compared efficiency of classical conditioning and early exposure. *Anim. Behav.* **59**, 1025–1034.

43 Brandes, C. and Menzel, R. (1990). Common mechanisms in proboscis extension conditioning and visual learning revealed by genetic selection in honey bees (*Apis mellifera capensis*). *J. Comp. Physiol. A.* **166**, 545–552.

44 Suchail, S., Guez, D. and Belzunces, L.P. (2000). Characteristics of imidacloprid toxicity in two *Apis mellifera* subspecies. *Environ. Toxicol. Chem.* **19**, 1901–1905.

45 Pham-Delègue, M.H., Girard, C., Le Métayer, M., Picard-Nizou, A.L., Hennequet, C., Pons, O. and Jouanin, L. (2000). Long-term effects of soybean protease inhibitors on digestive enzymes, survival and learning abilities of honeybees. *Entomol. Exp. Appl.* **95**, 21–29.

46 Decourtye, A., Le Métayer, M., Renou, M. and Pham-Delègue, M.H. (2000). Effets de doses sublétales de pesticides sur le comportement de l'abeille domestique *Apis mellifera* L. *Actes Colloq. Insectes Soc.*, Tours France, pp. 105–113.

47 Nauen, R., Tietjen, K., Wagner, K. and Elbert, A. (1998). Efficacy of plant metabolites of imidacloprid against *Myzus persicae* and *Aphis gossypii* (Homoptera: Aphididae). *Pestic. Sci.* **52**, 53–57.

48 Decourtye, A., Genecque, E., Marsault, D., Charreton, M. and Pham-Delègue, M.H. (2000). Impact de l'imidaclopride, et de ses deux principaux métabolites, sur l'apprentissage olfactif chez l'abeille domestique *Apis mellifera* L. *13ème Colloq. Physiol. Insecte*, Versailles, France, p. 46.

49 Decourtye, A., Roger, B., Odoux, J.F. and Pham-Delègue, M.H. (2000). Les effets de pesticides sur l'apprentissage olfactif chez l'abeille domestique *Apis mellifera* L. *15ème Conf. Int. Tournesol*, Toulouse, France, pp. 11–17.

50 CEB (1996). Méthode d'évaluation, sous tunnel en plein air, des effets à court terme des produits phytopharmaceutiques sur l'abeille domestique *Apis mellifera* L. Méthode no. 129. ANPP, Paris, p. 12.

51 Menzel, R. and Greggers, U. (1992). Temporal dynamics and foraging behaviour in honeybees. In: *Biology and Evolution of Social Insects* (Billen, J., Ed.). Leuven University Press, Leuven, pp. 303–318.

52 Menzel, R. (1993). Associative learning in honey bees. *Apidologie* **24**, 157–168.

53 Decourtye, A., Le Métayer, M., Pottiau, H., Tisseur, M., Odoux, J.F. and Pham-Delègue, M.H. (2000). Impairment of olfactory learning performances in the honey bee after long term ingestion of imidacloprid. In: *Hazards of Pesticides to Bees* (Pélissier, C. and Belzunces L.P., Eds). IOBC wprs Bulletin, Avignon, p. 33.

54 Guez, D., Suchail, S., Maleszka, R., Gauthier, M. and Belzunces, L.P. (2000). Sublethal effects of imidacloprid on learning and memory in honeybees. In: *Hazards of Pesticides to Bees* (Pélissier, C. and Belzunces L.P., Eds). IOBC wprs Bulletin, Avignon, pp. 5.

55 Patte, F., Etcheto, M., Marfaing, P. and Laffort, P. (1989). Electroantennogram stimulus–response curves for 59 odourants in the honey bee *Apis mellifera. J. Insect Physiol.* **35**, 667–675.

56 Laloi, D., Sandoz, J.C., Picard-Nizou, A.L., Marchesi, A., Pouvreau, A., Taséi, J.N., Poppy, G. and Pham-Delègue M.H. (1999). Olfactory conditioning of the proboscis extension in bumble bees. *Entomol. Exp. Appl.* **90**, 123–129.

57 Kirchner, W.H. (1999). Mad-bee-disease? Sublethal effects of imidacloprid ("Gaucho") on the behavior of honey-bees. *Apidologie* **30**, 422.

6 Effects of imidacloprid on the neural processes of memory in honey bees

C. Armengaud, M. Lambin, and M. Gauthier

Summary

The cholinergic system in insects is the main target of insecticides. One class of molecules, the neonicotinoids, induces direct activation of the neuronal nicotinic acetylcholine receptors (nAChRs). In the honey bee these receptors are mainly distributed in the olfactory pathways that link sensory neurons to antennal lobes and mushroom bodies. These structures seem to play an important role in olfactory conditioning. We have previously shown that cholinergic antagonists injected in different parts of the brain impaired the formation and retrieval of olfactory memory. We then advanced the hypothesis that, through the activation of the nAChR, the neonicotinoid imidacloprid (IMI) would lead to facilitation of the memory trace.

To test this hypothesis, IMI was applied topically upon the thorax and the effects were tested on the habituation of the proboscis extension reflex induced by repeated sugar stimulation of the antennae. Animals treated with IMI to a dose that did not affect sensory or motor functions needed fewer trials than nontreated animals to show a reflex inhibition. This effect can be interpreted as a learning facilitation.

We developed a functional histochemistry of cytochrome oxidase (CO) to reveal the brain targets of the drug in the honey bee brain. Following IMI injection, a CO staining increase, probably linked to an increase in metabolism, was observed in the antennal lobes. In integrative structures, in particular the calyces of mushroom bodies, IMI exerted a facilitatory or inhibitory effect on neuronal metabolism depending on the dose. The brain targets of nicotinic ligands, including pesticides, can be compared by using this technique.

Introduction

Two of the three main classes of insecticides exert their neurotoxic effects through action on the cholinergic system. This is the case for the new class of neonicotinoids, which are known to act on the nicotinic acetylcholine

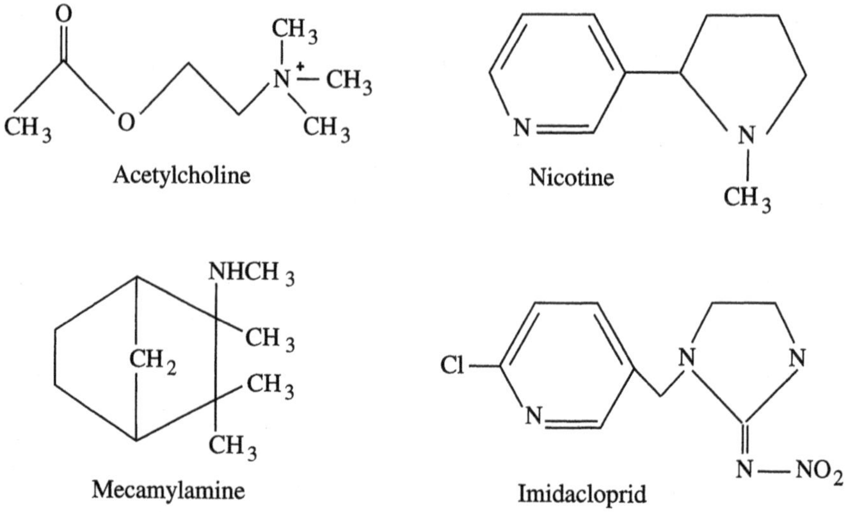

Figure 6.1 Chemical structures of acetylcholine and nicotinic cholinoceptor ligands used in this study.

receptor (nAChR) channel. Imidacloprid (IMI) {1–[(6-chloro-3-pyridinyl)methyl]-4,5-dihydro-N-nitro-1H-imidazol-2-amine} is one of these new molecules (Figure 6.1). According to the literature, IMI in insects acts at three pharmacologically distinct acetylcholine receptor sub-types inducing a dose-dependent depolarization [1]. Other electrophysio-logical effects of IMI have been described in different models. The single patch-clamp technique applied to the rat pheochromocytoma (PC 12) cells showed that the molecule may have both agonist and antagonist effects on the nAChR [2–4]. Binding experiments of [^{3}H]IMI to membranes from different species showed high-affinity binding sites in house fly head [5], and high- and low-affinity binding sites in the aphid *Myzus persicae* [6]. The nicotinic receptor subunit composition seems to exert a profound influence upon IMI binding affinity [7]. This brief review of the literature underlines the complex action of IMI on the nAChR.

The neurotransmitter acetylcholine (ACh) is distributed largely in the honey bee brain [8]. Acetylcholinesterase and ACh receptors have been identified in the antennal lobes and in the mushroom bodies (MBs), particularly in the calycal part [9, 10]. In addition, Kenyon cells, which fill the calyces, express functional nAChRs *in vitro* [11]. The involvement of the cholinergic system in memory processes in the honey bee has been demonstrated by intracranial injections of cholinergic antagonists using a classical conditioning procedure [12–14]. Local brain injections have shown that the nAChR antagonist mecamylamine impaired the recall or the formation of the memory trace depending on the brain site injection and the moment of the injection relative to the conditioning trial. From these experiments, we postulated that ACh, as in vertebrates, exerts a facilitatory effect on memory processes. We made the hypothesis that activation of the cholinergic pathways with agonists like those molecules belonging to the neonicotinoids would facilitate the formation and/or the recall of memory.

To test this hypothesis, we submitted honey bees to the habituation of the proboscis extension reflex (PER). This nonassociative learning paradigm can be easily used to detect the behavioral effect of different kinds of molecules. The PER is induced by antennal sucrose stimulation and involves activation of motor neurons situated in the subesophageal ganglion and driving the mouthpart muscles. The repetition of this non-noxious stimulation leads to a decrease in the response occurrence. We postulated that IMI could reduce the number of stimulations needed to observe the response decrease. However, given the neurotoxic action of IMI, the absence of the PER could indicate a problem of gustatory perception or a central motor disruption. In preliminary experiments, we defined the IMI dose that did not induce modifications of the gustatory threshold or a perturbation of motor activity.

The involvement of mushroom bodies in memory processes is well established in insects. Consequently imidacloprid brain targets were investigated using cytochrome oxidase (CO) histochemistry. CO activity is commonly used in vertebrates as an endogenous metabolic marker of neuronal activity. Energy demand due to neuronal activity increases the production of oxidative energy [15]. Classically, CO histochemistry is used in vertebrates to identify a pathological modification [16, 17] or the effect of chronic surgical and pharmacological treatments [18–20]. We attempted to develop a functional histochemistry of CO in honey bee brain that allowed the analysis of the short-term effect of cholinergic ligands including IMI on the metabolism of the different brain structures [21].

Materials and methods

Worker honey bees *(Apis mellifera)* were caught at the hive entrance and maintained with food and water *ad libitum* in small Plexiglas boxes until the beginning of the experiments. To evaluate the gustatory threshold and for learning and metabolism experiments, honey bees were immobilized individually in small plastic tubes with a drop of wax–collophane mixture laid between the dorsal part of the thorax and the tube. The head and the prothoracic legs were free to move, allowing the honey bee to clean its antennae from the repeated sucrose stimulations. Honey bees underwent a 2-hour starvation period before the beginning of the experiments.

Imidacloprid (Cluzeaux, France; molecular weight: 255.7; degree of purity 98 percent) was dissolved in dimethyl sulfoxide (DMSO; Sigma) to obtain a 10^{-1} M solution. Lower concentrations were obtained with successive dilutions in saline. Control groups were treated with DMSO dissolved in saline (vehicle) in the same proportions. For behavioral experiments, the drug or vehicle was used in topical applications (1 µl) to the thorax. Doses ranging from 1.25 to 5 ng/bee were used which were below the DL50 value (10 to 20 ng/bee, defined for thoracic application to *Apis mellifera* at 24 h; unpublished observations from L.P. Belzunces). For CO experiments, we tested the effect of intracranial injection of saline or IMI on worker honey bees receiving an injection of 0.5 µl saline or IMI (10^{-4}, 10^{-6}, or 10^{-8} M) at the brain surface. Nicotine (10^{-8}, 10^{-6}, and 10^{-4} M) and mecamylamine (10^{-2} M) were also tested as nAChR agonist and antagonist, respectively.

Behavioral tests

Gustatory threshold

The aim of this experiment was to study the effect of IMI on the gustatory perception. The gustatory threshold was defined as the lowest concentration of a sucrose solution applied to the antennae able to elicit a proboscis extension. The threshold was defined twice for each honey bee: first before any treatment and then after IMI or vehicle application. Several doses of IMI were used with several time-intervals between the application of the drug and the test.

The gustatory threshold was determined as follows. Fasted honey bees were submitted to antennal stimulations (1-minute intertrial interval) with increasing concentrations of sucrose solutions ranging from M/1024 to 4 M and following a geometric progression (M/1024, M/512, M/256, etc.). The range of increasing sucrose concentrations was applied twice separated by a 5-minute interval. The lowest concentration of sucrose that elicits the PER was defined as the gustatory threshold. Honey bees that failed to respond to one of the sucrose solutions were discarded. The remaining honey bees were fed with two drops of 50 percent (w/v) sucrose solution and fasted for 2 hours. This was done to ensure that the gustatory threshold determination under IMI application was made under the same motivational state. The thoracic application of vehicle or IMI at a dose of 1.25, 2.50, or 5 ng/bee was performed during the starvation period, 15 min, 30 min or 60 min before the second gustatory threshold determination. This second determination was done like the first one. For data quantification, any modification of the gustatory threshold between the two determinations from one sucrose concentration to the one immediately lower or higher was respectively quantified as -1 or $+1$ arbitrary unit.

Locomotion

Locomotion of honey bees was tested in an open-field-like apparatus consisting of a white Plexiglas box ($30 \times 30 \times 4$ cm) with a glass front for observation. The back surface was divided into 5-cm² squares and the box was illuminated from above. The box did not allow the honey bees to fly. A hole made in the bottom right-hand side of the box allowed the introduction of a single honey bee for a 5-min observation period. The position of the honey bee was recorded every 5 s. The duration of successive 5-s periods in the same square was reported as immobility as the locomotor activity of the honey bee in the same square was very low, if nonexistent. Otherwise, the honey bee was moving around. The effect of the drug on locomotor activity was studied 15, 30, and 60 min after application of a dose of 1.25, 2.50, and 5 ng/bee and was compared to the effect of vehicle.

Habituation

Fasted honey bees were stimulated repeatedly with a 50 percent (w/v) sucrose solution applied to one antenna at 1-min intervals. The habituation criterion was defined as three consecutive sucrose stimulations without proboscis extension. When this criterion was reached, the sucrose solution was applied to the controlateral antenna to rule out the eventuality of motor tiredness. Honey bees that did not respond to the 50 percent sucrose solution and to the restoration test of the reflex were discarded. IMI was applied at 1.25 ng/bee and the drug effect was tested after 15 min, 30 min or 1 hour in three independent groups. A group receiving no treatment and a solvent-treated group were also added.

CO histochemistry

Thirty minutes after injection of the drug, the animals were killed by rapid decapitation. Frontal sections (16 µm) from the whole brain were prepared for CO histochemistry, according to Wong-Riley [20]. Quantification of staining was performed by computer-aided densitometry of CO histochemistry intensity. We focused our investigations on antennal lobes, calyces, and α-lobes of MBs because it was previously shown that 30 min after an injection of AChR antagonists in these structures, memory processes were impaired [12, 13]. IMI was tested at concentrations of 10^{-8}, 10^{-6}, and 10^{-4} M, corresponding to doses ranging from 1.28 pg to 12.8 ng per honey bee. At higher doses IMI induced toxic and lethal effects.

Data analysis

Data sets were analyzed using a two-population independent two-tailed *t*-test or an analysis of variance (ANOVA). Figures show means ± s.e.m. In all cases, *P*-values less than 0.05 were considered as significant.

Results

Behavioral tests

Gustatory threshold

An increase in the gustatory threshold was observed between the first and the second determinations whatever the treatment (Figure 6.2). A very slight increase of less than one-half unit was found for the vehicle and for the lowest doses of IMI (1.25 and 2.50 ng/bee). Animals treated with the vehicle (controls) were not different from those receiving no treatment (data not shown). Groups that received 1.25 and 2.50 ng IMI were not different from controls, so in subsequent habituation experiments, both doses could have been used. A loss of sensitivity was noticed for the dose of 5 ng after 1 hour. This delayed effect seems not to be related to the time needed by imidacloprid to reach the brain from the thoracic application

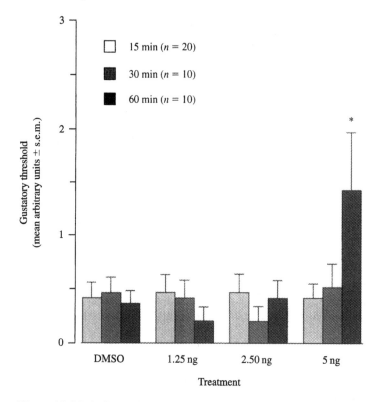

Figure 6.2 Variations of the gustatory threshold (arbitrary units) 15 min, 30 min, or 1 h after thoracic application of DMSO ($n = 20$ for each time) or imidacloprid at different doses (1.25, 2.5, 5 ng/bee). The number of imidacloprid-treated animals is indicated on the graph. *$P < 0.05$.

site since the high dose of 20 ng induces the same sensitivity loss as soon as 15 min after application (data not shown).

Locomotion

Opposite effects of IMI on motor displacements were observed depending on the dose (Figure 6.3). Compared to the vehicle, the lowest dose of IMI (1.25 ng/bee) induced an increase in displacements independently of time (shown as a decrease in immobility in Figure 6.3). A significant increase in locomotion was also observed for 2.5 ng/bee at 15 min. With 5 ng, IMI induced a decrease in the honey bee displacements in the box as soon as 30 min after application. The decrease in displacements was explained by a loss of motor coordination. The honey bees fell down on their backs, showing leg movements and body and wing shaking. Additional observations up to 2 hours after drug application showed that there was no behavioral recovery.

Unlike the previous experiment on gustatory perception, we did not observe a dose–effect relationship in this experiment, as there were more

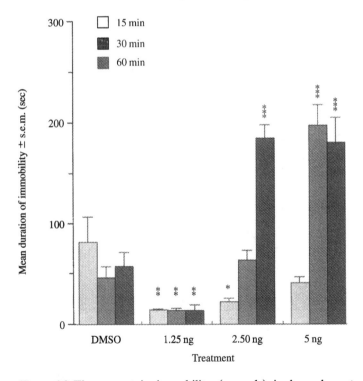

Figure 6.3 Time spent in immobility (seconds) in honeybees treated with DMSO or imidacloprid at different doses (1.25, 2.5, 5 ng/bee), 15 min, 30 min, or 1 h before the test. $n = 10$ in each group. *$P < 0.05$; **$P < 0.01$; ***$P < 0.001$.

numerous displacements at the lowest dose (1.25 ng/bee). This dose was retained to test the effect of IMI on habituation.

Habituation

Under IMI treatment (1.25 ng/bee), honey bees needed fewer trials to display PER habituation than honey bees receiving the vehicle or receiving no treatment (statistics highly significant in both cases, see Figure 6.4). There was no effect of time on the facilitating effect. This observation is closer to the enhancing effect of 1.25 ng of IMI on displacements, which is also independent of time. Dilute DMSO induced a slight but significant reduction in the number of trials compared to the groups receiving no treatment (statistics shown in Figure 6.4).

CO histochemistry

Histological modifications induced by IMI were of weak amplitude but very reproducible: for example in the antennal lobe, IMI 10^{-4} M induced a

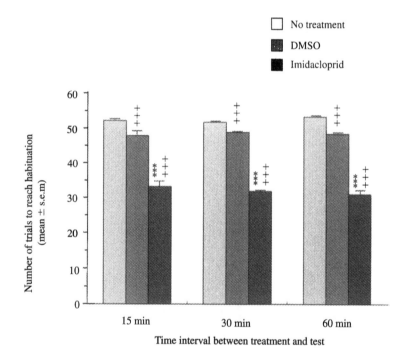

Figure 6.4 Number of trials required to reach habituation in animals receiving no treatment or animals treated topically with DMSO or imidacloprid (1.25 ng/bee) 15 min, 30 min, or 1 h before learning. $n = 20$ in each group. +++comparison to 'No treatment', $P < 0.001$. ***comparison to DMSO, $P < 0.001$.

staining increase in all the experiments performed. The intensity of staining was analyzed in the cortical layer and in the internal area of the
glomeruli. At the concentrations of 10^{-4}, 10^{-6}, and 10^{-8}M, a significant
increase in staining was obtained for the two regions of the glomeruli. The
increment ranged from +8 percent to +17 percent. A dose–response
effect was observed for this structure (Figure 6.5A).

The greatest modifications of CO labeling induced by IMI were
observed in the α-lobe stratification, corresponding for the dorsal layer B1,
to +23 percent of the saline group labeling (Figure 6.5B). For 10^{-4}M the
increment was significant in the dorsal, intermediate and ventral layers
(B1, B2, and B3).

In the calyces the 10^{-8}M IMI injection induced a significant reduction
in the labeling (Figure 6.5C). In the basal ring the mean gray level of the
10^{-6}M group was significantly lower compared to the saline group.

In the 10^{-4}M IMI-treated group, the CO staining in the upper (UD)
and lower (LD) divisions of the central body was significantly greater than
that of the saline group; the opposite effect was observed for 10^{-8}M
(Figure 6.5D).

In subsequent experiments other nAChR ligands were tested. CO was
stimulated by nicotine in a dose-dependent manner in many brain
regions (data not shown). In particular, the internal part of the glomeruli
exhibited significant increases of 19 percent at 10^{-4}M nicotine (Figure
6.6A).

The effects of nicotine were statistically significant in the B1, B2, and
B3 layers of the α-lobe (Figure 6.6B). The greatest stimulation by 10^{-4}M
nicotine administration was obtained for the B3 layer (+23 percent).
Moreover, for the ventral layer a significant increase was still present after
10^{-8}M (data not shown).

In calyces, whatever the concentration of nicotine tested, no significant
differences were found between the saline and nicotine groups whereas
10^{-4}M IMI induced an increase in labeling. Moreover, 30 min after 10^{-8}M
IMI injection, a decrease in brain metabolism was observed in the central
body, calyces, and α-lobe which was not observed with nicotine injection
to the same concentration and at the same interval.

Changes in the metabolic activity of the honey bee brain were examined following nAChR antagonist (mecamylamine) administration to high
concentration (10^{-2}M) inducing an impairment of retrieval processes [13].
Comparison between saline- and antagonist-treated brains indicates that
mecamylamine induced a significant decrease in neural metabolism in the
α-lobe (Figure 6.6B) and no effect in the other structures (Figure 6.6A, C,
D). Like IMI and nicotine, mecamylamine has a significant effect on the
α-lobe.

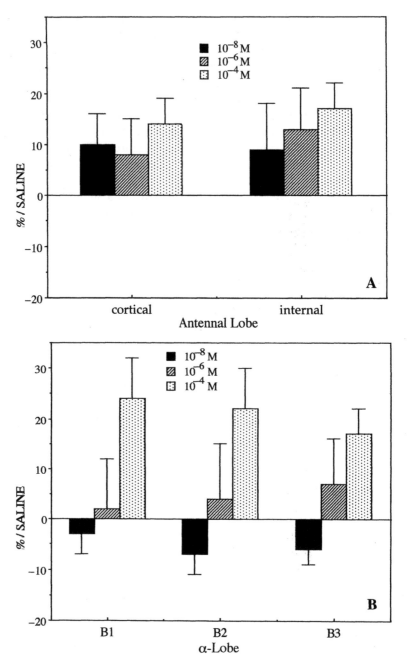

Figure 6.5 Relative variation of CO histochemistry induced by imidacloprid. (A) Antennal lobe: glomeruli cortical area, glomeruli internal area. (B) α-Lobe: B1, dorsal layer; B2, intermediate layer; B3, ventral layer. (C) Calyces: lip area, basal ring area. (D) Central body: UD, upper division of central body; LD, lower division

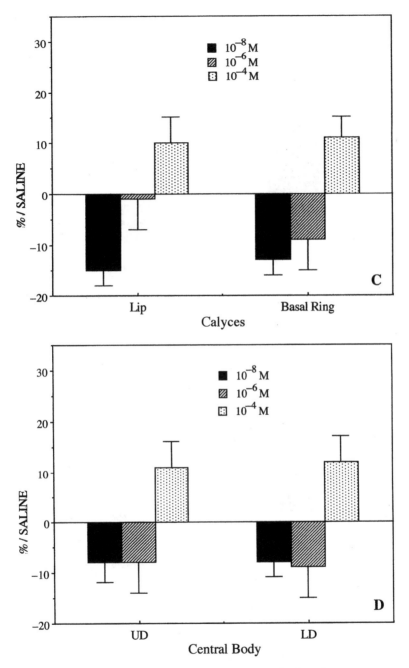

of central body. Each imidacloprid-treated brain value was expressed as $(\text{treated} \times 100/\text{saline}) - 100$. 0% represents optical density of saline group. Mean $\pm$ s.e.m was obtained by averaging the brain percentage variation of $n = 21$ (10^{-4}M), $n = 8$ (10^{-6}M), $n = 8$ (10^{-8}M) honey bees.

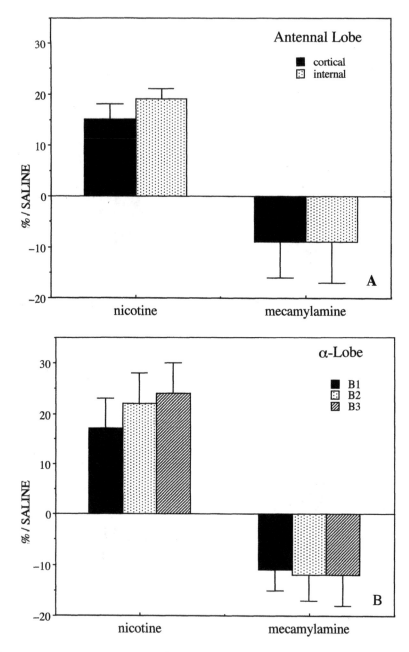

Figure 6.6 Effect of nicotine (10^{-4}M) and the nicotinic AChR antagonist mecamylamine (10^{-2}M) on cytochrome oxidase histochemistry in antennal lobe
(A), α-lobe (B), calyce (C), and central complex (D). Abbreviations
and result expression as in Figure 6.5.

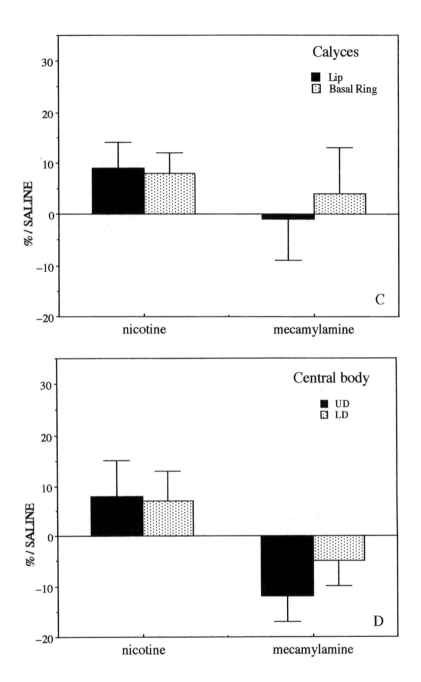

Discussion

Our hypothesis that IMI at a low dose (1.25 ng/bee corresponding to a micromolar concentration) could facilitate a simple form of learning is verified.

The experiments show a hierarchical effect of IMI applied topically. This effect depends on the physiological function tested. It seems that gustatory perception is less sensitive to the insecticide than motor function or learning processes. The first manifestation of the drug on the gustatory threshold appeared at 5 ng/bee after 1 hour whereas a strong effect of the drug at the dose of 1.25 ng/bee was observed on the other functions after the shortest interval. We do not retain the possibility that the drug needs more time to diffuse from the thorax to the brain than to the ventral nerve cord. The high doses of 10 and 20 ng induced a large increase in the gustatory threshold as soon as 30 min after application (results not shown).

An interesting finding in this work is the dual effects of IMI on motor displacements and brain metabolism depending on the dose. The locomotor activation observed at the lowest dose could indicate the specific effect of the drug linked to nicotinic activation whereas the higher doses would induce a nonspecific toxic effect. The complex effect of IMI on neuronal metabolism has also been observed after intracranial injection of the insecticide in the honey bee. All of the 10 regions analyzed showed a significant staining increase for the highest concentration (10^{-4}M). For the lower concentrations (10^{-6} and 10^{-8}M) a regional sensitivity and specificity were observed. The effect of IMI in the antennal lobes, the first relay of the olfactory information, was always an increase in metabolism. In integrative structures, in particular the calyces, the action of IMI is more complex. Low concentrations induced inhibition of CO histochemistry whereas high concentrations resulted in activation of CO. In a preliminary experiment brain injections of the nAChR antagonists mecamylamine and hexametonium only induced a decrease in the labeling. Hence, we were waiting for the agonist stimulating action of IMI. The electrophysiological effects of IMI are also complex. Modulation of the nAChR channel by IMI has been demonstrated in pheochromocytoma cells, corresponding to both multiple agonist and antagonist effects on acetylcholine-induced currents [2, 4]. Finally, the dual effect of IMI can be explained by the presence in the central nervous system of the honey bee of two types of nicotinic receptors as shown in the cockroach nervous system [1]. Stimulating effects, such as depolarization of the cockroach cercal afferent giant interneuron and inward currents with activation of the hybrid nAChR, were obtained with concentration up to 10^{-6}M [1, 22]. At 100 nM, IMI reversibly reduced the amplitude of the ACh responses recorded on SADβ2 hybrid receptors expressed in *Xenopus* oocytes [22]. Using [³H]IMI, high- and low-affinity nAChR-like binding sites have been characterized in the aphid *Myzus periscae* [6]. The dual agonist/antagonist effects of IMI on CO histochemistry could be linked to the presence of different subtypes of receptors in the different neuropiles. It is not out of the question that a desensitization of nAChRs would be induced with the concentration of 10^{-4} M IMI; desensitization was described in the cockroach in response to 10^{-5} M IMI [1]. In our experiment the specific effect of IMI would be observed with 10^{-8} M and corresponds to an increase in metabolism in the antennal lobes and a decrease in metabolism in integrative structures. However, agonist action (nicotine-like) of IMI, observed with the highest concentration (10^{-4} M), may contribute to the toxicity of this insecticide.

In conclusion, these results support our hypothesis that the cholinergic system is involved in learning processes in insects. We have previously shown that cholinergic antagonists impair the formation and recall of olfactory memory [12–14]. We show now that activation of the cholinergic system with IMI facilitates a reflex habituation. However, to demonstrate that the facilitatory effect is general and not limited to nonassociative learning, we have to submit IMI-treated honey bees to different tasks such as, for example, discriminative olfactory conditioning between reinforced and nonreinforced odors. However, this facilitatory effect is not so easy to understand at the cellular level. A simple hypothesis to advance is that the dose used in behavioral experiments (corresponding to a micromolar concentration) induces neuronal activation. This is not what is observed with CO experiments. At the concentration of 10^{-6} M, IMI induces a slight or no effect on brain metabolism. Our findings suggest that IMI exerts multiple actions at nAChRs. A new approach is currently being developed. Using the patch-clamp technique, ACh-ionic currents and their IMI-induced modifications are recorded in cultured Kenyon cells to see whether IMI behaves as an agonist or an antagonist on the honey bee nAChR.

References

1 Buckingham, S.D., Lapied, B., Le Corronc, H., Grolleau, F. and Satelle, B. (1997). Imidacloprid actions on insect neuronal acetylcholine receptors. *J. Exp. Biol.* **200**, 2685–2692.

2 Nagata, K., Aistrup, G.L., Song, J.H. and Narahashi, T. (1996). Subconductance-state currents generated by imidacloprid at the nicotinic acetylcholine receptor in PC 12 cells. *NeuroReport* **7**, 1025–1028.

3 Nagata, K., Iwanaga, Y., Shono, T. and Narahashi, T. (1997). Modulation of the neuronal nicotinic acetylcholine receptor channel by imidacloprid and cartap. *Pestic. Biochem. Physiol.* **59**, 119–128.

4 Nagata, K., Song, J.H. and Narahashi, T. (1998). Modulation of the neuronal nicotinic acetylcholine receptor-channel by the nitromethylene heterocycle imidacloprid. *J. Pharmacol. Exp. Ther.* **285**, 731–738.

5 Liu, M.-Y. and Casida, J.E. (1993). High affinity binding of [³H]imidacloprid in the insect acetylcholine receptor. *Pestic. Biochem. Physiol.* **46**, 40–46.

6 Lind, R.J., Clough, M.S., Reynolds, S.E. and Early, F.G.P. (1998). [³H]Imidacloprid labels high- and low-affinity nicotinic acetylcholine receptor-like

binding sites in the aphid *Myzus periscae* (Hemiptera: Aphididae). *Pestic. Biochem. Physiol.* **62**, 3–14.

7 Lansdell, S.J. and Millar, N.S. (2000). The influence of nicotinic receptor subunit composition upon agonist, alpha-bungarotoxin and insecticide (imidacloprid) binding affinity. *Neuropharmacology* **39**, 671–679.

8 Breer, H. (1987). Neurochemical aspects of cholinergic synapses in the insect brain. In: *Arthropod Brain. Its Evolution, Development, Structure and Functions* (Gupta, A.P., Ed.), Wiley, New York, pp. 415–437.

9 Kreissl, S. and Bicker, G. (1989). Histochemistry of acetylcholinesterase and immunocytochemistry of an acetylcholine receptor-like antigen in the brain of the honey bee. *J. Comp. Neurol.* **286**, 71–84.

10 Scheidler, A., Kaulen, P., Brüning, G. and Erber, J. (1990). Quantitative autoradiographic localization of ^{125}I α-bungarotoxin binding sites in the honeybee brain. *Brain Res.* **534**, 332–335.

11 Goldberg, F., Grünewald, B., Rosenboom, H. and Menzel, R. (1999). Nicotinic acetylcholine currents of cultured Kenyon cells from the mushroom bodies of the honeybee *Apis mellifera. J. Physiol.* **514**, 759–768.

12 Gauthier, M., Cano Lozano, V., Zajoual, A. and Richard, D. (1994). Effects of intracranial injections of scopolamine on olfactory conditioning retrieval in the honey bee. *Behav. Brain Res.* **63**, 145–149.

13 Cano Lozano, V., Bonnard, E., Gauthier, M. and Richard, D. (1996). Mecamylamine-induced impairment of acquisition and retrieval of olfactory conditioning in the honey bee. *Behav. Brain Res.* **81**, 215–222.

14 Cano Lozano, V. and Gauthier, M. (1998). Effects of the muscarinic antagonists atropine and pirenzepine on olfactory conditioning in the honey bee. *Pharmacol. Biochem. Behav.* **59**, 903–907.

15 Wong-Riley, M.T.T. (1998). Cytochrome oxidase: An endogenous metabolic marker of neuronal activity. *Trends Neurosci.* **12**, 94–101.

16 Brines, L.M., Tabuteau, H., Sundaresan, S., Kim, J., Spencer, D.D. and de Lanerolle, N. (1995). Regional distribution of hippocampal Na^+,K^+-ATPase, cytochrome oxidase, and total protein in temporal lobe epilepsy. *Epilepsia* **36**, 371–383.

17 Mutisya, E.M., Bowling, A.C. and Flint Beal, M. (1994). Cortical cytochrome oxidase activity is reduced in Alzheimer's disease. *J. Neurochem.* **63**, 2179–2184.

18 Cada, A., Gonzalez-Lima, F., Rose, G.M. and Bennett, M.C. (1995). Regional brain effects of sodium azide treatment on cytochrome oxidase activity: A quantitative histochemical study. *Metab. Brain Dis.* **10**, 303–320.

19 Rubio, S., Begega, A., Santin, L.J. and Arias, J.L. (1996). Ethanol- and diazepam-induced cytochrome oxidase activity in mammillary bodies. *Pharmacol. Biochem. Behav.* **55**, 309–314.

20 Wong-Riley, M.T.T (1979). Changes in the visual system of monocular sutured or enucleated cats demonstrable with cytochrome oxidase histochemistry. *Brain Res.* **171**, 11–28.

21 Armengaud, C., Aït-Oubah, J. and Gauthier, M. (1998). Functional staining of cytochrome oxidase activity in the honey bee brain. *J. Eur. Neurosci.* **10 (Suppl. 10)**, 260.

22 Matsuda, K. Buckingham, S.D., Freeman, J.C., Squire, M.D., Bayliss, H.A. and Sattelle, D.B. (1998). Effects of the α subunit on imidacloprid sensitivity of recombinant nicotinic acetylcholine receptors. *Br. J. Pharmacol.* **123**, 518–524.

7 Impact of agrochemicals on non-*Apis* bees

J.N. Tasei

Summary

Only few reports have been published on the reduction and recovery of native non-*Apis* bee populations, measured after temporary or permanent agrochemical pest control in North America. Small species were found to be the most sensitive. The assessment of pesticide toxicity and hazards to non-*Apis* bees has been practiced for about 50 years through various laboratory, semi-field, and field methods. Researches were conducted mainly on three species: *Nomia melanderi* (alkali bee), *Megachile rotundata* (alfalfa leafcutting bee), and *Bombus terrestris* (bumble bee). Toxicity tests performed in standardized conditions on adults and larvae showed that the intrinsic susceptibility of non-*Apis* bees measured by oral and topical LD_{50} or by LC_{50} varied to a great extent between species and also from *Apis mellifera*. Laboratory and semi-field tests have been used to assess the risks of sprays, field-weathered residues, or systemic compounds in nectar and pollen. The effects of several organophosphates, pyrethroids, neonicotinoids, and a carbamate, are discussed. Sublethal effects of deltamethrin, fenvalerate, trichlorfon, and imidacloprid have also been investigated. It has been shown that biochemical data from studies on detoxification in *M. rotundata* did not agree with toxicological parameters and risk assessment in the field.

Introduction

Many wild and cultivated plants are visited not only by honey bees (*Apis mellifera* in particular) but also by non-*Apis* bees which facilitate their fruit and seed setting. These Hymenoptera are represented by more than 20000 species throughout the world, belonging to nine families: *Colletidae, Oxaeidae, Halictidae, Andrenidae, Melittidae, Fideliidae, Megachilidae, Anthophoridae,* and *Apidae* [1]. This fauna is a natural resource which often sustains a prominent role in the pollination of crops and the maintenance of floral diversity, especially when honey bees are absent or not efficient. Many researchers have long emphasized the contribution of non-*Apis* bees, also called wild bees, to pollen transfer in cultivated and wild plants. Several authors have surveyed the bee diversity in Europe and compared the efficacy of several species in case studies [2, 3]. Other scientists have extensively investigated the different biological traits of solitary and social non-*Apis* species which proved to be highly variable [4–6]. Some species dig burrows into the ground, while others nest in twigs, timber, soil cavities, etc., and use all sorts of materials to protect their brood cells, such as wax, mud, leaf cuttings, wool, resin, and so on.

In areas where agricultural efficiency has been increased through the destruction of hedges and adventitious flowering plants, the cutting out of waste lands, and the reduction of crop diversity, the population of native non-*Apis* bees has been depleted. Generally, in the same areas, an additional factor in this depopulation may be the misuse of insecticides on crops. The importance of non-*Apis* bees for seed setting in wild and cultivated plants and the frequent shortage of pollinators on various crops have encouraged scientists to domesticate and multiply several non-*Apis* bees. Since the Second World War, the most popular has certainly been the alfalfa leafcutting bee, *Megachile rotundata (Megachilidae)*, for which the management techniques have been constantly improved [7, 8].

This solitary bee can nest gregariously in tunnels of wooden or plastic shelters established temporarily in alfalfa fields grown for seed. For the same purpose, in the USA and New Zealand, artificial nesting beds have been created close to alfalfa seed crops, to increase the population of the ground-nesting *Halictidae* bee *Nomia melanderi* [9, 10]. Several "mason bees" (*Megachilidae*) are propagated commercially in tunneled domiciles to improve fruit production in various countries in Asia, America, and Europe: *Osmia lignaria* [11], *Osmia cornifrons* [12], and more recently, *Osmia cornuta* [13]. Since the 1990s, the social bee *Bombus terrestris* has been reared *en masse* mostly in Europe to pollinate vegetables (initially, the tomato) in greenhouses [14]. Now other bumblebee species are also being produced in Asia and North America.

Owing to the biological differences among non-*Apis* species and honey bees, one can predict that their population may not be affected similarly by agrochemicals. Thus, for example, the death of a solitary female bee in charge of a nest means the end of reproductive activity, while in social bees deficits following spraying may be compensated by workers and also by new bees emerging from the brood. Moreover, except for the species cited above, native non-*Apis* bees live in natural habitats that cannot be removed from hazardous sites. Despite technical difficulties, some researchers have investigated the impact of large-scale insecticide applications on non-*Apis* populations. In addition, the economic importance of the domesticated non-*Apis* bee has favored laboratory and field studies on the toxicity and hazards of pesticides to the three main species, *M. rotundata*, *N. melanderi*, and *B. terrestris*. The availability of individuals now produced *en masse* enables advances in methodology often inspired from honey bee studies, and comparisons of sensitivity between *Apis* and non-*Apis* bees are now possible.

Historical background of pesticide risk assessment on non-*Apis* bees

The earliest scientific article mentioning concern about the effect of a pesticide on non-*Apis* bees appeared in 1946 in a Canadian journal [15]. The author, from the Massachusetts State College, reported very briefly on the mortality rates of unidentified solitary bees and bumble bees collected from apple flowers and introduced in cages to be exposed to DDT, "dusted lightly through the screen covering the top of the cage." He assessed mortality several times within 48 hours in a first test and 60 hours in a second experiment where honey bees were treated in the same way. The conclusions were that the number of experimental insects – 10 solitary bees, 5 to 10 bumble bees and 10 honey bees – were too low to enable definite conclusions to be made. It was also recognized that laboratory and field tests may not always agree. However, both solitary bees and honey bees seemed equally affected while the death of the bumble bees was retarded, which means they were presumably more tolerant. Further studies on DDT effects on wild pollinators were performed in the UK [16] and published in 1948 and in the USA [17, 18] in 1949 and 1950. The UK authors collected *B. terrestris*, *B. agrorum*, and solitary bees (*Andrena flavipes*), and then performed laboratory tests, using various concentrations of toxic material which was spread on a glass wall placed in boxes for contact tests or diluted in sucrose solution for feeding tests. They also used sprayed blooms in a third kind of contact laboratory test and made field observations on treated apple blossoms on which they collected several *Andrena* species and *Osmia rufa*. They concluded that the susceptibility of worker bumble bees to DDT was comparable to that of honey bees and that queens and drones of *B. agrorum* and *B. terrestris* were more resistant. The authors also insisted on the environmental impact of bumble bee queen losses which entail hundreds of workers that would not be produced. One paper from the USA reported on laboratory tests on several solitary species collected in a field, belonging to the genera *Nomia*, *Megachile*, *Melissodes*, *Anthidium*, and *Agapostemon* [18]. Experimental bees were exposed to dry DDT residues on screens which had been previously immersed in a DDT solution at different concentrations. The comparison with honey bees showed that solitary bees were more resistant than *Apis mellifera* at the same concentration and exposure duration. It was also found that females were more resistant than males. The other article by US authors [17] described an experimental procedure to evaluate the hazards to *N. melanderi* of a DDT spray on an alfalfa field in bloom. DDT was applied before the bees started foraging. Before and after the spray, the bee density on flowers was assessed by sweeping alfalfa plants with a net. Pollen loads were also collected from female hind legs to measure the rate of exposed insects. Since this species lives in aggregations close to cultivated alfalfa, dead bees could be counted at the nesting site as well as the number of active nests. Despite the inaccuracy of the method and the absence of statistical interpretation, the results evidenced a moderate repellency for a few hours and a toxicity of the residues estimated at 15 percent of nests becoming inactive.

The first calculation of LD_{50} was published in 1963 [19]. The American authors used leafcutting bees as test insects. They applied acetone dilutions of various compounds on the abdomen of anesthetized bees. They found that *M. rotundata* was more susceptible to three of the tested pesticides than honey bees, and less susceptible to two of them, including carbaryl.

The earliest research using biochemical techniques for studying the toxic action of insecticides was presented in a PhD thesis in 1972 [20]. The author estimated the effect of drugs such as the synergist piperonyl butoxide on the insecticide action of carbaryl by measuring *in vitro* the microsomal enzyme activity in *M. rotundata*. He found drug absorption enhanced this activity up to 4–5 times and reduced bee lipid content by 20–30 percent. At the same time two other American scientists [21] compared the effect of trichlorfon and carbophenothion on acetylcholinesterase in the leafcutting bee and the honey bee. In the solitary bee, enzyme inhibition was stronger with trichlorfon, and with carbophenothion enzyme recovered 10 minutes after application. Changes in enzyme activity were similar in *M. rotundata* and *A. mellifera*.

In 1975 the first estimation of the impact of an insecticide applied on a large scale was published [22]. The author compared the diversity and abundance of native pollinators of Canadian lowbush blueberries in a control area and in areas contaminated with fenitrothion sprayed on forests of New Brunswick. Pollinators were mainly *Bombus* spp., *Andrenidae*, and *Halictidae*. Data of the population census were interpreted through statistical analyses which evidenced that the lowest diversity and abundance index was in areas close to treated forests. Moreover, carcasses found in these areas showed the highest residue rates. Both results corroborated the crop failures reported by blueberry growers of the province.

If we consider that insecticide repellency, mentioned in early studies, is not a typically sublethal effect, the first report of a consequence of low doses of insecticide on non-*Apis* bees appeared in 1981 [23]. The authors, comparing two pyrethroids and organophosphates on the leafcutting bee in laboratory tests, found that a high percentage of comatose bees recovered. This was observed only with the pyrethroids fenvalerate and decamethrin (deltamethrin) which caused a strong "knock-down" effect within the first hour after application.

The earliest study on the possible effects of systemic compounds on wild bees appeared in 1972 [24]. The American scientists tested the hazards of soil application of insecticide solutions of aldicarb, oxydemeton-methyl, and metasystox. They used sweet clover plants cultivated in pots and visited by *M. rotundata* and found no mortality at the recommended dosages, while the application of 10 times the recommended dose of aldicarb resulted in significant mortality of females, revealing that some active substance was transferred to the nectar.

The first test of an insect growth regulator (IGR) on a non-*Apis* bee was presented during a symposium in 1993 [25]. The authors observed adult mortality and brood development in bumble bee colonies (*B. terrestris*) maintained in cages. Forage plants were treated during activity hours with the IGR fenoxycarb. It was concluded that the IGR did not present a negative action on adult bumble bees but that a larval test had to be developed for an adequate assessment of the brood mortality.

Survey of testing methods

Among the various testing methods described by authors, a large number was aimed at measuring mortality rates in standardized laboratory conditions. Some procedures permitted the calculation of the LD_{50} or the LC_{50} of compounds through contact or feeding tests. They supplied data enabling comparisons of acute toxicity of pesticides between non-*Apis* bees and between honey bees and non-*Apis* bees.

Other kinds of tests were performed in a cage, tunnel, or greenhouse and their objectives were to assess the consequences of sprays or residues on bees exposed to compounds in more or less standardized conditions. In these tests, scientists did not expect to estimate the reaction of an insect to a measured substance intake or deposit but to assess risk in practice. The advantage of keeping bees in such enclosures is to ensure the permanent exposure of the insects but the counterpart is an overestimate of the hazards.

The third kind of assessment was hazard testing conducted in the field either by using domesticated non-*Apis* bees maintained in artificial domiciles or by monitoring native populations in their natural habitat. The drawback of these methods is that standardization is not possible since the exposure of experimental insects to test compounds is not controlled. In the case of native population monitoring, the main difficulty is the interpretation of data due to the number of factors involved in population changes during seasons and years.

The first category will be referred below to as "laboratory tests," the second as "semi-field tests," and the third as "field tests."

Laboratory tests

Median lethal dose assessment

Laboratory procedures for estimating LD_{50} were first applied to *M. rotundata* and *N. melanderi* in the USA because of their commercial importance in areas where alfalfa seed was produced. The detailed description of topical tests among the early articles [26] indicated that test leafcutting bees were obtained from incubated cells and then immobilized at 10°C. The treatment was a drop of 1.7 µl applied to the thoracic scutum with a micro-injector. Twenty leafcutting bees were necessary for each dosage–mortality test. After treatment, bees were placed into screened boxes with feeders containing a 20 percent honey solution. The boxes were maintained in a micro-biotron illuminated at 27°C. Mortality was recorded every 24 hours for 4 days following the treatment. The LD_{50} was established at 72 hours for leaf-cutting bees but at 48 hours for *N. melanderi*. Other scientists who studied the intrinsic susceptibility of solitary bees used similar techniques [21] or adopted variations such as the reduction to 48 hours of the duration of the mortality check in leafcutting tests [27]. This duration dropped to 24 hours for *N. melanderi* and *M. rotundata* [28]. These authors used six concentrations of the test solution falling between the limits of the expected value, and their data were analyzed with the probit analysis method. A device designed for accurate measurement of the consumption of pesticide solutions by any bee species was described in 1973 [29] but it seems it was not used in practice by other authors who established the oral LD_{50} of aldicarb in *N. melanderi* and *M. rotundata* with a more simple feeding system [30].

The first approach to studying acute contact toxicity of pesticides to bumble bees was derived from the method described for honey bees [31]. After a gap of 29 years, a detailed method to determine acute contact and oral toxicity in bumble bees was presented in a symposium [32] and completed later [33]. For the contact test the authors recommended collecting workers of average size and age then using five concentrations per replicate and performing two replicates. The 1-µl drop of pesticide solution in acetone was deposited on the ventral part of the thorax and the mortality recorded every 24 hours for 3 days. A negative control with acetone and a positive one with either dimethoate or parathion were also recommended. The method for oral toxicity derived from the European guidelines for honey bees was modified in order to be adapted to bumble bees which have no trophallaxis. The principle of the test was to cage individually 30 bumble bees per concentration, maintained in the dark at 25°C. The test substance was dissolved in a sucrose solution and the mean weight of the bees determined. Mortality checks and controls were similar to those of the contact test. Some variations in these guidelines appeared in other articles [33], in particular the use of water as solvent and the duration of 10 days for mortality recording.

The LD_{50} of several compounds was established in the social bee *Trigona spinipes* using a topical test protocol similar to the one for bumble bees [34].

Tests on residues

Even when no LD_{50} was calculated, other kinds of laboratory procedures have been used to test pesticide toxicity comparatively. The first comparisons of the susceptibility of bumble bees, solitary bees, and honey bees to DDT were performed in cages with glass or screened walls previously sprayed or immersed and dried [16, 18]. A standardized method was proposed to test pesticide effects on *M. rotundata*, using filter papers soaked in test solutions, dried, and placed on the bottom of screened boxes containing 15 individuals which emerged within 48 hours. The boxes were kept at 27°C under constant light. No food was supplied and mortality was recorded every 1 or 2 hours for 1 day. Identical boxes were used for testing the effect of sprays. In this case, after the treatment, leafcutting bees were introduced into clean boxes and placed in standardized conditions [35]. In further articles [23, 36] it was suggested that each filter paper should absorb the amount of solution corresponding to the field recommended dosage.

Several Canadian and American scientists have developed laboratory bioassays for risk assessment when non-*Apis* bees were exposed to treated plants. In the earliest studies, foliage was sampled from alfalfa plots previously treated with the test insecticide and placed in cages where 10 leafcutting bees or *N. melanderi* could walk and feed on a sugar solution. They recorded mortality after 24 or 48 hours [37, 38] and tested various ages of residues to establish the RT 25, that is the residual time required to obtain a bee mortality of 25 percent after a test exposure to field-weathered spray deposits [39]. This method was extended to the bumble bees *B. centralis* and *B. rufocinctus* and to honey bees [40]. In further research on field-weathered residual exposure tests, the authors, who followed the main guidelines, standardized the method. They sampled the upper 15-cm portion of test plants and placed about $500\,cm^2$ of foliage in screened cages 45 cm long with plastic Petri dishes as top and bottom. Twelve to 30 test bees were introduced into each cage maintained at 26–29°C. Mortality was recorded after 24 hours and each treatment was replicated four times [41–43]. A Canadian scientist used a "tube chamber" constructed of clear plastic sheets forming a tube 14.5 cm in diameter and 49 cm high. This exposure chamber was separated by a screen partition into a top and a bottom section to test vapor and residue hazards, respectively. Potted test plants were sprayed at the field rate then dried and moved to a climate room at 28°C and a 16 L:8 D photoperiod. A tube chamber was positioned over each plant and 10 test bees were introduced into each of the sections. Mortality was recorded after 24 hours [44].

Tests on sublethal effects

Laboratory tests were used to assess the effects of low doses or concentrations on bumble bees. To estimate whether a 0.01–0.02 µg topical application of deltamethrin affected the longevity of bumble bees, 32 workers per treatment were kept in disposable boxes each containing eight insects, maintained in the dark at 20°C. Mortality was recorded daily [45]. A more recent article reported on a feeding test conducted on queenless micro-colonies of three workers of *B. terrestris* to study sublethal effects of low concentrations of imidacloprid in the food on worker survival, brood size and larval development [46].

Testing insect growth regulators

Queen right *B. terrestris* colonies were used in the laboratory for testing the toxicity of IGRs on brood when the substance was ingested by workers for 24 hours. The technique proposed was to photograph the brood every day during the week before the 24-day feeding period and over the next 5 weeks [47]. A standardized larval test was described to evaluate IGR hazards to *B. terrestris* brood. Brood cells with 10 young larvae each had to be placed in small rearing boxes at 28°C and 50 percent HR. They were attributed to groups of three nurse workers and fed with sucrose syrup and pollen dough. After pupation, workers were removed and the brood was reared until the adults emerged. The test substance had to be dissolved in the food and fed to the test groups for 24 hours. It was to be applied to larvae of different ages. For each larval age and each test substance three replications were necessary [48].

Semi-field tests

All these tests were conducted in field or greenhouse cages and also in greenhouse compartments, i.e. under nearly natural climatic conditions and permanent exposure. Sometimes, parallel experiments were conducted in greenhouse and in field cages to determine whether both situations gave similar results.

Greenhouse cages and compartments

Potted *Melilotus alba* was often used as a test plant in greenhouses because of its abundant flowering. The effects of three systemic compounds were estimated by applying converted field dosage to sweetclover testing pots placed in cages. Ten leafcutting bee were introduced into each cage and mortality was recorded every day [24]. The relative repellent effect of two pyrethroids on nesting leafcutting bee females was assessed in greenhouse compartments where treated and control sweetclover plants were placed together [23]. The sublethal effect of low doses of deltamethrin on the fecundity of females was studied by using marked leafcutting bees nesting in the same compartment and foraging sweet-clover [27]. Comparison of the effect of imidacloprid seed dressing on the visiting rate of sunflower heads by bumble bees (*B. terrestris*) was performed by cultivating treated and control potted sunflower in a compartment [49]. For studying the effects of IGRs on bumble bee larvae, Tornier [50] suggested rearing in greenhouse compartments queen right colonies of *B. terrestris* with 30 marked workers each and a similar amount of brood. The entrance of the hives should be equipped with a dead bee and larvae trap. A picture of the brood should be taken before and after the test period [50]. Results on IGR effects on *B. terrestris* were obtained in 5×3 m greenhouse cubicles where tomato plants had been treated [51].

Field cages

For testing residues, 6×6 m alfalfa field plots were covered with cages, each containing 50 females and 50 males of *M. rotundata*, which were released a week after application. Dead females were counted by examining the straws in which they nested and larval mortality was assessed by splitting the straws lengthwise once nesting was completed [52]. The effects of a systemic granular side dressing of alfalfa on leafcutting bee mortality were also studied in cages covering 3.6×6 m plots. *M. rotundata* could nest in straws or laminated boards in insulated shelters. These devices enabled cell extraction and residue analysis of provisions [53–55]. Similar experiments were conducted in alfalfa fields, using cages ranging from 1.2×1.2 m to 6×6 m and various leafcutting bee densities in relation to the amount of forage [30–44].

Field cages were also used to test the effects of sprays on bees. For this purpose, Heller *et al.* [56] reported a comparative trial with three replications where *M. rotundata* was reared in screened tunnels 17×6 m, partitioned into three sections where *Sinapis alba* was grown as a test plant. The spray was applied during the foraging period. A field cage method was presented for testing IGR effects on *B. terrestris*, using *Phacelia* as a test plant in 3×4 m cages [48]. Before introduction into the test cage and IGR application, test colonies containing 50–70 workers were attributed standardized egg cells and brood with larvae of known age ranging from 1 to 6 days. The cage period lasted 2–3 weeks, then colonies were returned to the laboratory until adult emergence [48].

Smaller removable screened cages containing test bees were used for a standardized exposure to experimental sprays on alfalfa [19] or fenitrothion aerial spray for forest protection [57]. The first authors removed the cages after the spray and placed them in a holding room where bees could feed on a sucrose solution and they assessed the 24-hour mortality. The other authors used $5 \times 7.5 \times 3.5$ cm individual screened compartments where bumble bee queens were introduced before the spray and fed twice daily. The caged queens were left in place for 4 hours following the treatment, before being moved to the laboratory for a 7-day observation period. "Krome-Kote" cards placed adjacent to boxes enabled a good estimate of exposure by counting the insecticide droplets on $5 \times 1\,cm^2$ per card.

Field tests

Tests with leafcutting bees

A few articles have reported on field tests on *M. rotundata* using shelters in alfalfa fields. In the earliest one the author placed two shelters ($86 \times 50 \times 39\,cm$) in two distant parts of a crop 1 km long and 45 m wide. One half of the field was sprayed with naled, the second half with trichlorfon during a calm evening. Shelters were supplied with boards drilled with hundreds of holes each accommodating paper soda straws 0.5 cm in diameter and 6.5 cm long where leafcutting bees were nesting. Fifty nests were marked and monitored in each field before and after the treatment. Straws were extracted and examined at night. When the nesting period was completed, boards were returned to the laboratory at 26°C for 1 month, then straws containing the marked nests were hibernated for 4 months prior to an incubation period of 20 days at 26°C. When incubation was completed, marked nests were dissected to record larval mortality [58]. Later authors also comparing two compounds preferred testing sprays on six alfalfa plots ranging from 200 to 1600 m², each being at least 300 m away from others. On each plot they placed a small leafcutting bee shelter where 114 to 237 females established their nests in polystyrene grooved boards 4 cm deep. Females were counted early morning before they began to fly. Treatments were applied during foraging hours when the number of females in nesting tunnels had been stable for 3 consecutive days. Two plots were used as control, two were sprayed with phosalone, and two with deltamethrin. Female numbers were assessed seven times during the 3 weeks following treatments. The exposure of foragers was estimated by analyzing pollen samples from brood cells provisioned by female bees at $t-1$ and $t+1$. At the end of flight activity nesting boards were moved to the laboratory and left until the larval development was completed. Then cells were extracted and samples of 600–800 cells per treatment were incubated after a 2-month hibernation. When adult emergence was finished, closed cells were opened to examine their content [59]. In another experiment conducted with similar lay-out and material, the authors used coded colored marks on every leaf plug as soon as a nest was completed to assess the larval mortality in relation to the date of cell provisioning. Samples of plug leaves, pollen provisions, and live larvae enabled residue analysis for the two compounds tested, alphamethrin and phosalone [36].

Tralomethrin was tested for bee hazard in alfalfa fields pollinated by leafcutting bees and treated by airplane or helicopter. In a first test, the authors observed the fate of females nesting in shelters placed in separate plots of a large field, which received applications at different rates in the evening. In a second test they preferred to use separate fields to compare the effects of treatments. Evaluation of hazards were done by pre-application and post-application records of the number of active females per 5-second scan per nesting unit. This count was replicated 10 times. The number of females in 13 nest tunnels was also assessed (25 replications) at night before the application and 2 days after the application [28].

Tests with bumble bees

For studying the possible effect of a systemic dressing of sunflower seeds on homing behavior and nest development of *B. terrestris,* 20 colonies of approximately 50 individuals were prepared and all the workers were marked on the thorax the day they were moved to fields, i.e. at the beginning of flowering. Ten colonies were placed in a large treated field surrounded by more than 400 ha of treated sunflower. The other 10 colonies were in a control field, 20 km away in a large nontreated zone. Exposure to residues in sunflower nectar and pollen was estimated by identifying pollen grains carried by a total of 241 nectar and pollen gatherers collected at the hive entrance. After a 9-day field period, the 20 hives were removed to the laboratory after sunset. They received identical food until new queens emerged, then marked and unmarked workers were counted [49]. An attempt to establish a standardized field test for IGRs was not satisfactory. The authors placed six small colonies of *B. terrestris* (less than 50 workers) near a 2400 m^2 *Phacelia* plot and applied triflumuron 3 days after colony introduction. They recorded the forager density on 5×1 m^2 spots, the flight activity for 10 min every day at the hive entrance, the origin of the pollen collected by the workers, and the larval mortality by counting dead larvae inside and outside the colony and also by counting the number of larvae, egg cells, and cocoons from pictures taken every day. Counting dead larvae was almost impossible and the authors suggested that a special trap to assess larval loss should be devised. Data interpretation was difficult due to the kind of colony development which is unpredictable in bumble bees [60].

*Monitoring populations of native non-*Apis *bees*

The impact of chemical control of North American forest moths is of great concern for scientists and fruit growers close to treated areas and various methods have been used to assess the consequences of aerial sprays on native pollinators. Short-term effects were studied by observing 25 "sight units," each unit being a small blooming plot of about 0.8 m^2. The 25 units were on the same plant species. Each sighting conducted on warm hours of the day lasted 10 s. Weekly observations preceded and followed closely pesticide applications [61]. A quite different method was described for assessing bumble bee density. Twenty line transects were selected along road-sides. All these sample areas were classified in categories related to their "spray history." Each site was visited at least once by an observer walking at a constant pace. The caste, sex, and species of bumble bee were recorded and divided by the forage quantity in each transect thus giving a "bees per forage-mile" estimate. The forage quantity was calculated by measuring the length of each stand of the dominant visited plants [57]. Sampling wild bee population with a net was used for assessing the impact of fenitrothion on blueberry pollinators. One hundred sweeps were taken in selected flowering blueberry crops, during the warmest hours of 3 days of sampling [22].

Comparative toxicity and hazards of pesticides to non-*Apis* bees

Acute and chronic toxicity

Data on acute toxicity have been gathered in Tables 7.1 to 7.3. Table 7.1 shows that the median lethal dose (LD_{50}) in the leafcutting bee varied from 0.0003 to 30 μg/bee depending on the test substance [19, 26]. For *M. rotundata* the most toxic insecticides in topical tests were malathion and dicrotophos, whereas the least was carbaryl [19, 26], *N. melanderi* was also most susceptible to dicrotophos but less susceptible to fipronil than *M. rotundata* [26, 43]. Dieldrin toxicity was the highest to *T. spinipes* while that of carbaryl was the lowest [34]. The least susceptible species to carbaryl was *M. rotundata*, and the most, *T. spinipes*, with a ratio of about 41 [19, 34], the honey bee being intermediate. Deltamethrin was 76 times more toxic to *M. rotundata* females than to *B. terrestris* workers [27, 63]. This pyrethroid showed a similar toxicity to leafcutting bees and honey bees [27, 64]. With deltamethrin, trichlorfon, and carbophenothion, female leafcutting bees were about twice as tolerant as males [21, 27] Immature stages may be more susceptible to insecticides than adults. For example, after topical application, aldicarb was seven times more toxic to third instar larvae of *M. rotundata* than to adults [30]. The LD_{50} can also be expressed in μg/g of bee which is considered by some authors as a better approach to the intrinsic toxicity of a substance to bees [21, 30, 63]. According to several authors, the mean weights of *M. rotundata*, *N. melanderi*, *A. mellifera*, and *B. terrestris* are 0.036, 0.100, 0.118, and 0.190 g, respectively. In the case of deltamethrin the new LD_{50} in the female leafcutting bee, *B. terrestris*, and *A. mellifera* is 0.33, 4.8, and 0.08 μg/g (if we take 0.01 μg/bee as the LD_{50} for honey bees [64]). This means that honey bees are less tolerant to deltamethrin than leafcutting bees and bumble bees.

In Table 7.2 the feeding test with two IGRs revealed a much greater susceptibility of *B. terrestris* larvae to diflubenzuron than to fenoxycarb.

Table 7.1 Acute toxicity* of pesticides to four non-Apis bee species and honey bees (topical LD$_{50}$ (µg/bee))

Year/Ref.	Compound	Megachile rotundata	Nomia melanderi	Bombus terrestris	Trigona spinipes	Apis mellifera
1988/27	Deltamethrin	0.0052 (♂) 0.012 (♀)				
1987/63	Deltamethrin			0.91		
1982/64	Deltamethrin					0.01
1991/28	Tralomethrin	0.011				
1984/30	Aldicarb	0.4308				
1973/21	Trichlorfon	8.975 (♂) 18.488 (♀)				3.374
	Carbophenothion	0.154 (♂) 0.22 (♀)				1.491
1973/26	Trichlorfon	0.136	0.0465			0.0240
	Dieldrin	0.0036	0.0023			0.0006
	Parathion	0.0157	0.0015			0.0030
	Oxydemeton methyl	0.133	0.0082			0.0030
	Dicrotophos	0.0003	0.0010			0.0010
	Malathion	0.0005	0.0036			0.0020
1999/43	Fipronil	0.004	1.130			0.0130
1963/19	Carbaryl	30.50				1.2700
1989/34	Parathion				0.0956	
	Carbaryl				0.7472	
	Malathion				0.2649	
	Dieldrin				0.0289	
	Dicrotophos				0.1685	
1999/33	Imidacloprid			2.5		

Note
*If no indication of sex is given, the chemical has been tested against females.

Table 7.2 Oral toxicity of two IGRs to *Bombus terrestris* larvae (after Gretenkord and Drescher, 1996) [48]

Compound	Age of larvae (days)	LD_{50} (ng/larva)
Fenoxycarb	1	>650
	4	>1740
	6	>3710
Diflubenzuron	1	7.7
	4	52.9
	6	5112.0

Table 7.3 Oral toxicity of insecticides to three non-*Apis* bees and honey bees (oral LD_{50} (μg/bee))

Year/Ref.	Compound	Megachile rotundata	Nomia melanderi	Bombus terrestris	Apis mellifera
1999/33	Imidacloprid			0.04	
1984/30	Aldicarb	0.398 ($\male$)	0.41		0.071
		0.244 ($\female$)			
1993/62	Deltamethrin			0.6	
	Oxydemeton methyl			0.75	
	Pirimicarb			8.5	
	Phosalone			60.0	

Another striking difference was the rapid decrease of toxicity of diflubenzuron between 4-day-old and 6-day-old larvae [48].

Table 7.3 reveals that oral toxicity has not been investigated as much as topical toxicity. In *B. terrestris* the oral toxicity of phosalone was 1500 times lower than that of imidacloprid [33, 62]. The topical toxicity of imidacloprid in *B. terrestris* was 62 times lower than the oral toxicity [33]. *M. rotundata* and *N. melanderi* were less susceptible to aldicarb than honey bees. Contrary to the topical toxicity of other compounds [21, 27], the oral toxicity of aldicarb was lower in male leafcutting bees than in females [30].

A chronic feeding test was performed for 21 days with aldicarb, which showed medium lethal concentration values of 1.6, 2.0, and 3.9 mg/kg for honey bees, *N. melanderi*, and *M. rotundata*, respectively [30]. With IGR insecticides, LC_{50} estimated on young bumble bee larvae was higher for diflubenzuron than for fenoxycarb while the converse was observed for honey bees [48].

Susceptibility of bees to residues

Tests with contaminated paper

Through tests on paper it was possible to classify several pesticides used on blooming alfalfa, according to their hazards to male leafcutting bees. In a first experiment it was shown that toxicity decreased according to the following ranking: endosulfan > trichlorfon > phosalone > oxydemeton-methyl > pirimicarb. The last substance did not affect bees whereas residues of the other pesticides were still active after 5 days [65]. The mortality curves of the residual action of deltamethrin and fenvalerate did not have the typical aspect of those related to nonpyrethroid insecticides (Figure 7.1). This was an indication of the "knock-down" effect which was more marked in fenvalerate, a larger proportion of male leafcutting bees recovering, than in deltamethrin [23]. Alphamethrin residues at the field rate of 10 g/ha were less hazardous to *M. rotundata* males than phosalone at the rate of 1000 g/ha; after a 4-hour exposure the mortality rate recorded at 24 hours was 12 and 47 percent, respectively [36].

Tests with contaminated leaves

Acidified residues of trichlorfon were more efficient against pest insects. They were tested on alfalfa-treated leaves kept in Petri dishes and proved to be no more hazardous to leafcutting bees than the non-acidified compound. Conversely, mortality in honey bees was twice as much as that with

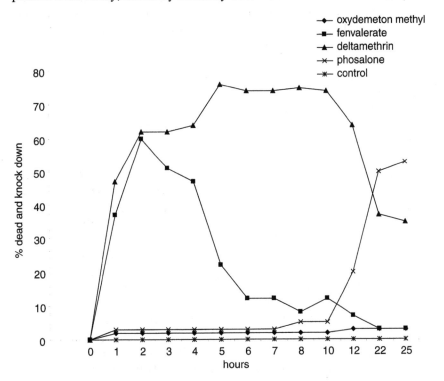

Figure 7.1 Mortality rate and knock-down effect of four insecticides against *Megachile rotundata* males exposed to residues on paper (after Tasei and Dinet, 1981) [23].

Table 7.4 Residues of naled and oxydemeton methyl recovered in alfalfa leaves, pollen, nectar and pollen ball of leafcutting bees (after George and Rincker, 1985) [54]

	Sampling interval (days following spray)	Leaves	Pollen	Nectar	Pollen ball
Naled/Dichlorvos* (mg/kg)	0	0.32/8.44	−/−	−/−	−/−
	1	0.63/1.37	nd**/3.99	0.20/nd**	−/−
	5	nd**/0.02	nd**/nd**	nd**/nd**	−/−
	13	nd**/0.34	−/−	−/−	nd**/nd**
Oxydemeton methyl (mg/kg)	0.5	56.4			−
	3	16.6			0.1
	14	1.2			0.3

Notes
*Dichlorvos is a metabolite of naled.
**not detected.

trichlorfon alone [38]. After a 24-hour contact with residues on alfalfa leaves, trichlorfon was more hazardous to *M. rotundata* than deltamethrin and methoxychlor. Males were more affected than females [44]. Residues of acephate and naled applied on alfalfa foliage with the stickers "Sur-tix®" and "Bond®" were less hazardous than residues without the stickers, whereas malathion caused 100 percent mortality even with the stickers [41]. Residues were measured in the leaf, pollen, and nectar of alfalfa treated with naled and oxydemeton methyl and pollen-nectar balls extracted from nests of leafcutting bees foraging the caged test flowers. More residues of the second insecticide were recovered in the leaves. A metabolite of naled (dichlorvos) was recovered in the pollen and leaves 1 day and 13 days after application. No residue could be detected in the pollen balls. Oxydemeton methyl residues were determined in pollen balls (Table 7.4). No adverse effect was observed on bees in the cages [54]. The residual toxicity of endosulfan, carbaryl, and trichlorfon was assessed on alfalfa foliage 3 hours after application. The mortality of the test insects with trichlorfon was 31, 5, and 17 percent, in *N. melanderi*, *M. rotundata*, and honey bees, respectively, which was considered as a low level, whereas with endosulfan the proportions were: 100, 71, and 11 percent. With carbaryl, the mortality rate of the three species was higher than 91 percent [37], which is consistent with a study reporting that female *M. rotundata* was affected when foraging alfalfa sprayed with carbaryl before bloom (Figure 7.2) [52]. The "Residual Time 25," which is the age of residues causing 25 percent mortality among the tested bees, was used by several authors to classify insecticides according to their hazards and recommend for late-evening sprays those with a RT 25 less than 8 hours. RT 25 estimation of field-weathered residues on alfalfa showed that tralomethrin was not hazardous to *M. rotundata* and

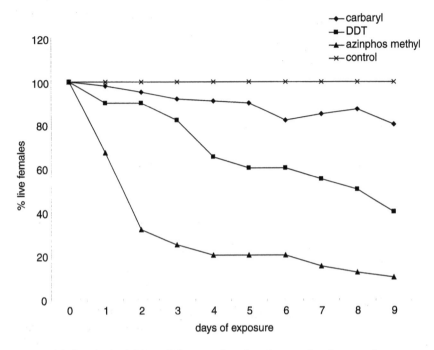

Figure 7.2 Survival of *Megachile rotundata* females nesting in greenhouse cages and exposed to *Medicago sativa* treated with three insecticides (after Waller, 1969) [52].

N. melanderi if applied late evening [28]. The same procedure testing the hazards of naled, imidacloprid, endosulfan, esfenvalerate, and oxydemeton methyl to *M. rotundata*, *N. melanderi*, and *B. occidentalis* revealed that the only insecticide with an RT 25 less than 8 hours in the three species was oxydemeton methyl [42]. It has been suggested that in contact testing and in the field practice, "insecticide pick-up" by pollinators, which is a parameter of hazards to bees, may be correlated with insect size. The pick-up, which is the ratio "weight of insecticide/weight of bee body" increases as the ratio "bee surface/bee volume." We can say that the larger the insect, the lower the pick-up since the volume and thus weight increase more rapidly than the surface [40]. According to this author, small bees such as *M. rotundata* (26 mg) and *N. melanderi* (87 mg) are more sensitive than large ones such as *B. centralis* (221 mg), honey bees being intermediate at 128 mg. The "surface/volume" ratios are: 1.0, 1.3, and 2.0 for *A. mellifera*, *N. melanderi*, and *M. rotundata*, respectively [39].

Susceptibility of larvae to contaminated food

Phosalone and alphamethrin were applied at 1000 and 10 g/ha on two experimental alfalfa fields where *M. rotundata* shelters were established.

Pollen ball samples were extracted from nests 5, 10, and 27 days following sprays, for residue analysis. No residues of the pyrethroid could be detected and phosalone concentration decreased from 1 to 0.1 mg/kg within the 3-week sampling period. Larval mortality was very stable in the four cell samples collected when the larval development was completed: before treatment 3.5 and 4.8 percent of larvae died in alphamethrin and phosalone samples versus 3.5 and 4.8 percent after treatment, respectively. No residues were detected in live larvae, which means that both molecules were metabolized [36]. Residues of deltamethrin were determined in leaf-cutting bee provisions collected in a field shelter placed in an alfalfa crop sprayed at the recommended rate. The maximum concentration was 0.01 mg/kg. A laboratory feeding test with pollen artificially contaminated with 1 mg/kg deltamethrin resulted in 55 percent larval mortality while no mortality occurred when the contamination rate was 0.1 mg/kg. It was concluded that the recommended dose of deltamethrin, 7.5 g/ha, was not hazardous to *M. rotundata* larvae [27]. Deltamethrin sprayed at 12.5 g/ha on rape, *Brassica napus oleifera*, was determined in anthers, nectar, bumble bee foragers and honey pots provisioned by a *B. terrestris* colony. In 1-day-old samples, residues were 0.2, 0.02, 0.15, and 0.005 mg/kg, respectively. A chronic feeding test using sugar solution contaminated at 0.01 and 0.2 mg/kg demonstrated that even the high-level concentration did not affect bumble bee larvae, which means that a dose twice as much as the recommended rate was nonhazardous to *B. terrestris* larvae [45]. After application at the recommended rates of aldicarb, dimethoate, carbofuran, and trichlorfon to alfalfa plots visited by leafcutting bees, pollen balls were sampled from their nests within various periods following treatment. No residues of the first three substances were detected whereas the maximum rate determined was 5 mg/kg for trichlorfon, and no larval mortality was observed in the plots treated with the three systemic insecticides whereas trichlorfon resulted in 22 percent dead larvae [53]. Two IGRs, diflubenzuron and fenoxycarb, were sprayed on caged *Phacelia* at the rate of 300 g/ha (recommended rate) and 1200 g/ha (double rate), respectively. Residues of diflubenzuron in pollen collected by *B. terrestris* ranged from 62 to 2 mg/kg within the 7-day period following application. During the same period, the figures for fenoxycarb varied from 217 to 7.5 mg/kg. Two days after application, diflubenzuron killed almost all the larvae except the old ones which was consistent with previous laboratory studies indicating an LC_{50} of 1.18 mg/kg and an LD_{50} 664 times higher in 6-day-old larvae than in 1-day-old ones [48]. In addition, during the whole period in the cage, colonies were not able to rear new brood even though queens continued to lay eggs. It was suggested that diflubenzuron had an ovicidal effect on queen ovaries. Normal eggs were laid when colonies were returned to the laboratory and fed with noncontaminated pollen. Fenoxy-carb was totally nonhazardous to *B. terrestris* whereas it is harmful to honey bees, and diflubenzuron which was safe to bumble bees is classified

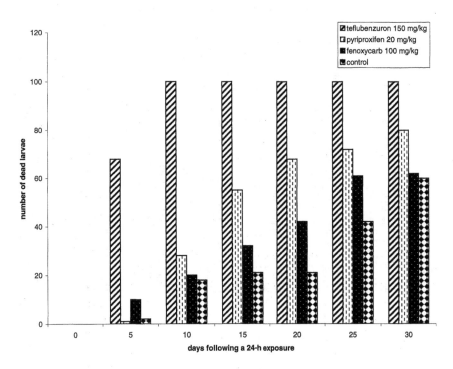

Figure 7.3 Mortality in *Bombus terrestris* larvae exposed to three IGR compounds (after de Wael *et al.*, 1995) [47].

as nonhazardous to *A. mellifera* [48]. The effects of the three IGRs, fenoxycarb, pyriproxyfen, and teflubenzuron, were compared in a laboratory study, using concentrations in bumble bee food based on the standard application rate in greenhouse, that is: 100, 20, and 150 mg/kg, respectively. After a 24-hour exposure, mortality records in larvae populations revealed no negative effects of fenoxycarb and pyriproxyfen, whereas teflubenzuron killed all the larvae that were ejected by *B. terrestris* workers (Figure 7.3). This substance also arrested egg development and no developing brood appeared for 5 weeks in the treated colony [47].

Susceptibility of non-Apis bees to field applications of pesticides

Few experiments in field conditions have been reported. The earliest one showed that a population of *M. rotundata* females reared in an alfalfa field was not reduced significantly after a treatment with trichlorfon in late evening. However, the number of cells completed per day was reduced during the post-treatment period (Figure 7.4) and the number of dead immature individuals was a maximum the day following application. It was concluded that trichlorfon was a short residual substance [58]. In a similar

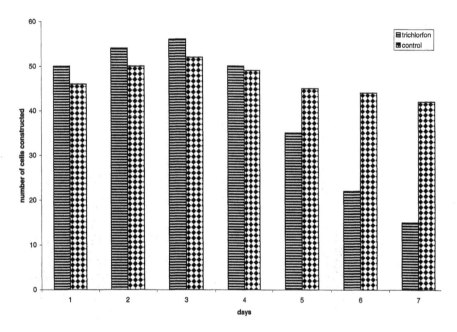

Figure 7.4 Egg-laying activity of *Megachile rotundata* in 50 nests in each of two experimental alfalfa plots, one of which was treated on Day 3 with trichlorfon (after Torchio, 1983) [58].

study deltamethrin and phosalone applications on four alfalfa plots resulted in low but significant losses of leafcutting bee females, compared to two control plots (Figure 7.5). From pollen analysis it was estimated that 70 and 60 percent of females, respectively, were exposed to the spray. When nests and larval development were completed the authors found a significant increase in the number of dead old larvae when bees were exposed to test compounds (Table 7.5) [59]. Leafcutting bee females were more affected by alphamethrin sprays applied to alfalfa fields at 10 g/ha.

Table 7.5 Mortality rates of different stages of *Megachile rotundata* progeny (after Tasei and Carre, 1985) [59]

	Deltamethrin	Phosalone	Control
Sample size	640	618	775
% Eggs and young larvae	2.8	5.2*	1.6
Prepupae	17.3*	12.8*	7.5
Pupae	1.9	0.5	0.9
Adults	0.6	0.5	0.5
Total	22.6*	18.9*	10.5

Note
*Mortality significantly higher than in control.

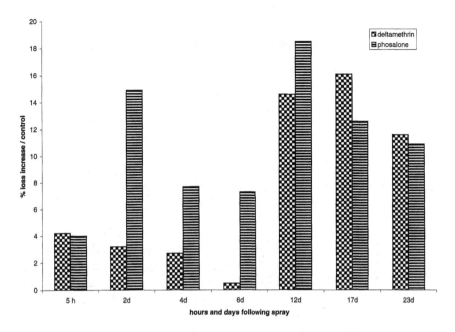

Figure 7.5 Increase of female losses in *Megachile rotundata* compared to a control
population, when shelters were placed in alfalfa fields treated with
deltamethrin and phosalone (after Tasei and Carre, 1985) [59].

The day following application 20 percent of nesting females had disappeared compared to 5.5 percent in the field treated with phosalone and 0.8
percent in the control. At the time of application, exposure rates of foragers were estimated at 82 and 74 percent. Larvae were slightly affected in
treated fields during the 3 days post-treatment, mortality reaching 4.8 and
5.4 percent, respectively, while it was 1.2 percent in the control [36].
Observations of *B. terrestris* colonies moved to two sunflower fields, one
treated with imidacloprid as a seed dressing, the other being nontreated,
showed that 60 percent of foragers in the treated field and 50 percent in
the control visited sunflowers during the 9-day test period. Losses of
workers accounted for 33 and 23 percent of the population marked the day
of introduction into fields, but the difference was not significant. It was
concluded that *B. terrestris* was not affected by the systemic properties of
imidacloprid seed dressing [49].

Short- and long-term impact of agrochemicals on native populations

Because of difficulties in determining the loss of bees in the native non-*Apis* bee population, the impact of agrochemicals on native pollinators and the consequences on crops and long-term effects on vegetation have not been well documented [66]. The main case story is that presented in 1979 on the effects of fenitrothion used for protecting Canadian forests. During a 3-year period beginning in 1971 this compound was sprayed on a large scale, then its use was discontinued. Monitoring studies over 8 years were conducted on the impact of forest sprays on bees and on the pollination of blueberries, the main crop in the spray areas of New Brunswick. After the depopulation following the 3 years of treatment, a slow recovery appeared where sprays were stopped. Migration, mostly for bumble bees, and resident reproduction, mostly for solitary bees, were the two causes of this recovery [67]. During the first period, a survey of pollinators in blueberry fields cultivated at various proximities to fenitrothion sprays showed that the diversity and abundance index in nearby crops was 5–6 times less than in distant ones. Collected bees were bumble bees, solitary ground nesting species belonging to *Andrenidae* and *Halictidae* families. The author presumed that blueberry production failures might be attributed in part to fenitrothion sprays which reduced blueberry pollinators [22]. A later study in the same Canadian province showed that bumble bees exposed in cages to fenitrothion sprays were strongly affected. The negative effects were observed at least 150 m from the flight path of the spray aircraft, thus indicating serious spray drift. A survey of the bumble bee population demonstrated that reduction in their densities was associated with fenitrothion sprays. This reduction persisted for at least 2 years after these treatments were discontinued, and the author presumed that 3 to 4 years were necessary for a total recovery. Apparent recovery may be due to movement of queens from unsprayed areas or local individuals emerging later [57]. Other compounds such as carbaryl, trichlorfon, acephate, and diflubenzuron were used in North America to control the spruce budworm and Douglas-fir tussock moth in Maine, Montana, and Pacific Northwest forests. In Maine, carbaryl sprayed at 0.84 kg/ha sometimes resulted in more than 50 percent mortality of native bees of *Andrena*, *Dialictus*, and *Nomada* genus. In addition, in sprayed areas the bee population depletion was associated with a lower fruit set of *Viburnum cassanoides*. In Montana, wild bee densities were not affected by carbaryl and trichlorfon sprays at 1.12 kg/ha but there was a significant reduction in the proportion of native bees of small size, belonging to the families *Megachilidae*, *Andrenidae*, and *Halictidae* [66]. In the Pacific Northwest, carbaryl and acephate depressed foraging populations of wild bees observed on flowering "sight units" and fruit production of bluebells (*Mertensia paniculata*) was significantly reduced in an area treated with acephate. Conversely, diflubenzuron at 0.275 kg/ha did not affect the native bee population and was not hazardous to the honey bee brood. This IGR was thus recommended for moth control [61].

Sublethal effects

Very few authors have investigated the sublethal effects of pesticides on non-*Apis* bees. Reported studies deal with repellency, knock-down, fecundity, longevity, lifespan, food uptake in adults, and growth rate of larvae.

Repellency

This symptom was first reported in 1948 on *Andrena flavipes* which was repelled by a DDT application [16]. Further assessments were those on *M. rotundata* [23] and *B. terrestris* [25], which reacted negatively to residues of the three pyrethroids: fenvalerate, deltamethrin, and lambda-cyhalothrin. Visits of plants treated with pyrethroids by *M. rotundata* females were reduced by 50 percent compared to control. Generally, bees approached the flowers but did not touch them, or if they did, the contact was very brief. This repellency lasted more than 1 hour with fenvalerate and more than 3 hours with deltamethrin [23].

Knock-down

One hour following a fenvalerate spray at 50 g/ha, more than 80 percent of *M. rotundata* males were "knocked down" and counted dead, but 62 percent of the treated population recovered within 14 hours. This was not the case for males sprayed with deltamethrin at 7.5 g/ha which were all killed. When males were maintained on dry residues of both compounds, the knock-down effect was also observed and recovery affected 44 and 59 percent of males treated with deltamethrin and fenvalerate, respectively [23].

Fecundity

The fecundity of *M. rotundata* females foraging in a field sprayed with trichlorfon was significantly affected, since 4 days after application the number of cells completed per female dropped by 66 percent compared to control [58].

Longevity

When applied to male leafcutting bees a dose of deltamethrin equal to $0.04 \times LD 1$ reduced their survival rate by 50 percent after 6 days. According to the authors, females reacted similarly [27]. In a chronic feeding test where sucrose solution and pollen dough contained 10 and 6 µg/kg imidacloprid, the longevity of *B. terrestris* workers was affected only during the first month of the 3-month trial and the survival rate was reduced by 10 percent [46]. Conversely, the lifespan of bumble bees was prolonged considerably by chronic ingestion of deltamethrin at 0.2 mg/kg [45].

Food consumption

Topical application of 0.01–0.02 μg deltamethrin per worker resulted in a significant increase in sucrose solution intake, whereas a reduced consumption was observed when workers were fed solutions containing 0.1–0.2 mg/kg deltamethrin [45].

Growth rate of larvae

In *M. rotundata*, the duration of the larval development was 2 days longer if pollen balls were contaminated with 0.1 mg/kg deltamethrin [27].

Metabolism and toxicity

Synergism and detoxification processes in non-*Apis* bees have been investigated by several scientists. One of the earliest articles discussed the selective toxicity of trichlorfon to honey bees and *M. rotundata*. This compound was 18–34 times more toxic to *A. mellifera* than to leafcutting bee females and it was hypothesized that this differential toxicity could be associated with the pH of the body fluid which is 6.0 and 6.8 in *A. mellifera* and *M. rotundata*, respectively, and induces a greater stability of the molecule in the honey bee body [21]. The speed of penetration in *M. rotundata* was investigated with radioactive carbaryl which showed a 14 percent penetration after 5 minutes, 24 percent after 1 hour, and 41 percent after 8 hours. In addition, eight metabolites were recovered in the organosoluble fraction of the leafcutting bees [68]. Carbaryl served as a model for some studies that aimed at understanding variations in the toxicity to *M. rotundata*. When leafcutting males aged, their susceptibility increased rapidly, the LD_{50} for 1-, 2-, 3-, 4-day-old males being 240, 166, 109, and 51 μg/g bee, respectively. In both sexes the lipid content and microsomal enzyme activity decreased with aging. Three drugs were tested on leafcutting bees prior to carbaryl application. The first one, piperonyl butoxide, resulted in a strong synergy since LD_{50} dropped from 245 to 11 μg/g bee for 1-day-old females. The synergist ratio (LD_{50} of carbaryl alone/LD_{50} of carbaryl + piperonyl) decreased when females aged. With the second drug, chlorcyclizine, the LD_{50} for males was doubled whereas the third, aminopyrine, reduced the LD_{50} for females. The LC_{50} of carbaryl was 81.8 mg/kg for the compound alone and 47.5 mg/kg when piperonyl was added [20, 69, 70]. Experiments with radioactive carbaryl revealed a maximum persistence of the molecule when aminopyrine was used while chlorcyclizine reduced persistence and phenobarbital had no effect on it [71]. Chlorcyclizine modified the midgut structure and increased the susceptibility of *N. melanderi* to parathion [72]. More recent biochemical studies have shown that *M. rotundata* possesses seven enzymes susceptible to organophosphate inhibition [73]. Since serine esterases are the major target of organophosphate insecticides, they were used to establish the kinetics of cytosolic esterases of *M. rotundata* females and to estimate the effects of four organophosphorus compounds: naled, trichlorfon, oxydemeton methyl, and paraoxon. The method was based on the measurement of hydrolysis of *p*-nitrophenylacetate by cytosolic preparations of leafcutting bees. It was demonstrated that a mixed mechanism of inhibition was involved and that the order of toxicity, based on inhibition constants, was: naled > paraoxon > trichlorfon > oxydemeton methyl [74]. The similarity of LD_{50} for trichlorfon and oxydemeton methyl [26] suggests that these compounds have different penetration speeds and, in addition, other enzyme systems such as polysubstrate mono-oxygenases or glutathione *S*-transferases, which may metabolize the insecticides before they reach their target, have been removed [74].

Conclusion

The use of domestication techniques has made it possible to assess the toxicity of some compounds used for pest control to a restricted number of non-*Apis* bee species. Standardized laboratory tests have enabled comparative studies to be performed which demonstrate that susceptibility to a compound can vary according to species; for example, *B. terrestris* is 60 to 90 times more tolerant to deltamethrin than *A. mellifera*. In addition, large differences in toxicity to a bee species appear within the same insecticide category. An example among IGRs is the toxicity of diflubenzuron to bumble bees which is 85 times higher than that of fenoxycarb. Owing to deficiencies in method harmonization, authors who used the same bee species have not always agree with one another when attributing toxicity ranks to identical series of pesticides.

A great variety of materials and procedures have been used for estimating hazards of pesticides to bees by either the semi-field or the field method. Cages and greenhouses were generally preferred because exposure rates cannot be controlled in the field.

Studies on detoxification in *M. rotundata* did not agree with previous toxicological data and thus expected hazards in the field. It was assumed that before a pesticide reaches its biochemical target several factors of major importance intervene in the contamination process. In particular, one should pay attention to the following:

* Insecticide pick-up depends on the insect size and is related to the ratio "surface/volume." Therefore, small bees are more sensitive than large ones.
* Penetration speed through the cuticle may be variable.
* Aged bees are more susceptible than callow individuals.
* Males are less tolerant than females.
* Degradation of pesticide in bees may depend on the pH of insect fluid, which may vary between species.

Low doses of several compounds, deltamethrin, trichlorfon, and imida-cloprid, tested on *M. rotundata* and *B. terrestris*, resulted in various sub-lethal effects: repellency, knock-down, reduced fecundity, longevity or food consumption, and prolonged larval development.

Although the assessment of ecological consequences of temporary or permanent pest control by insecticides met technical difficulties, it has been shown that a population of small bees was more likely to be depleted than that of large species (bumble bees). Recovery was also more rapid with bumble bees, due to migration of queens from untreated areas, while solitary species recovered mostly through local reproduction. Reduction of fruit sets in some crops pollinated by native bees was associated with depression of pollinator population.

Additionally, exposure profiles of honey bees, bumble bees, and soli-tary bees differ significantly, due to respective flight activity hours, flight seasons, foraging habits, and nesting behavior, which result in different ecological impacts of pest management by agrochemicals.

References

1 Michener, C.D. (1974). *The Social Behavior of the Bees. A Comparative Study.* The Belknap Press of Harvard University Press, Cambridge, MA, p. 404.

2 Williams, I.H. (1996). Aspects of bee diversity and crop pollination in the European Union. In: *The Conservation of Bees* (Matheson, A., Buchmann, S.L., O'Toole, C., Westrich, P. and Williams, I., Eds). Academic Press, pp. 63–80.

3 Richards, K.W. (1996). Comparative efficacy of species for pollination of legume seed crops. In: *The Conservation of Bees* (Matheson, A., Buchmann, S.L., O'Toole, C., Westrich, P. and Williams, I., Eds). Academic Press, pp. 81–104.

4 Malyshev, S.I. (1935). The nesting habits of solitary bees. A comparative study. *Eos* **II**, 201–309 (plates III–XV).

5 Stephen, W.P., Bohart, G.E. and Torchio, P.F. (1969). *The Biology and Exter-nal Morphology of Bees.* Agricultural Experimental Station, Oregon State Uni-versity, Corvallis, p. 140.

6 Alford, D.V. (1975). *Bumblebees.* Davis-Pointer, London, p. 352.

7 Stephen, W.P. (1962). *Propagation of the Leaf-cutter Bee for Alfalfa Seed Pro-duction.* Station Bulletin 586, Agricultural Experimental Station, Oregon State University, Corvallis, p. 16.

8 Krunic, M.D., Tasei, J.N. and Pinzauti, M. (1995). Biology and management of *Megachile rotundata* Fabricius under European conditions. *Apicoltura* **10**, 71–97.

9 Stephen, W.P. (1960). Artificial bee beds for propagation of the Alkali bee, *Nomia melanderi. J. Econ. Entomol.* **53**, 1025–1030.

10 Donovan, B.J. and Macfarlane, R.P. (1984). Bees and pollination. In: *New Zealand Pests and Beneficial Insects* (Scott, R.R., Ed.). Lincoln University College of Agriculture, New Zealand, pp. 247–270.

11 Torchio, P.F. (1991). Use of *Osmia lignaria propinqua* (Hymenoptera:

Megachilidae) as a mobile pollinator of orchard crops. *Environ. Entomol.* **20**, 590–596.

12 Sekita, N. and Yamada, M. (1993). Use of *Osmia cornifrons* for pollination of apples in Aomori prefecture, Japan. *Jarq* **26**, 264–270.

13 Bosch, J. (1994). Improvement of field management of *Osmia cornuta* (Latreille) (Hymenoptera, Megachilidae) to pollinate almond. *Apidologie* **25**, 71–83.

14 van Heemert, C., de Ruijter, A., van den Eijnde, J. and van der Steen, J. (1990). Year-round production of bumble bee colonies for crop pollination. *Bee World* **71**, 54–56.

15 Shaw, F.R. (1946). Some observations on the effect of a 5 percent DDT dust on bees. *Can. Entomol.* **58**, 110.

16 Way, M.J. and Synge, A.D. (1948). The effects of DDT and of benzene hexachloride on bees. *Ann. Appl. Biol.* **35**, 94–109.

17 Bohart, G.E. and Lieberman, F.V. (1949). Effect of an experimental field application of DDT dust on *Nomia melanderi*. *J. Econ. Entomol.* **42**, 519–522.

18 Linsley, E.G., McSwain, J.W. and Smith, R.F. (1950). Comparative susceptibility of wild bees and honey bees to DDT. *J. Econ. Entomol.* **43**, 59–62.

19 Johansen, C., Jaycox, E.R. and Hutt, R. (1963). *The Effect of Pesticides on the Alfalfa Leafcutting Bee Megachile rotundata*. Stations Circular 418, Washington Agricultural Experiment Stations, Institute of Agricultural Sciences, Washington State University, p. 12.

20 Lee, R.M. (1972). Xenobiotic detoxication potential and drug induced changes in carbaryl toxicity in the alfalfa leafcutter bee, *Megachile rotundata* (Fabricius). *Diss. Abst.*, Int. serie B. 33, no. 1.

21 Ahmad, Z. and Johansen, C. (1973). Selective toxicity of carbophenothion and trichlorfon to the honey bee and the alfalfa leafcutting bee. *Environ. Entomol.* **2**, 27–30.

22 Kevan, P.G. (1975). Forest application of the insecticide fenitrothion and its effects on wild pollinators (Hymenoptera: Apoidea) of lowbush blueberries (*Vaccinium* spp.) in southern New Brunswick, Canada. *Biol. Conserv.* **7**, 301–309.

23 Tasei, J.N. and Dinet, P. (1981). Effets comparés de deux pyréthrinoides de synthèse et de trois insecticides organophosphorés sur les mégachiles (*Megachile rotundata* F. = *pacifica* Pz.). *Apidologie* **12**, 363–376.

24 Mizuta, H.M. and Johansen, C.A. (1972). *The Hazard of Plant-Systemic Insecticides to Nectar-Collecting Bees*. Technical Bulletin 72, Washington Agricultural Experiment Station, College of Agriculture, Washington State University, p. 8.

25 Gretenkord, C. and Drescher, W. (1993). Development of a cage test method for the evaluation of pesticide hazards to the bumble bee *Bombus terrestris* L. In: *Proceedings of the Vth International Symposium on The Hazards of Pesticides to Bees* (Harrison, E.G., Ed.). Oct. 26–28, 1993, Wageningen, Netherlands, Appendix 20, pp. 151–154.

26 Torchio, P.F. (1973). Relative toxicity of insecticides, to the honey bee, alkali bee and alfalfa leafcutting bee (Hymenoptera: Apidae, Halictidae, Megachilidae). *J. Kansas Entomol. Soc.* **46**, 446–453.

27 Tasei, J.N., Carre, S., Moscatelli, B. and Grondeau, C. (1988). Recherche de la DL 50 de la deltamethrine (Decis) chez *Megachile rotundata* F. abeille

pollinisatrice de la luzerne (*Medicago sativa* L.) et des effets de doses infraléthales sur les adultes et les larves. *Apidologie* **19**, 291–306.

28 Mayer, D.F., Lunden, J.D. and Williams, R.E. (1991). Tralomethrin insecticide and domesticated pollinators. *Am. Bee J.* **131**, 461–463.

29 Torchio, P.F. and Youssef, N.N. (1973). A method of feeding bees measured amounts of insecticides in solution. *Can. Entomol.* **105**, 1011–1014.

30 Johansen C.A., Rincker, C.M., George, D.A., Mayer, D.F. and Kious C.W. (1984). Effects of aldicarb and its biologically active metabolites on bees. *Environ. Entomol.* **13**, 1386–1398.

31 Stevenson, J.H. and Racey, P.A. (1967). Toxicity of insecticides to bumble bees. In: *Report of Rothamsted Experiment Station for 1966.* (AFRC Institute of Arable Crops Research, Rothamsted Experimental Station, Harpenden, UK), p. 176.

32 van der Steen, J.J.M., Gretenkord, C. and Schaefer, H. (1996). Method to determine the acute contact LD_{50} of pesticides for bumble bees (*Bombus terrestris* L.). In: *Proceedings of the VIth International Symposium on Hazards of Pesticides to Bees* (Lewis, G., Ed.). Sept. 17–19, 1996, Braunschweig, Germany, Appendix 28.

33 Bortolotti, L., Grazioso, E., Porrini, C. and Sbrenna, G. (2001). Effect of pesticides on bumblebee, *Bombus terrestris* L., in the laboratory. In: *Hazards of Pesticides to Bees*, Vol. 98 (Belzunces, L., Pelissier, C. and Lewis, G., Eds). INRA, pp. 217–225.

34 Macieira, O.J.D. and Hebling-Beraldo, M.J.A. (1989). Laboratory toxicity of insecticides to workers of *Trigona spinipes* (F., 1793) (Hymenoptera, Apidae). *J. Apic. Res.* **28**, 3–6.

35 Tasei, J.N. (1977). Méthode de test de toxicité des insecticides applicable aux abeilles solitaires et plus particulièrement à *Megachile pacifica* Pz. (Hymenoptera Megachilidae). *Apidologie* **8**, 129–139.

36 Tasei, J.N. and Carre, S. (1987). Effects of the pyrethroid insecticide, WL 85871 and phosalone on adults and progeny of the leaf-cutting bee *Megachile rotundata* F. pollinator of lucerne. *Pestic. Sci.* **21**, 119–128.

37 Johansen, C. and Eves, J. (1967). *Toxicity of Insecticides to the Alkali Bee and the Alfalfa Leafcutting Bee.* Washington Agricultural Experiment Station, College of Agriculture, Washington State University, Circular 475.

38 Johansen, C. and Eves, J. (1972). Acidified sprays, pollinator safety, and integrated pest control on alfalfa grown for seed. *J. Econ. Entomol.* **65**, 546–551.

39 Johansen, C.A., Mayer, D.F., Eves, J. and Kious, C.W. (1983). Pesticides and bees. *Environ. Entomol.* **12**, 1513–1518.

40 Johansen, C.A. (1972). Toxicity of field-weathered insecticide residues to four kinds of bees. *Environ. Entomol.* **1**, 393–394.

41 Mayer, D.F., Johansen, C.A., Lunden, J.D. and Rathbone, L. (1987). Bee hazard of insecticides combined with chemical stickers. *Am. Bee J.* **127**, 493–495.

42 Mayer, D.F., Patten, K.D., Macfarlane, R.P. and Shanks, C.H. (1994). Differences between susceptibility of four pollinator species (Hymenoptera: Apoidea) to field weathered insecticide residues. *Melanderia* **50**, 24–28.

43 Mayer, D.F. and Lunden, J.D. (1999). Field and laboratory tests of the effects of fipronil on adult female bees of *Apis mellifera*, *Megachile rotundata* and *Nomia melanderi. J. Apic. Res.* **38**, 191–197.

44 Charnetski, W.A. (1988). Toxicity of insecticides to the alfalfa leafcutter bee, *Megachile rotundata* F. (Hymenoptera: Megachilidae), established by three evaluation methods. *Can. Entomol.* **120**, 297–305.

45 Tasei, J.N., Sabik, H., Pirastru, L., Langiu, E., Blanche, J.M., Fournier, J. and Taglioni, J.P. (1994). Effects of sublethal doses of deltamethrin (Decis Ce) on *Bombus terrestris. J. Apic. Res.* **33**, 129–135.

46 Tasei, J.N., Lerin, J. and Ripault, G. (2000). Sub-lethal effects of imidacloprid on bumblebees, *Bombus terrestris* (Hymenoptera: Apidae), during a laboratory feeding test. *Pest. Manag. Sci.* **56**, 784–788.

47 De Wael, L., de Greef, M. and van Laere, O. (1995). Toxicity of pyriproxifen and fenoxycarb to bumble bee brood using a new method for testing insect growth regulators. *J. Apic. Res.* **34**, 3–8.

48 Gretenkord, C. and Drescher, W. (1996). Laboratory and cage test methods for the evaluation of the effects of growth regulators (Insegar®, Dimilin®) on the brood of *Bombus terrestris* L. In: *Proceedings of the VIth International Symposium on Hazards of Pesticides to Bees* (Lewis, G., Ed.). Sept. 17–19, Braunschweig, Germany, Appendix 29.

49 Tasei, J.N., Ripault, G. and Rivault, E. (2001). Effects of Gaucho® seed coating on bumblebees visiting sunflower. In: *Hazards of Pesticides to Bees*, Vol. 98 (Belzunces, L., Pelissier, C. and Lewis, G., Eds). INRA, pp. 207–212.

50 Tornier, I. (2001). Side effects of an insect growth regulator on bumble-bees and honey-bees. In: *Hazards of Pesticides to Bees*, Vol. 98 (Belzunces, L., Pelissier, C. and Lewis, G., Eds). INRA, p. 299.

51 Thompson, H. and Barrett, K. (2001). Assessing the effects of glasshouse application of a novel insect growth regulator on bumble bee colonies. In: *Hazards of Pesticides to Bees*, Vol. 98 (Belzunces, L., Pelissier, C. and Lewis, G., Eds). INRA, pp. 227–228.

52 Waller, G.D. (1969). Susceptibility of an alfalfa leafcutting bee to residues of insecticides on foliage. *J. Econ. Entomol.* **62**, 189–192.

53 George, D.A. and Rincker, C.M. (1982). Residues of commercially used insecticides in the environment of *Megachile rotundata. J. Econ. Entomol.* **75**, 319–323.

54 George, D.A. and Rincker, C.M. (1985). Results and conclusions of using pesticides with the alfalfa leafcutting bee in the production of alfalfa seed. *J. Agric. Entomol.* **2**, 93–97.

55 Rincker, C.M. and George, D.A. (1985). Effect of pesticide residues in alfalfa pollen and nectar on the foraging and reproduction activities of alfalfa leafcutting bees *Megachile rotundata. J. Appl. Seed Prod.* **3**, 33–37.

56 Heller, J.J., Baumeister, R., Mattioda, H. and Tasei, J.N. (1990). A new method to test the hazards of insecticide spraying on *Megachile rotundata* F.: Trials in tunnels. In: *Proceedings of the IVth International Symposium on The Harmonization of Methods for Testing the Toxicity of Pesticides to Bees* (Research Institute of Apiculture at Dol, Czechoslovakia, Ed.). May 15–18, Rez, Czechoslovakia, pp. 41–45.

57 Plowright, R.C., Pendrel, B.A. and McLaren, I.A. (1978). The impact of aerial fenitrothion spraying upon the population biology of bumble bees (*Bombus* Latr.: Hym.) in South Western New Brunswick. *Can. Entomol.* **110**, 1145–1156.

58 Torchio, P.F. (1983). The effects of field applications of naled and trichlorfon on the alfalfa leafcutting bee, *Megachile rotundata* (Fabricius). *J. Kansas Entomol. Soc.* **56**, 62–68.

59 Tasei, J.N. and Carre, S. (1985). Effets du traitement de la luzerne en fleurs (*Medicago sativa* L.) avec de la deltamethrine et de la phosalone sur l'abeille solitaire: *Megachile rotundata* F. (Hym., Megachilidae). *Acta Oecol./Oecol. Appl.* **6**, 165–173.

60 Schafer, H. and Muhlen, W. (1996). First experiences to test side-effect of alsystin on bumble bees (*Bombus terrestris* L.) in the field. In: *Proceedings of the VIth International Symposium on Hazards of Pesticides to Bees* (Lewis, G., Ed.). Sept. 17–19, Braunschweig, Germany, Appendix 30.

61 Robinson, W.S. and Johansen, C.A. (1978). Effects of control chemicals for Douglas-fir Tussock moth *Orgia pseudotsugata* (McDonnough) on forest pollination (Lepidoptera: Lymantriidae). *Melanderia* **30**, 9–56.

62 Gretenkord, C. and Drescher, W. (1993). Effects of 4 pesticides (Decis, Metasystox, Pirimor, Rubitox) on the bumblebee *Bombus terrestris* L.: Determination of the oral LD_{50} and preliminary results with semi-field tests. *Apidologie* **24**, 519–521.

63 Tasei, J.N., Carre, S., Grondeau, C. and Hureau, J.M. (1987). Effets d'applications insecticides à l'égard d'apoides pollinisateurs autres que l'abeille domestique (*Megachile rotundata* F. et *Bombus terrestris* L.). In: *Proceedings of the International Conference on Pests in Agriculture* (Association Nationale de Protection des Plantes, Ed.). Dec. 1–3, Paris, pp. 127–136.

64 Bos, C. and Masson, C. (1982). Toxicity and repellent effect of the synthetic pyrethroids on bees: Methodological aspects. In: *Proceedings of the IInd International Symposium on The Harmonization of Methods for Testing the Toxicity of Pesticides to Bees* (Viel, J.A. and Debray, Ph., Eds). Sept. 21–23, Hohenheim, Germany, Appendix 9.

65 Tasei, J.N., Capou, J. and Michaud, D. (1977). Action de quelques insecticides sur une abeille solitaire: *Megachile pacifica* Panz. (Hym., Megachilidae). *Apidologie* **8**, 111–127.

66 NRCC (1981). *Pesticide–Pollinator Interactions.* Environmental Secretariat, publication no. 18471, NRCC/CNRC, Ottawa, Canada.

67 Kevan, P.G. and LaBerge, W.E. (1978). Demise and recovery of native pollinator populations through pesticide use and some economic implications. In: *Proceedings of the IVth International Symposium* on *Pollination* (Caron, D.M., Ed.). Oct. 11–13, College Park, MD, USA, pp. 489–508.

68 Guirguis, G.N. and Brindley, W.A. (1975). Carbaryl penetration into and metabolism by alfalfa leafcutting bees, *Megachile pacifica*. *J. Agric. Food Chem.* **23**, 274–279.

69 Guirguis, G.N. and Brindley, W.A. (1974). Insecticide susceptibility and response to selected pollens of larval alfalfa leafcutting bees, *Megachile pacifica* (Panzer) (Hym.: Megachilidae). *Environ. Entomol.* **3**, 691–694.

70 Lee, R.M. and Brindley, W.A. (1974). Synergist ratios, EPN detoxication, lipid and drug-induced changes in carbaryl toxicity in *Megachile pacifica*. *Environ. Entomol.* **3**, 899–907.

71 Guirguis, G.G. and Brindley, W.A. (1976). Effect of chlorcyclizine, aminopyrine, or phenobarbital on carbaryl metabolism in alfalfa leafcutting bees. *Environ. Entomol.* **5**, 590–594.

72 Moradeshaghi, M.J., Brindley, W.A. and Youssef, N.N. (1974). Chlorcyclizine and SKF 525A effects on parathion toxicity and midgut tissue structure in alkali bees, *Nomia melanderi*. *Environ. Entomol.* **3**, 455–463.

73 Frohlich, D.R., Burris, T.E. and Brindley, W.A. (1989). Characterization of glutathione *S*-transferases in a solitary bee, *Megachile rotundata* (Fab.) (Hym.: Megachilidae), and inhibition by chalcones, flavone, quercitin and tridiphanediol. *Comp. Biochem. Physiol.* **94B**, 661–665.

74 Frolhlich, D.R., Boeker, E.A. and Brindley, W.A. (1990). The kinetics and inhibition of *p*-nitrophenylacetate-hydrolysing esterases in solitary bee, *Megachile rotundata* (Fab.). *Xenobiotica* **20**, 481–487.

8 Honey bees as indicators of radionuclide contamination

A truly useful biomonitor?

T.K. Haarmann

Summary

The concept of using honey bees as indicators of the presence of environmental contaminants continues to receive much deserved attention around the globe. Many studies have demonstrated that honey bees can be used successfully to sample an area for environmental contaminants. Honey bees are currently being used to monitor a variety of environmental pollutants including many trace elements and radionuclides. Information collected from these monitoring programs can support the ongoing attempts to assess the influences of contaminants on living systems and their impacts to ecosystems. In addition, comparing the concentration of contaminants in the hive and bees to the known concentrations in the surrounding area is useful in modeling the redistribution of contaminants through ecosystems. Understanding the dynamics of the interactions between honey bees and contaminants becomes a critical component in interpreting the data collected as part of a monitoring program. In particular, incorporating honey bees into an environmental monitoring program designed to examine radionuclides presents unique issues and problems. While honey bees can be indicators of radionuclide contamination, how truly useful are they? This chapter describes a series of field experiments designed to examine some of the pros and cons of using honey bees in this capacity.

Introduction

Background

Many facilities around the world are actively involved in the research and development of nuclear-related materials and the production of nuclear energy. Inherent in the many processes involved in this type of work is the production of radioisotopes. Unfortunately, some of these radionuclide waste products have found their way into surrounding natural areas. Historically, sampling for environmental contaminants has been done on the various abiotic components (i.e. water and soil) of an ecosystem and has often excluded the sampling of many of the biotic components. The ongoing interest in assessing the influences of contaminants on living systems has generated questions on how best to incorporate sampling data into ecological risk assessment models. The primary concerns involve determining which methods are best to monitor these contaminants and how to analyze the influences these contaminants have on biological systems. How might we integrate sampling of both biotic and abiotic components of an ecosystem?

One innovative sampling method incorporates insects – honey bees (*Apis mellifera*) – as monitors of environmental contamination. Using honey bees as indicators of radionuclide contamination is an inexpensive form of environmental monitoring, especially considering the numerous sampling points the foraging bees visit. Sampling at one location (the hive) can provide information from various points across a landscape relative to the distribution and bioavailability of contaminants. Comparing the concentration of contaminants in the hive products or the honey bees to the known concentrations in the surrounding area can be useful in modeling the redistribution of contaminants through ecosystems. The nature of honey bee ecology makes them an excellent living system from which to monitor the presence of contaminants and explore their impacts.

Past research has demonstrated that honey bees are useful indicators of environmental contamination [1–3]. Honey bees can be thought of as mobile samplers that efficiently cover a large sample area and then return to a central location [4]. Honey bees forage in an area with a radius as large as 6 km and often cover a total area up to 100 square km [5, 6]. Each hive contains thousands of bees, most of whom will forage for nectar, water, pollen, and plant resins, which are all brought back into the hive. During these foraging flights, bees inadvertently contact and accumulate a wide array of pollutants, some of which are brought back to the colony [7]. These contaminants often become incorporated into the bee tissue, the wax, the honey, or the hive itself [8]. Honey bees have been used in the past to monitor the presence and distribution trace elements, including fluoride [9, 10], lead [11], zinc [12], nickel [13], and potassium [14], and the bioavailability of radionuclides [15–17], including cesium [17, 18], tritium [19, 20], and plutonium [21].

Unfortunately, there are still many gaps in our knowledge concerning the use of honey bees as indicators of contamination. Specifically, there are many unanswered questions concerning the dynamics of radionuclide redistribution through ecological systems. One question is often asked – Do we understand enough about honey bees as indicators of radionuclides to successfully incorporate them into an environmental monitoring or surveillance program?

This chapter will explore the issue of using honey bees as monitors by reviewing several recent studies conducted at the United States

Department of Energy's Los Alamos National Laboratory (LANL). LANL, which is located in north-central New Mexico, has been involved in the research and development of nuclear-related materials for the past five decades and is an excellent location to conduct this type of research.

Experimental questions

A series of field experiments were conducted to investigate various aspects of using honey bees as monitors. The goal of this research was to understand the feasibility, including the limitations, of using honey bees in this capacity. The experiments were designed to include research into some basic issues, such as comparing the consistency of analytical sample results collected from similar bee colonies, to more complex questions addressing the dynamics of radionuclide redistribution through an ecosystem. Specifically, as part of these field experiments, the following questions were explored:

- Do bee tissue samples taken from the same colony yield the same results?
- Do bee tissue samples taken from similar colonies under similar conditions yield the same results?
- Is there an accumulation of radionuclides within colonies over time?
- Might the proportion of forager bees to nurse bees in a particular sample influence the radionuclide contaminant levels found in that sample?
- How does the radionuclide concentration in flowers influence the levels of contaminants found in the bees?
- What is the primary source of contamination in the study site: water or nectar?
- Are the levels of contaminants in the bees, flowers, and water correlated, and do they demonstrate similar trends over time?
- Is there an observable bioaccumulation of radionuclides within bees or flowers?

Field experiments

This section of the chapter will briefly review the LANL field studies and the results of these studies. The significance of each of these experiments will be examined in the Discussion section of this chapter. Field research was conducted at LANL during 1994, 1995, and 1996. The study site was located adjacent to a 7-million-liter, radioactive waste lagoon that contained known bioavailable contamination including tritium, cobalt-56, cobalt-60, manganese-54, sodium-22, and tungsten-181. The lagoon was the nearest source of water for the colonies in the experiment.

Variability study

The primary focus of this study was to address the basic question – How consistent are the radionuclide concentrations in bee samples? If one of the primary objectives is to eventually use data collected from honey bees as part of an environmental monitoring program, or more importantly, as input into an ecological risk assessment model, then one would hope there is a certain degree of consistency between samples. In other words, if 25 samples were collected from a beehive, and each one was analyzed for tritium, one would assume there would be relative consistency between the radiochemical analytical results. A large disparity in the concentrations of tritium in bee samples would make the results suspect. In this study, first the consistency of bee samples collected from a single colony was examined. Second, the consistency of samples collected from several colonies in the same location was assessed.

As part of this experiment, a series of honey bee samples was collected from colonies located at the LANL study site near the radioactive lagoon, and analyzed for concentrations of radionuclides (gamma-emitting nuclides, uranium, and tritium). There were two groups of colonies used in the experiment. One group had been located at the study site for 4 months, the other group for several years. A detailed description of this experiment is described in Haarmann [22]. Table 8.1 shows an example of the data that were collected as part of this study.

The results indicated that generally a low variability in radionuclide concentrations existed between samples collected within the same colony. Furthermore, results indicated that a higher variability existed between samples that were collected from adjacent colonies.

Accumulation study

In the past, there have been various environmental surveillance programs that have used honey bees as monitors of radionuclide contamination. Typically, beehives are placed around a facility or particular region, and samples are collected on a regular basis. The hives used in this type of monitoring program are often located at the site year after year. Often, the scientists in charge of these monitoring programs have contaminant/honey bee data dating back several years, if not decades. As an example, let us suppose that one of these scientists is interested in using these long-term data to estimate the concentration of radionuclides in the environment based on the levels of radionuclides in the bees? If the bee samples were collected from the same hive for several years in a row, are the results reflective of what is really environmentally bioavailable to honey bees, or simply a reflection of the accumulation of contaminants within that particular hive? The accumulation study was designed to examine data collected at the study to address the question – Is there an accumulation of radionuclides within colonies over time?

Table 8.1 An example of the data collected during the LANL variability study

Colony	Sample	Tritium (pCi/ml)	Analytical uncertainty	Cobalt-57 (pCi/g)	Analytical uncertainty	Cobalt-60 (pCi/g)	Analytical uncertainty	Manganese-54 (pCi/g)	Analytical uncertainty	Sodium-22 (pCi/g)	Analytical uncertainty
New 1	1	176.55	3.05	29.75	7.37	1.67	0.38	1.50	0.52	7.69	0.92
	2	171.79	2.97	30.33	7.93	1.71	0.38	1.65	0.72	7.41	0.92
	3	173.10	2.99	28.86	7.39	<0.17*	NA	<0.28	NA	7.25	0.90
	4	168.35	2.91	29.75	8.15	1.38	0.38	1.50	0.43	6.58	0.82
	5	171.30	2.96	28.07	7.30	1.30	0.32	1.51	0.52	6.83	0.87
New 2	1	141.50	2.49	32.16	8.24	1.77	0.29	1.57	0.51	5.71	0.70
	2	150.78	2.64	29.17	7.36	1.76	0.37	1.43	0.45	5.79	0.72
	3	148.62	2.62	29.18	7.39	1.55	0.30	1.49	0.56	5.97	0.73
	4	149.00	2.62	31.74	8.19	1.63	0.31	1.94	0.63	6.12	0.77
	5	147.40	2.59	26.90	6.50	1.93	0.34	1.98	0.59	6.49	0.81
Old 1	1	400.74	6.73	119.57	32.14	4.28	0.75	<0.65	NA	10.26	1.28
	2	396.79	6.66	99.19	25.64	4.61	0.66	2.93	0.94	11.19	1.36
	3	401.95	6.75	108.93	28.36	4.69	0.68	2.71	0.72	10.71	1.30
	4	407.69	6.85	114.74	29.30	5.27	0.75	2.95	0.67	11.26	1.36
	5	405.56	6.81	90.95	22.26	4.69	0.68	3.02	0.98	11.68	1.40
Old 2	1	693.43	11.56	58.96	13.75	3.40	0.53	2.24	0.46	12.68	1.48
	2	702.34	11.71	9.32	1.25	2.97	0.39	1.15	0.35	14.02	1.37
	3	692.59	11.53	56.04	13.28	3.25	0.52	2.08	0.69	12.85	1.50
	4	690.47	11.50	46.72	10.46	3.20	0.49	2.22	0.69	13.45	1.55
	5	714.46	11.91	49.03	10.80	3.89	0.59	2.26	0.66	14.07	1.60

Note
*<signifies a below detection limit value.

To explore this issue, bee samples from colonies that had been located at the study site for several years were compared to bee samples that had been collected from colonies located at the site for 4 months (Table 8.1). A detailed description of the experiment and results is described in Haarmann [22]. The results indicated that there was a significant difference between radionuclide samples taken from different aged colonies. Colonies that had been in the study site more years had consistently higher levels of radionuclides than newer colonies. Thus, it appears that over time, there is a measurable accumulation of radionuclides within a colony.

Caste study

Commonly, a sample of bees used for radiochemical analysis comprises up to 1200 individual bees. Some protocols for collecting these samples suggest collecting foragers at the front of the hive as they are returning, while other protocols suggest opening the beehive and collecting bees directly off the frames. In the latter case, depending on which part of the beehive the samples are collected from, the sample may consist of mostly foragers, mostly nurse bees, or a combination of both. Do forager bees contain higher concentrations of radionuclides than nurse bees? Might the proportion of forager bees to nurse bees in a particular sample influence the radionuclide concentrations found in that sample? The caste study was designed to explore these questions.

Separate nurse bee samples and forager samples were collected from colonies located at the study site and analyzed for concentrations of radionuclides (gamma-emitting nuclides and tritium). Figure 8.1 shows a series of boxplots of the forager and nurse bee sample radionuclide concentrations. Detailed results from these experiments are reported in Haarmann [23]. While a statistical analysis indicated that there were no significant differences between the contaminant levels in forager and nurse bees, some insight into the differences in radionuclide concentrations between the two castes emerged. This issue will be addressed further in the Discussion section below.

Flower study

Imagine that an organization or facility is interested in establishing an environmental monitoring program with plans to include bees as indicators of radionuclides in the environment. Based on the experiments described above, they would have a better understanding of the influences that something as simple as sample collection might have on radiochemical analytical results. Once sampling protocols were established, they would need to examine other factors that might influence the concentrations found in bee samples. One of these factors is nectar. If nectar contains radionuclides that are gathered by the bees during foraging, is all nectar

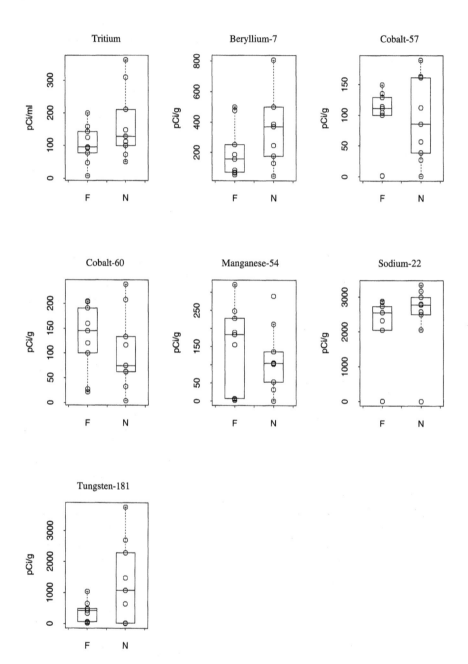

Figure 8.1 Boxplots of the concentrations of radionuclides in samples of forager (F) and nurse (N) bees. Each boxplot graphs the individual sample results, the median (shown as the middle horizontal line of the box), interquartile range (enclosed in the box), and twice the interquartile range (whiskers extend to twice the interquartile range).

considered equal? Do the flowers of different plant species have different concentrations of radionuclides that might influence the concentrations in the bees?

Flowers of the three main forage plants in the study site were collected and analyzed for radionuclides (gamma-emitting nuclides and tritium). These flowers came from salt cedar (*Tamarix ramosissima*), white sweet clover (*Melilotus albus*), and rabbit brush (*Chrysothamnus nauseosus*). Results from this study indicated that there were no significant differences in the amounts of radionuclides found in the flowers of these three plants. Figure 8.2 shows a series of boxplots of the floral sample concentrations. Detailed results from these experiments can be found in Haarmann [23].

Redistribution study

Yet another field experiment was initiated as part of this ongoing study. The purpose of this study was to investigate the redistribution of contaminants within the study site as the contaminants move from the source, in this case a radioactive waste lagoon, to the honey bees. This experiment was designed to explore several questions: (1) Do the bees take up the majority of contaminants from the lagoon or from nearby flowers? (2) Are the levels of contaminants in the bees, flowers, and water correlated, and do they demonstrate similar trends? (3) Is there an observable bioaccumulation of contaminants within the bees or flowers? A detailed summary of this experiment and results are published in Haarmann [24].

In this study, samples of water, flowers, and honey bees were collected from the contaminated study site for two consecutive years. The samples were analyzed for radionuclides (tritium and gamma-emitting nuclides), and the results were compared using rank sum, correlation, and trend analysis. The results were then used to assess the redistribution pathway of radionuclides within the site. Table 8.2 lists the radiochemical analytical results. The results indicated that honey bees received the majority of their contamination directly from the source – the radioactive waste lagoon. The amount of contamination the bees received from flowers during nectar collection appeared to be insignificant compared to the amount received during water collection. The results did not demonstrate significant patterns of correlation or trends between the lagoon, bees, or flowers. Sample results showed a significant bioaccumulation of cobalt-60 and sodium-22 within the honey bees, but no significant bioaccumulation within the flowers.

Discussion

This section will address the significance of the aforementioned studies as they relate to the use of honey bees as part of an environmental monitoring program. In addition, some recommendations will be made for using

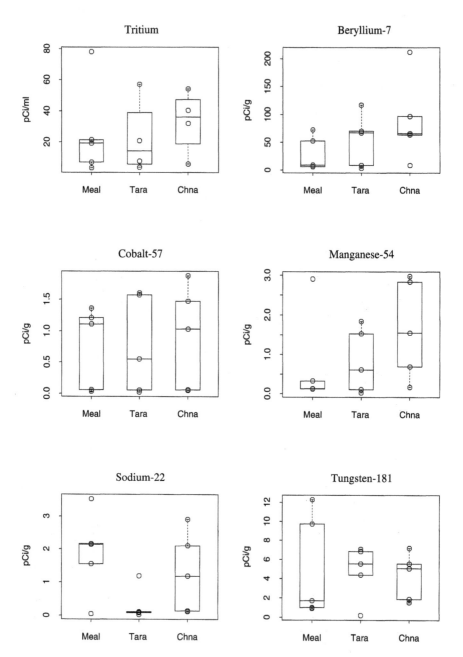

Figure 8.2 Boxplots of the concentrations of radionuclides in flower samples of three plants (*Melilotus albus* [Meal], *Tamarix ramosissima* [Tara], and *Chrysothamnus nauseosus* [Chna]). Each boxplot graphs the individual sample results, the median (shown as the middle horizontal line of the box), interquartile range (enclosed in the box), and twice the interquartile range (whiskers extend to twice the interquartile range).

Table 8.2 Level of radionuclides in samples collected at the LANL study site as part of the redistribution study

Sample type	Sample number	Tritium (pCi/ml)	Analytical uncertainty	Co-56 (pCi/g)	Analytical uncertainty	Co-60 (pCi/g)	Analytical uncertainty	Mn-54 (pCi/g)	Analytical uncertainty	Na-22 (pCi/g)	Analytical uncertainty	W-181 (pCi/g)	Analytical uncertainty
Lagoon	1	4849.00	2.00	0.03*	0.1	0.3*	0.03	0.2*	0.02	101.5*	8.4	71.2*	6.8
	2	3740.00	132.00	<6.56*	NA²	5.2*	0.6	4.4*	0.8	122.0*	11.0	67.0*	17.0
	3	2546.00	101.00	<4.1*	NA	6.0*	0.7	11.0*	1.0	132.0*	12.0	82.0*	18.0
	4	2555.00	102.00	9.5*	3.2	21.0*	2.0	76.0*	7.0	170.0*	16.0	215.0*	34.0
Floral	1	12.04	246.00	1.3	0.7	<0.3	NA	0.2	NA	1.7	1.3	5.0	1.8
	2	67.55	338.00	4.2	0.9	1.5	0.5	1.5	0.5	1.8	0.6	8.4	3.3
	3	26.54	0.25	<0.4	NA	<0.3	NA	0.8	0.2	<0.1	NA	8.1	1.5
	4	13.01	0.20	<1.4	NA	<0.3	NA	<0.3	NA	<0.4	NA	9.8	3.6
	5	29.28	0.25	<0.5	NA	<0.3	NA	1.0	0.2	<0.1	NA	8.6	1.6
Bees	1	0.14	0.14	14.6	5.5	48.6	5.4	62.0	7.7	2031.0	181.0	183.0	50.0
	2	171.82	0.49	<14.1	NA	62.3	7.0	37.7	6.5	2722.0	242.0	164.0	51.0
	3	480.38	0.77	27.0	11.4	163.0	17.0	53.7	8.4	4392.0	389.0	335.0	73.0
	4	77.90	0.36	<13.2	NA	115.0	12.0	64.1	9.8	3158.0	2832.0	242.0	68.0
	5	445.90	0.74	23.9	8.7	154.0	16.0	383.0	38.0	2489.0	223.0	311.0	67.0
	6	164.38	0.41	<11.0	NA	78.0	9.4	39.0	7.3	2815.0	251.0	267.0	56.0
	7	318.64	0.64	<17.9	NA	340.0	35.0	154.0	19.0	5253.0	466.0	1046.0	159.0
	8	629.14	0.87	36.8	14.1	553.0	53.0	523.0	51.0	4559.0	403.0	849.0	125.0

Notes
1 pCi/g measurements are ash weight. These numbers were converted to wet weight when appropriate for certain statistical tests.
2 NA = not applicable.
*values are given in picocuries per milliliter (pCi/ml).
< signifies a below detection limit value.

bees in this capacity, and will include several suggestions for future studies.

Variability study

The results of this study verify that the issue of sample consistency should be examined when using bees as monitors of radionuclides. Generally, samples taken from a colony in the same general area (i.e. honey frames) displayed small variability. However, samples taken from different colonies were more variable. Interestingly, the concentrations of tritium and sodium-22 found in bee samples taken from similar colonies during the same period were inconsistent, while levels of cobalt-57, cobalt-60, and manganese-54 were consistent. Any scientist collecting environmental monitoring data should examine these inconsistencies carefully before drawing any conclusions about the data. Within honey bees, why are certain radionuclide concentrations more variable than others?

Tritium and sodium-22 samples are among the radionuclides that demonstrate sample inconsistency. These inconsistencies are likely to be a result of the dynamics of tritium and sodium-22 in the bee's body. Both hydrogen and sodium are actively involved in several physiological processes and are readily transported through the bee's body. Hence, tritium and sodium-22 are transported as well. It is probable that, within an individual bee, the concentrations of these radionuclides fluctuate continually. The fluctuations would be influenced by, among other things, temperature regulation, spatial and temporal foraging patterns, energy expenditure, and flight activity. In environments where the rate of exposure to colonies is consistent, there may be greater differences in the concentrations of those elements that are active in physiological processes than in concentrations of elements that are less active.

Accumulation study

Past research has shown that radionuclides can be found in bee tissue, honey, pollen, and wax [8, 25]. One would assume that in the case of those radionuclides with a half-life that exceeds one year, the contaminants potentially remain within a colony for several years. Thus, the longer a colony remains in a contaminated area, the greater the accumulation of radionuclide contaminants. Subsequently, bee tissue samples from older colonies would be expected to have higher levels of radionuclides. In older colonies, contaminants are likely to be passed to young bees via trophallaxis and direct contact before any foraging activities. Thus, when an individual bee begins to forage, it may already contain elevated levels of radionuclides. Furthermore, during the winter months, bees in these older colonies are feeding on contaminated honey.

This "precontamination" of foragers may result in tissue samples that show higher levels of radioisotopes than are in fact bioavailable to the bees during foraging. It is not hard to imagine that this fact would have ramifications for an environmental monitoring program that is interested in studying the bioavailability of radionuclides.

Is it therefore safe to assume that all older colonies have higher concentrations of radionuclides? In this experiment, one of the newer colonies had higher levels of tritium than one of the older colonies. Obviously, there are many variables that can influence the levels of contaminants in a colony, bioavailability being only one of them. The fact that a newer colony would have a higher concentration of radionuclides than an older colony suggests that there is a complicated interplay between these variables.

Caste study

This experiment found no significant statistical difference between forager bees and nurse bees. However, one might expect the forager bees to have higher levels of contamination because (1) they are older than the nurse bees and have had the longest time exposure to the contamination and (2) they continually come in direct contact with the contamination sources while foraging. It is possible that equilibrium is reached between the levels of radionuclides in foragers and nurse bees. In classic experiments with radioactive nectar, Free [26] demonstrated that over 75 percent of foragers involved in food exchange contained the radioactive nectar after 24 hours. Using colored and radioactively labeled nectar, Nixon and Ribbands [27] showed that over 50 percent of a colony's workers contained the tracer nectar only 24 hours after 10 foragers had brought it into the colony. Assuming that contamination is spread through the colony very quickly, equilibrium between the levels of radionuclides in the foragers and nurse bees should be achieved within a short period of time.

The data showed that nurse bees tend to have slightly higher concentrations of beryllium-7, sodium-22, tungsten-181, and tritium. As counterintuitive as this may seem, there may be a good reason for this. Radionuclides tend to follow pathways similar to the nutrient analog [28]. The nurse bee samples with slightly higher levels of contaminants seen in this experiment support the accumulation study, suggesting that radioisotopes of physiologically important elements, such as hydrogen and sodium, are readily transported through the honey bee's body. Forager bees possibly expel sodium-22 or tritium via respiration during activities that require increased metabolic activity (i.e. flight). The increased metabolic activities of foragers may ultimately contribute to slightly lower levels of certain contaminants in forager bees than in nurse bees.

Flower study

Theoretically, a variation in floral contaminant levels might influence the levels in bees that forage on those flowers. However, the experiment verified that there were no significant differences in the levels of contaminants in the flowers of the three main forage plants. Therefore, the species of flower the bees had visited probably had little influence on the concentrations of radionuclides found in the bees themselves. In addition, the uptake of contaminants via flowers may have contributed little to the overall levels in the honey bees, since there was a radioactive waste lagoon nearby that contained much higher concentrations of radionuclides. Because bees collected water from the lagoon, they conceivably accumulated most of their contaminants from the water rather than from the nectar of surrounding flowers. Although the particular species of flowering plants used as forage in the study did not appear to have significantly influenced the radionuclide concentrations found in the bees, there were some notable graphical trends.

Although the salt cedar plant is halophytic (e.g. grows in saline soil), the concentration of sodium-22 in the salt cedar flowers was very low. Like its nutrient analog, sodium-22 is probably absorbed by salt cedar and accumulated in the leaves. Salt cedar increases surface soil salinity by transporting salts to the leaves and subsequently releasing these salts back into the surrounding soils when the leaves are shed [29], thus giving it a competitive advantage over non-halophytic plants [30]. It is likely that the majority of sodium-22 is being partitioned into the leaves rather than the flowers.

Rabbit brush, on the other hand, had the highest levels for three of the six contaminants (manganese-54, beryllium-7, and tritium). This is consistent with studies conducted by Fresquez *et al.* [31], which demonstrated that rabbit brush tends to readily take up radionuclides (strontium-90 and uranium) in contaminated sites. While salt cedar and rabbit brush are perennials and sweet clover is an annual, there did not appear to be a clear correlation between the accumulation of contaminants in these plants and their life cycle. Again, this study emphasizes the importance of taking into account all the factors that might influence the radionuclide concentrations within a honey bee.

Redistribution study

Previous studies at LANL have investigated the redistribution of radionuclide contaminants within the environment. Hakonson and Bostick [21] measured the contaminant levels of tritium, cesium-137, and plutonium in bees, honey, surface water, and vegetation. The authors concluded that tritium levels in bees appear to equilibrate with the source. Cesium-137 and plutonium concentrations were low or undetectable in the bees during this study, and therefore difficult to use in the analysis. The authors suggested that because there appeared to be several locations from which the bees received the radionuclides, it was difficult to interpret the data and understand patterns of redistribution.

In this study, because the lagoon was the only major source of tritium, the redistribution of tritium within the study site is easier to understand. Because the levels detected in the flowers were consistently less than those present in the bees, and because the lagoon levels were consistently higher than the levels in the bees, the bees were receiving the majority of their tritium from the lagoon, with much less being contributed by the flowers. In areas with lower source levels, the redistribution patterns would certainly be different, including the possibility that the flowers would be a significant contributor of tritium to the bees.

Consistently, the floral samples contained the lowest levels of all contaminants. The levels were all significantly lower than those observed in either the lagoon or the bees. These results are to be expected because the majority of plants in the study site were not taking up the contaminants directly from the lagoon water; and therefore, the redistribution of contaminants to the plants in the area was somewhat limited.

The levels of cobalt-60 and sodium-22 detected in the bee samples were significantly higher than the levels in the lagoon samples. As part of an ongoing LANL surveillance program, air, water, soil, and foodstuffs were monitored in the study site [32]. These studies indicated that the only major source of cobalt-56, cobalt-60, manganese-54, sodium-22, and tungsten-181 near the study site was the waste lagoon. Because the bees were only receiving cobalt-60 and sodium-22 from the lagoon, and because the levels found in the bees were significantly higher than those at the source, it is apparent that bioaccumulation of sodium-22 and cobalt-60 was occurring within the honey bees. There was no significant bioaccumulation of radionuclides within the floral samples.

While a correlation analysis of the data did not detect statistical significance, one should not rule out a relationship between the levels of contaminants in the lagoon and those in the flowers and bees. Analyses indicating "no significant correlation" in the contaminant levels may simply be a result of the small sample size and the difficulties associated with detecting correlations of data sets with small sample sizes. The strongest positive correlation appears to be between the levels of contaminants in the lagoon and the bees. This is in agreement with the findings of the statistical analysis that indicated that the lagoon is the primary source of contamination for the bees. Similarly, Fresquez *et al.* [20] examined 17 years of data on the tritium levels in honey and bees at LANL, and found no significant correlation between the levels in the bees and the honey.

A trend analysis indicated that, for the most part, upward trends were seen in the lagoon and the bees for all the contaminants. This further supports the hypothesis that the bees were receiving the majority of their contamination from the lagoon. The floral samples showed a variety of trends. The first-year tritium lagoon and flower trends showed upward trends, while the next year showed opposite trends. In fact, for most cases the flowers and lagoon showed opposite trends.

In conclusion, while trend and correlation analyses did not result in statistically significant findings, the bioaccumulation of certain radionuclides within the honey bees was apparent. Nonetheless, this study is helpful in understanding which point sources significantly contribute to the levels of contamination within the bees, as well as the issue of bioaccumulation of certain radionuclides within the honey bees. As part of any contaminant monitoring program, if we hope to get the most out of the data collected from honey bees, the redistribution of contaminants within the study area will certainly need to be taken into account.

Bees as indicators: are they truly useful?

The results of the experiments described in this chapter confirm the findings of many other studies demonstrating that honey bees are good indicators that contamination is bioavailable [4, 20, 33]. At a fundamental level, bees are useful indicators of radionuclide contamination. However, it is apparent that an effective environmental monitoring program would have to do more than simply collect samples of honey bees and use those data at face value. The findings of the experiments presented in this chapter suggest that there is a complicated interplay of many physical and chemical factors that influence the radionuclide concentrations within an individual honey bee.

The data collected in these experiments could be useful in (1) the planning and study design of projects that will use honey bees as monitors of environmental contamination, (2) ideas for the management of honey bee colonies when used in monitoring projects (i.e. how long to leave a colony in a particular area), and (3) the development of protocols for sample collection.

Based on the finding of these studies, when designing and implementing an environmental monitoring program for radionuclides that uses honey bees, one should consider the following:

- Because intracolony sample variability is small, a single sample from each colony adequately represents the levels of contaminants within that colony for that point in time.
- If bee sample results from two locations are to be compared, it is best to avoid subsampling of colonies. Rather than collecting several samples from one colony, it would be preferable to take one sample from each of several colonies. Locations that are to be monitored should contain as many beehives as possible.
- Because intercolony variability is low for some radionuclides (cobalt-

57, cobalt-60, and manganese-54) and higher for others (tritium and sodium-22), depending on the radionuclide in question, it cannot be assumed that colonies in the same area that are exposed to similar conditions will yield consistent sample results. Sufficient quantities of samples must be collected to compensate for this inconsistency in sample variability. It is recommended that when sampling for tritium or sodium-22, samples be collected from several different hives within the study area. The samples can then be treated in one of two ways: (1) all samples can be combined into a composite sample or (2) the analytical results from the samples collected from all the hives can be averaged together to calculate the mean level of contaminants within the study area.

- Because there is a temporal contaminant accumulation within colonies located in a contaminated area, monitoring programs should use colonies of the same age. It would be preferable to replace colonies on an annual basis.

- Although the studies discussed in this chapter did not demonstrate a statistically significant difference between levels of contaminants in forager or nurse bees, it is still recommended that all samples be collected from the same temporal caste, since forager bee samples showed an overall lower level of radionuclides. Sampling from one temporal caste will eliminate any bias that may be introduced by sampling different castes.

- The particular species of plant used as forage by honey bees is an issue that might need to be addressed when interpreting the sample results. While there is no evidence that the levels of contaminants in flowers are significantly different in the LANL study site, this may not be true for all areas.

- Bioaccumulation of certain radionuclides occurs in the honey bees. Bioaccumulation is an important component of understanding the redistribution of contaminants within a biological system. Therefore, the propensity for bioaccumulation of certain radionuclides should be factored into the analysis and interpretation of results.

- Redistribution pathway dynamics need to be understood to accurately interpret sample results. This includes successful identification of the primary source(s) of contamination.

- It is often difficult to demonstrate a significant correlation or trend between the levels of contaminants in the source and those seen in the bees. Therefore, one cannot assume that high contaminant source levels will automatically equate to high levels in the bees. Again, an understanding of redistribution pathways plays a crucial role in data analysis.

Future studies

This chapter has stressed the importance of teasing apart the physical and chemical factors in an ecosystem that might influence radionuclide concentrations in honey bees. We have a long way to go in understanding the dynamics of these interactions. It would be helpful to establish long-term, large-scale projects that investigate the interactions between honey bees and radionuclides in the environment. Additionally, data collected as part of these studies should be incorporated more often into ecological risk assessment models to help predict xenobiotic impacts to ecosystems.

Conclusion

As discussed in this chapter, the findings of the experiments verify that honey bees are indeed good indicators of radionuclide contamination when it is present in the environment. In addition, the data provide insight into those factors that contribute to the overall levels of contaminants detected in the honey bees. These factors include temporal contaminant accumulation, the type of plant species used as forage, and the redistribution of contaminants within ecosystems.

At present, one of the challenges we face is the incorporation of these types of sampling data into ecological risk assessment models. How good are the data? Can we interpret the analytical results meaningfully? Are honey bees a good species to use? These are but a few of the issues we will struggle with if we want to successfully employ honey bees as indicators of environmental contamination.

References

1 Cesco, S., Barbattini, R. and Agabit, M.F. (1994). Honey bees and bee products as possible indicators of cadmium and lead environmental pollution: An experience of biological monitoring in Portogruaro city (Venice, Italy). *Apicoltura* **9**, 103–118.

2 Bromenshenk, J.J. (1988). *Regional monitoring of pollutants with honey bees.* In: *Progress in Environmental Specimen Banking* (Wise S., Zeisler R. and Golstein G.M., Eds). National Bureau of Standards Special Publication, 740. Section 18, pp. 156–170.

3 Konopacka, Z., Pohorecka, K., Syrocka, K. and Chaber, J. (1993). The contents of cadmium, lead, nitrates, and nitrites in pollen loads collected from different sures in the vicinity of Poland. *Pszczelnicze Zesz. Nauk.* **37**, 181–187.

4 Bromenshenk, J.J. (1992). Site-specific and regional monitoring with honey bees. In: *Ecological Indicators*, Vol. 1. Proceedings of the International Symposium on Ecological Indicators, Fort Lauderdale, FL. 16–19 Oct. 1990 (McKenzie D.H., Hyatt D.E. and McDonald V.J., Eds). Elsevier Science, London, UK.

5 Leita, L., Muhlbachova, G., Cesco, S., Barbattini, R. and Mondini, C. (1996). Investigation of the use of honey bees and honey bee products to assess heavy metals contamination. *Environ. Monit. Assess.* **43**, 1–9.

6 Visscher, P.K. and Seeley, T.D. (1982). Foraging strategy of honeybee colonies in a temperate deciduous forest. *Ecology* **63**, 790–801.

7 Bromenshenk, J.J., Carlson, S.R., Simpson, J.C. and Thomas, J.M. (1985). Pollution monitoring of Puget Sound with honey bees. *Science* **227**, 800–801.

8 Wallwork-Barber, M.K., Ferenbaugh, R.W. and Gladney, E.S. (1982). The use of honey bees as monitors of environmental pollution. *Am. Bee J.* **122**, 770–772.

9 Bromenshenk, J.J., Cronn, R.C., Nugent, J.J. and Olbu, G.J. (1988). Biomonitoring for the Idaho National Engineering Laboratory: Evaluation of fluoride in honey bees. *Am. Bee J.* **128**, 799–800.

10 Mayer, D.G., Lunden, I.D. and Weinstein, L.H. (1988). Evaluation of fluoride levels and effects on honey bees (*Apis mellifera* L.). *Fluoride* **21**, 113–120.

11 Migula, P., Binkowska, K., Kafel, A. and Nakonieczy, M. (1989). Heavy metal contents and adenylate energy charge in insects from industrialized regions as indices of environmental stress. In: *Proceedings of the 5th International Conference, Bioindicatores Deteriorisationis Regionis, II* (Bohac, J. and Ruzicka, V., Eds). Institute of Landscape Ecology, Ceske Budejovice, Czechoslovakia. pp. 340–349.

12 Bromenshenk, J.J., Gudatis, J.L., Cronn, R.C., Nugent, J.J. and Olbu, G.J. (1988). Uptake and impact of heavy metals to honey bees. *Am. Bee J.* **128**, 800–801.

13 Balestra, V., Celli, G. and Porrini, C. (1992). Bees, honey, larvae and pollen in biomonitoring of atmospheric pollution. *Aerobiologia* **8**, 122–126.

14 Barbattini, R., Frilli, F., Iob, M., Giovani, C. and Padovani, R. (1991). Transfer of cesium and potassium by the "apiarian chain" in some areas of Friuli NE Italy. *Apicoltura* **7**, 85–87.

15 Gilbert, M.D. and Lisk, D.J. (1978). Honey as an environmental indicator of radionuclide contamination. *Bull. Environ. Contam. Toxicol.* **19**, 32–34.

16 Morse, R.A., Van Campen, D.R., Getenmann, W.H. and Lisk, D.J. (1980). Analysis of radioactivity in honeys produced near Three-Mile Island Nuclear Power Plant. *Nutr. Rep. Int.* **22**, 319–321.

17 Tonelli, D., Gattavecchia, E., Ghini, S., Porrini, C., Celli, G. and Mercuri, A.M. (1990). Honey bees and their products as indicators of environmental radioactive pollution. *J. Radioanal. Nucl. Chem.* **141**, 427–436.

18 Bettoli, M.G., Sabatini, A.G. and Vecchi, M.A. (1987). Honey produced in Italy since the Chernobyl incident. *Apitalia* **14**, 5–7.

19 White, G.C., Hakonson, T.E. and Bostick, K.V. (1983). Fitting a model of tritium uptake by honey bees to data. *Ecol. Model.* **18**, 241–251.

20 Fresquez, P.R., Armstrong, D.R. and Pratt, L.H. (1997). Radionuclides in bees and honey within and around Los Alamos National Laboratory. *J. Environ. Sci. Health* **A32**, 1309–1323.

21 Hakonson, T.E. and Bostick, K.V. (1976). The availability of environmental radioactivity to honey bee colonies at Los Alamos. *J. Environ. Qual.* **5**, 307–309.

22 Haarmann, T.K. (1997). Honey bees as indicators of radionuclide contamination: Exploring sample consistency and temporal contaminant accumulation. *J. Apic. Res.* **36**, 77–88.

23 Haarmann, T.K. (1998). Honey bees as indicators of radionuclide contamination: Comparative studies of contaminant levels in forager and nurse bees and in the flowers of three plant species. *Arch. Environ. Contam. Toxicol.* **35**, 287–294.

24 Haarmann, T.K. (1998). Honey bees (Hymenoptera: Apidae) as indicators of radionuclide contamination: Investigating contaminant redistribution using concentrations in water, flowers, and honey bees. *J. Econ. Entomol.* **91**, 1072–1077.

25 Kirkham, M.B. and Carey, J.C. (1977). Pollen as an indicator of radionuclide pollution. *J. Nucl. Agric. Biol.* **6**, 71–4.

26 Free, J.B. (1954). The transmission of food between worker honeybees. *Anim. Behav.* **5**, 41–47.

27 Nixon, H.L. and Ribbands, C.R. (1952). Food transmission within the honey-bee community. *Proc. R. Soc. London* (B) **140**, 43–50.

28 Whicker, F.W. and Shultz, V. (1982). *Radioecology: Nuclear Energy and the Environment*, Vol. I. CRC Press, Inc., Boca Raton, FL. 212pp.

29 Baum, B.R. (1978) *The Genus Tamarix*. Israel Academy of Sciences and Humanities, Jerusalem. 209pp.

30 Brotherson, J.D. and Winkel V. (1986). Habitat relationships of saltcedar (*Tamarix ramosissima*) in central Utah. *Great Basin Naturalist* **46**, 535–541.

31 Fresquez, P.R., Foxx, T.S. and Naranjo, L. (1995). Strontium concentrations in chamisa (*Chrysothamnus nauseousus*) shrub plants growing in a former liquid waste disposal area in Bayo Canyon. Los Alamos National Laboratory report LA-13050-MS.

32 Los Alamos National Laboratory (1996). Environmental surveillance at Los Alamos during 1995. Los Alamos National Laboratory report LA-13210-ENV.

33 Debackere, M. (1972). Industrial air pollution and apiculture. *Air Pollut. Apicult.* **6**, 145–155.

9 Cesium-134 and Cesium-137 in French honeys collected after the Chernobyl accident

J. Devillers, N. Ben Ghouma-Tomasella, and J.C. Doré

Summary

French honeys collected since the Chernobyl accident in May 1986 were subjected to gamma spectrometry to estimate their radioactive contamination. ^{134}Cs and ^{137}Cs were used as markers of the artificial radioactivity. Differences were found according to the date of sampling, the department of sampling, and the type of honey. However, the results showed conclusively that the French honeys have been contaminated by radionuclides after this catastrophe.

Introduction

The explosion of the number four reactor at Chernobyl (Ukraine) on April 26, 1986, was the greatest peacetime industrial disaster of all time. In addition to massive radioactive contamination in the vicinity of the reactor [1], the explosion and the ensuing 10-day fire propelled an aerosol of radionuclides and particulates into the atmosphere. The amount of radioactive materials released during this accident totaled about 10^{19} Becquerels (Bq) [2]. While the releases contained numerous fission products (^{133}Xe, ^{131}I, ^{134}Cs, ^{137}Cs, ^{132}Te, ^{89}Sr, ^{90}Sr, ^{140}Ba, ^{95}Zr, ^{99}Mo, ^{103}Ru, ^{106}Ru, ^{141}Ce, ^{144}Ce, ^{239}Np, ^{238}Pu, ^{239}Pu, ^{240}Pu, ^{241}Pu, ^{242}Cm), the iodine, cesium, and strontium components were the most deleterious. Indeed, while ^{131}I preferentially fixed by the thyroid increases the risk of cancer of this gland, cesium and strontium are readily incorporated into biological tissues as homologues of potassium and calcium, respectively. During the Chernobyl accident, the release of these elements was estimated at approximately 1760, 54, 85, 115, and 10 petabecquerels (10^{15} Bq) for ^{131}I, ^{134}Cs, ^{137}Cs, ^{89}Sr, and ^{90}Sr, respectively [2].

Widespread distribution of radioactivity throughout the northern hemisphere was noted. A contributing factor was the variation in meteorological conditions and wind regimes during the period of release. Because of the duration of the release and the high altitude (about 1 km) it reached, the radioactivity transported by the multiple plumes from Chernobyl was detected not only in the former Soviet Union but also in northern and southern Europe, Canada, Japan, and the United States. Campaigns of measurements were performed in these countries and bioindicators were also used to assess the biological and ecological risks brought about this accident. While most of the governmental authorities provided information to protect their populations, in France a lack of transparency was adopted by the government and the nuclear lobby yielding an absence of published data to estimate correctly the levels of radioactive contamination and their consequences on living species and humans. Recently, a book [3] has been published, compiling interesting data recorded in 1986, after the Chernobyl accident, on concentrations of radionuclides in soils, crops, milk, fishes, fungi, wild animals, and so on. Unfortunately, data on radionuclide concentrations in French honey are scarce. Analysis of the literature published at that date yields the same conclusion [4, 5]. Consequently, in this chapter an attempt is made to fill this gap by presenting original data dealing with the level of contamination of French honey in 1986. In addition, for comparison purposes, some analytical results from recent measurements are also given.

Materials and methods

All the honeys were analyzed by gamma spectrometry to determine their contamination by ^{134}Cs and ^{137}Cs. The presence of ^{134}Cs ($T_{1/2} = 2.06$ years) in such samples, collected in the second part of 1986, was a good indication that the source of radioactive contamination resulted from the Chernobyl accident. Conversely, it is worth noting that ^{137}Cs ($T_{1/2} = 30.07$ years) contamination could have resulted from Chernobyl or from much earlier aboveground nuclear testing [6]. The honeys were homogenized and conditioned in 250-cm^3 jars. All samples were analyzed with high-purity Ge detectors associated with a multichannel analyzer (8000 channels). The energy range was 20 keV to 1.8 MeV.

A p-type detector was used for samples analyzed before 1995, while the most recent samples were analyzed with an n-type detector. All detectors presented a typical resolution of 1.7 keV for the Co peak at 1.33 MeV. Energy calibration was set with a Canadian (Canmet) reference sample. Efficiency calibration was set with aqueous europium 152 and barium 133 sources used in the same geometry as the unknown samples. Samples were counted from 20000 to 80000 seconds, depending on the mass and intensity of the radioactivity. ^{137}Cs was quantified from its 661.7 keV peak and ^{134}Cs by means of its 604.7 keV peak. Depending on the ^{134}Cs or ^{137}Cs activity, counting time, and mass of honey samples, the counting error was always less than 10 percent.

Results and discussion

Levels of radioactive contamination in French honeys between 1986 and 1989

The dates and location of sampling, the different types of French honeys, and the radionuclide concentrations of ^{134}Cs and ^{137}Cs (in Bq/kg to raw (wet) weight) found in the samples between 1986 and 1989 are given in Table 9.1. The department numbers cited in the third column of this table are reported in Figure 9.1. On this figure, the gradient of the radioactive

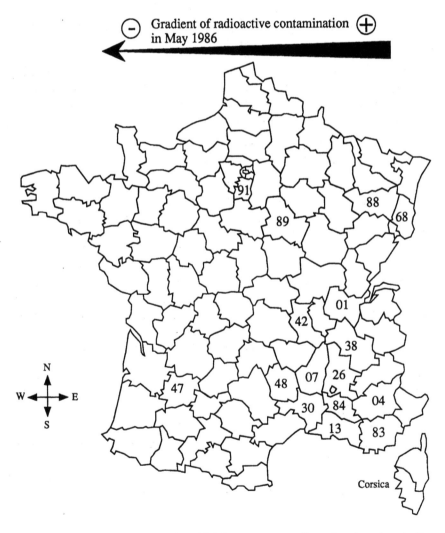

Figure 9.1 French departments in which honeys were collected and analyzed after the Chernobyl accident.

Table 9.1 ^{134}Cs and ^{137}Cs concentrations (Bq/kg) in various honeys collected in France between 1986 and 1989

Collection sampling	Date of analysis	Dpt.*	Type of honey	^{134}Cs	^{137}Cs
June-86	Oct-86	89	acacia	6	10
June-86	Oct-86	89	rape	5	13
Aug-86	Oct-86	89	multiflora	9	15
??-86	Oct-86	07	chestnut	20	53
??-86	Nov-86	84	mountain	14 ± 3**	29 ± 5
June-86	Jan-87	83	"garrigue"	33 ± 15	48 ± 16
July-86	Jan-87	07	thyme + lime tree + acacia	28 ± 9	86 ± 16
Aug-86	Jan-87	42	fir	42 ± 15	169 ± 35
June-86	Feb-87	01	acacia	10 ± 6	17 ± 7
??-86	June-87	26	mountain	10 ± 3	27 ± 6
Aug-86	June-87	01	rape	5 ± 3	19 ± 5
??-86	June-87	01	fir	22 ± 5	67 ± 12
Aug-86	June-87	83	lavender + thyme	14 ± 4	20 ± 6
June-87	July-87	38	acacia	<4	<4
??-86	July-87	84	lavender	<3	10 ± 4
??-86	July-87	84	lavender + other flowers from Provence	13 ± 5	35 ± 8
??-86	July-87	04	lavender	4 ± 3	8 ± 4
??-86	July-87	84	fir	60 ± 14	174 ± 28
July-87	July-87	26	honeydew	<2	6 ± 2
July-87	July-87	26	sainfoin + lime tree	<3	<3
June-86	Sep-87	38	dandelion + lime tree	<3	8 ± 3
Aug-87	Sep-87	38	mixed	8 ± 3	19 ± 4
June-86	Oct-87	13	honeydew	165 ± 20	425 ± 20
??-87	Oct-87	26	lavender	<4	<4
Aug-87	Oct-87	48	mountain	<1	14 ± 4
Aug-87	Oct-87	07	multiflora	7 ± 2	22 ± 5
Aug-87	Oct-87	01	multiflora + honeydew	<2	<2
Sep-87	Oct-87	26	lavender	<3	<2
Oct-87	Oct-87	13	multiflora	6 ± 2	14 ± 3
June-87	Nov-87	01	acacia	<3	<4
Aug-87	Nov-87	04	lavender	<3	7 ± 4
June-87	Dec-87	83	thyme	<2	6 ± 3
Aug-87	Dec-87	04	lavender	<2	3 ± 1
Aug-87	Dec-87	04	lavender	<2	<2
Aug-87	Dec-87	04	multiflora	<2	<3
July-87	Jan-88	30	chestnut	3 ± 1	17 ± 4
Sep-87	Mar-88	07	multiflora	3 ± 1	12 ± 3
Sep-87	Mar-88	26	multiflora	<2	9 ± 3
Oct-87	Mar-88	48	heather	7 ± 3	36 ± 7
Sep-87	Apr-88	07	mountain	2.4 ± 0.7	7.4 ± 1.6
??	Oct-88	88	fir	2.7 ± 1.7	24.5 ± 4.8
May-89	Aug-89	68	multiflora	<0.4	2.3 ± 0.9
June-89	Aug-89	68	multiflora	<0.3	1.5 ± 0.6
June-89	Aug-89	68	multiflora	<0.3	<0.4
??-89	Aug-89	68	multiflora + acacia	<0.2	1.4 ± 0.4
July-89	Aug-89	47	sunflower	<0.3	<0.4

Notes
*Dpt. = Department number (see Figure 9.1).
**2σ + 10%.

contamination is also indicated. Broadly speaking, the flux of contamination at the beginning of May 1986 was from the east to the west of the country.

The concentrations listed in Table 9.1 have to be interpreted with care. Most of the honeys analyzed by the CRIIRAD after the Chernobyl accident were spontaneously provided by beekeepers or individuals who wanted to have information on the level of contamination of their honey. Consequently, no standardized protocol was used to collect the honey samples. However, there is no doubt that all the samples have been harvested after the Chernobyl accident even if it is necessary, for interpreting the data, to account for the date of the analyses, especially for ^{134}Cs which presents a half-life of 2.06 years. In addition, the geographical origin and floral source of each honey sample are not in doubt. Consequently, we assert that interesting and useful (eco)toxicological information can be extracted from an analysis of Table 9.1.

Thus, the concentrations of ^{134}Cs and ^{137}Cs found in the different samples clearly reveal that the French honeys have been contaminated by radionuclides after the Chernobyl accident. As indicated previously, the concentrations of ^{134}Cs provide a good indication that the source of radioactive contamination resulted from the Chernobyl accident. ^{137}Cs could have resulted from the Chernobyl accident or from nuclear weapon tests [6]. However, the ratios of ^{134}Cs to ^{137}Cs calculated for the different samples in Table 9.1 (results not shown) reveal that the former hypothesis is the most likely. This table also shows that the level of contamination depended mainly on the type of honey. This finding is in agreement with numerous published studies (e.g. [7–11]). More specifically, the highest radioactive contamination was found in a honeydew honey. This sample was obtained in June 1986 and analyzed in October 1987 and contained 165 ± 20 Bq/kg of ^{134}Cs and 425 ± 20 Bq/kg of ^{137}Cs. The other two honeydew honeys, gathered later, presented a lower level of radionuclide contamination. Interestingly, in samples of honeydew honey collected in different regions of Italy in May–July 1986, Tonelli and co-workers [8] showed that the radionuclide activity of ^{137}Cs ranged from 31.4 to 362.7 Bq/kg. Similarly, Barisic and co-workers [11] revealed recently, that the radionuclide activity of ^{137}Cs in honeydew honey from Gorski Kotar (Croatia) ranged from 4.8 to 36.2 Bq/kg while that of meadow honey collected in the same geographical location ranged from 0 to 1 Bq/kg. As stressed by these authors [8, 11], it appears that the honeydew honey could be used as an indicator of cesium pollution, even a long time after the radioactive contamination.

The multiflora honeys present various concentrations of ^{134}Cs and ^{137}Cs in relation to the sampling date, geographical origin, and certainly soil characteristics. Xerophytic and aromatic plants (e.g. "garrigue," lavender, thyme), due to their anatomical and physiological characteristics, yield honeys often presenting a high level of radioactive contamination. In the same way, even if the heather honey was collected in 1987 and analyzed in the first part of 1988 (Table 9.1), its measured radioactivity (i.e. $^{134}Cs = 7 \pm 3$ Bq/kg and $^{137}Cs = 36 \pm 7$ Bq/kg) confirms that the heather plants are bioindicators of cesium pollution [7, 12, 13]. The radioactivity detected in the sunflower honey is low but no formal conclusion can be drawn from this analytical result (Table 9.1). In fact, only one sample of this type of honey was analyzed. In addition, this honey was collected and analyzed about 3 years after the Chernobyl accident (i.e. mid-1989). Finally, it is important to note that the sample was collected in a low contaminated area (Figure 9.1).

Comparison of the levels of radioactive contamination found in the honeys produced from the different species of trees also provides interesting information. Thus fir honey seems to highly concentrate radionuclides. The ^{134}Cs and ^{137}Cs concentrations in chestnut honeys are also generally high. Conversely, acacia honey appears as a weak indicator of radioactive pollution. Our results confirm those found by Tonelli and co-workers [8], who report that the mean concentrations of ^{137}Cs found in Italian chestnut honey and acacia honey in May–June 1986 were 70.2 ± 58.7 Bq/kg (22.2 to 180) and 27.3 ± 19.6 Bq/kg (5.1 to 65.5), respectively.

Levels of radioactive contamination in French honeys in 1999/2000

In order to determine the change in radioactive contamination in the French honeys, we measured the concentrations of ^{134}Cs and ^{137}Cs in 14 honeys collected and analyzed in 1999/2000 (Table 9.2). The Ardèche department (07) was principally selected to gather the samples because, first, it was highly contaminated after the Chernobyl accident; the ^{137}Cs concentration in the surface soil layer of this department ranged from 2065 to 12260 Bq/m^2 [3]. Second, in this department it was also possible to find various types of honeys, especially those obtained from xerophytes and aromatic plants which are known to accumulate radionuclides. The last sample in Table 9.2 was selected because it was obtained from an apicultural center of research (Bures-sur-Yvette, INRA) and because this honey was harvested in a rainy department for which, in 1986, no measurement of radioactivity was made, but for which measurements on pollens were available [4].

Because of the fairly short half-life of ^{134}Cs, it is not surprising to see only traces of this radionuclide in all the recent analyzed honey samples (Table 9.2). However, Table 9.2 shows that fairly different concentrations of ^{137}Cs are found in the honey samples. In general, the acacia honeys appear to be less contaminated than the other honeys of different botanical origins. However, because of the different concentrations recorded for the same type of honey, it is clear that specific topographical, climatical, and ecological factors have influenced the radionuclide contamination process of these honeys. Thus, as stressed previously, the high concentra-

Table 9.2 ^{134}Cs and ^{137}Cs concentrations (Bq/kg) in various French honeys collected and analyzed in 1999/2000

French department*	Type of honey	^{134}Cs	^{137}Cs
07	acacia	<0.27	<0.34
07	heather	<0.4	20.9 ± 3.3**
07	acacia	<0.15	1.04 ± 0.34
07	"garrigue"	<0.34	2.46 ± 0.97
07	heather	<0.56	1.06
07	flowers from Provence	<0.29	<0.37
07	lavender	<0.32	<0.31
07	acacia	<0.39	<0.5
07	acacia	<0.5	<0.61
07	"garrigue"	<0.42	<0.53
07	multiflora	<0.39	3.51 ± 1.1
07	"garrigue"	<0.26	<0.31
07	mountain	<0.25	3.57 ± 0.94
91	multiflora	<0.22	0.6 ± 0.43

Notes
*Location on Figure 9.1.
**$2\sigma + 10\%$.

tion of ^{137}Cs found in one of the heather honeys is certainly due to its botanical origin. In addition, it is worth mentioning that the corresponding apiary was located at a high altitude. This location could also explain the fairly high concentration of ^{137}Cs found in this honey and also that found in the mountain honey (i.e. 3.57 ± 0.94 Bq/kg).

The concentrations of ^{134}Cs and ^{137}Cs recently measured in the multiflora honey from Bures-sur-Yvette (91) are only <0.22 and 0.6 ± 0.43 Bq/kg, respectively. Conversely, it is interesting to note that pollen collected in the hives of this center of apicultural research between May 10 and May 25, 1986, was contaminated by 34 Bq/kg of ^{134}Cs and 97 Bq/kg of ^{137}Cs (certainly expressed on a wet mass basis) [4].

Concluding remarks

Whereas for numerous countries (e.g. Italy, Croatia, USA) it is possible to readily obtain data on the level of radioactive contamination of their honeys after the Chernobyl accident [e.g. 10, 13–16], this is the first time that similar information has been published on French honeys. Indeed, in the recent book by Renaud and co-workers [3] on the level of radioactive contamination in France after the Chernobyl accident, almost nothing is written about the radionuclide pollution of French honeys while figures are given for milk, crops, and fish. The authors only indicate [3, p. 114] that the concentrations of ^{137}Cs in French honeys after the Chernobyl accident ranged from 1 to 80 Bq/kg with a mean of 24 Bq/kg. Similarly, Vaillant [5] only provided three measurements of ^{134}Cs and ^{137}Cs in French honeys collected after the Chernobyl accident, but no information was given about their botanical origin.

Thus, as stressed throughout the text, although all our data must be interpreted with care, it is clear that they retrospectively allow us to draw a picture of the level of radioactive contamination of French honey after the Chernobyl accident and to have an idea of the current situation. It is unfortunate that it was not possible to also collect data on the levels of radioactivity in bees and other bee products since these indicators have also shown their relevance in quantifying environmental radioactive contaminations [6, 8, 17–27]. Finally, it is worth mentioning that data on Corsican honey are missing. This is particularly disappointing because it has been shown that the ^{137}Cs concentration in the surface soil layer of Corsica after the Chernobyl accident ranged from 970 to 31 760 Bq/m^2 [3]. Vaillant [5] indicated that the concentrations of ^{134}Cs and ^{137}Cs in Corsican honey, at that date, were equal to 34 and 224 Bq/kg, respectively, but the validity and reliability of these data are questionable. Consequently, we plan to perform a rational campaign of measurements in Corsica to estimate the current level of radioactive contamination of the honey.

References

1 Chesser, R.K., Sugg, D.W., Lomakin, M.D., van den Bussche, R.A., DeWoody, J.A., Jagoe, C.H., Dallas, C.E., Whicher, F.W., Smith, M.H., Gaschak, S.P., Chizhevsky, I.V., Lyabik, V.V., Buntova, E.G., Holloman, K. and Baker, R.J. (2000). Concentrations and dose rate estimates of ^{134,137}Cesium and ^{90}Strontium in small mammals at Chernobyl, Ukraine. *Environ. Toxicol. Chem.* **19**, 305–312.

2 Sweet, W. (1996). Chernobyl's stressful after-effects. *IEEE Spectrum* Nov. 27–34.

3 Renaud, P., Beaugelin, K., Maubert, H. and Ledenvic, P. (1999). *Les Retombées en France de l'Accident de Tchernobyl. Conséquences Radioécologiques et Dosimétriques.* EDP Sciences, Les Ulis, p. 146.

4 Vaillant, J. (1986). Répercussions de l'accident nucléaire de Tchernobyl sur les produits apicoles. *Santé de l'Abeille* **95**, 211.

5 Vaillant, J. (1986). Le césium reste. *Santé de l'Abeille* **96**, 251.

6 Ford, B.C., Jester, W.A., Griffith, S.M., Morse, R.A., Zall, R.R., Burgett, D.M., Bodyfelt, F.W. and Lisk, D.J. (1988). Cesium-134 and cesium-137 in honey bees and cheese samples collected in the U.S. after the Chernobyl accident. *Chemosphere* **17**, 1153–1157.

7 Bunzl, K. and Kracke, W. (1981). $^{239/240}$Pu, ^{137}Cs, ^{90}Sr, and ^{40}K in different types of honey. *Health Phys.* **41**, 554–558.

8 Tonelli, D., Gattavecchia, E., Ghini, S., Porrini, C., Celli, G. and Mercuri, A.M. (1990). Honey bees and their products as indicators of environmental radioactive pollution. *J. Radioanal. Nucl. Chem.* **141**, 427–436.

9 Giovani, C., Padovani, R., Frilli, F., Barbattini, R. and Iob, M. (1991). Il miele come indicatore della contaminazione radioattiva. *Apicoltura* **7**, 137–149.

10 Barišić, D., Lulić, S., Kezić, N. and Vertačnik, A. (1992). ^{137}Cs in flowers, pollen and honey from the Republic of Croatia four years after the Chernobyl accident. *Apidologie* **23**, 71–78.

11 Barišić, D., Vertačnik, A., Bromenshenk, J.J., Kezić, N., Lulić, S., Hus, M., Kraljević, P., Simpraga, M. and Seletković, Z. (1999). Radionuclides and selected elements in soil and honey from Gorski Kotar, Croatia. *Apidologie* **30**, 277–287.

12 Assmann-Werthmüller, U., Werthmüller, K. and Molzahn, D. (1991). Cesium contamination of heather honey. *J. Rad. Nucl. Chem. Art.* **149**, 123–129.

13 Molzahn, D. and Assmann-Werthmüller, U. (1993). Caesium radioactivity in several selected species of honey. *Sci. Total Environ.* **130/131**, 95–108.

14 Ropolo, R., Patetta, A. and Manino, A. (1987). Osservazioni su radioattività e miele nel 1986 in Piemonte. *Apicolt. Mod.* **78**, 11–15.

15 Ropolo, R., Manino, A. and Patetta, A. (1988). Radioattività e miele in Piemonte un anno dopo. *Apicolt. Mod.* **79**, 147–151.

16 Ropolo, R., Patetta, A. and Manino, A. (1996). Dieci anni da Chernobyl. *Apicolt. Mod.* **87**, 51–56.

17 Hakonson, T.E., Johnson, L.J. and Purtymun, W.D. (1973). The applicability of the honeybee as an indicator of environmental radiocontamination. *Health Phys.* **25**, 329–330.

18 Hakonson, T.E. and Bostick, K.V. (1974). The use of honeybee colonies as bio-indicators of cesium-137, tritium, and plutonium in the Los-Alamos environs. *Health Phys.* **27**, 632.

19 Hakonson, T.E. and Bostick, K.V. (1976). The availability of environmental radioactivity to honey bee colonies at Los Alamos. *J. Environ. Qual.* **5**, 307–310.

20 Kirkham, M.B. and Corey, J.C. (1977). Pollen as indicator of radionuclide pollution. *J. Nucl. Agric. Biol.* **6**, 71–74.

21 Eldridge, J.S., Oakes, T.W., Parsons, D.W. and Fell, R.D. (1982). Radionuclide concentrations in honey and bees near radioactive waste disposal sites. *Health Phys.* **43**, 159.

22 Ravetto, P., Cavaglia, D., Colombo, V. and Peila, D. (1987). Proposta per l'utilizzazione dell'ape mellifera come efficiente indicatore di contaminazioni radioattive. *Apicolt. Mod.* **78**, 187–195.

23 Fresquez, P.R. and Armstrong, D.R. (1996). Radionuclide concentrations in bees and honey in the vicinity of Los Alamos laboratory. *Health Phys.* **70 (Suppl 6)**, S69.

24 Fresquez, P.R., Armstrong, D.R. and Pratt, L.H. (1997). Radionuclides in bees and honey within and around Los Alamos National Laboratory. *J. Environ. Sci. Health Part A* **32**, 1309–1323.

25 Haarmann, T.K. (1997). Honey bees as indicators of radionuclide contamination: Exploring colony variability and temporal contaminant accumulation. *J. Apic. Res.* **36**, 77–87.

26 Haarmann, T.K. (1998). Honey bees (Hymenoptera: Apidae) as indicators of radionuclide contamination: Investigating contaminant redistribution using concentrations in water, flowers, and honey bees. *J. Econ. Entomol.* **91**, 1072–1077.

27 Haarmann, T.K. (1998). Honey bees as indicators of radionuclide contamination: Comparative studies of contaminant levels in forager and nurse bees and in the flowers of three plant species. *Arch. Environ. Contam. Toxicol.* **35**, 287–294.

10 The role of honey bees in environmental monitoring in Croatia

D. Barišić, J.J. Bromenshenk,
N. Kezić, and A. Vertačnik

Summary

The products of honey bees can be used as indicators and monitors of a variety of environmental pollutants because of the bees' ability to collect materials that reflect their immediate environmental conditions. The area covered by honey bees in their nectar- or honeydew-gathering process can be presented as a circle with a few kilometers radius. It seems that the honey could be a good random sample, representative of a broad area. Radionuclides, cations, and chemical compounds deposited as fallout due to global atmospheric pollution or as constitutive elements or trace elements of soil can migrate upwards by plant uptake. Concentrations of ^{137}Cs in various honey types during the 1990s in Croatia are presented in this report. The results of analyses of honey samples archived in Austria, Germany, and Slovenia from 1952 through 1995 provide an intriguing and unique history of ^{137}Cs pollution in Europe. The research also documents the levels of ^{137}Cs, ^{40}K, Ca, Fe, Rb, Sr, Cu, Zn, Pb, Ni, Mn, and Cr in soils, coniferous tree branches, and honey, and compares the transfer from soil into nectar honey, mixtures of nectar and honeydew honey, and honeydew honey in fir and spruce forests in Croatia. For all of the elemental concentrations investigated, no significant differences, at level $P < 0.05$, were found between honeydew honey and mixed honey, regardless of the soil type where the honey was collected from. Elemental transfer factors from soils into nectar honey were significantly lower than those for honeydew honey.

Honey bees in radioactive environmental monitoring

In many cases, the spread of environmental contaminants is related to air pollution. The first incidence of air pollution is lost in unrecorded history, but it certainly goes back to the time of the discovery of fire. Air pollution refers to the presence in the outdoor atmosphere of one or more contaminants, occurring in quantities, of a duration, and with characteristics that are known to be injurious to human, animal, and plant life, or to property, or that unreasonably interfere with the comfortable enjoyment of life and property [1]. Once released from sources into the atmosphere, pollutants can be transported large distances due to the global atmosphere circulation. Any factor that restricts the air movement will prevent the movement and dispersion of pollutants entering the atmosphere. In addition to large-scale effects of air movement, local air circulation in valleys and on the slopes of hills or mountains is very important from an air pollution viewpoint, especially during pollutant deposition processes.

In the past century, as well as nowadays, environmental pollution has been closely connected to human activities and industrial development. Developments in the field of atomic energy have introduced radioactive particles as a new and serious type of environmental pollution. Some of the radionuclides formed in nuclear reactions are the most potent poisons known. Moreover, there is no way, except radioactive decay by time, of neutralizing radionuclides. Additionally, radionuclides cannot be detected by human senses, and many members of the public are nervous of or frightened by any manifestation of radioactivity.

Radioactive pollution and random representative sample

Radionuclides, as well as heavy metals and trace elements, occur either as normal constituents of soils or as a result of dry or wet depositional processes due to global atmospheric contamination. The natural radionuclide ^{40}K is a normal constituent of soils, while the presence of ^{137}Cs in soils is an artifact of global atmospheric radioactive pollution. The cesium isotope ^{137}Cs was produced as a by-product of the atmospheric testing of thermonuclear weapons during the period extending from the 1950s to the 1970s. It was distributed globally within the stratosphere and deposited as wet fallout and/or during dry deposition processes. Since the 1970s, the main contributors of atmospheric radionuclides have been operational releases from nuclear power plants and nuclear reactor accidents. The last significant release of radioactive cesium that was deposited on the earth's surface occurred mainly in Europe, during and after the Chernobyl accident of 1986.

The contamination of Croatian territory following the Chernobyl incident is illustrated by the ^{137}Cs content ($kBq\,m^{-2}$) that was found in the first 25 cm of vertical soil profiles (Figure 10.1). Chernobyl-derived ^{137}Cs contamination of the Croatian landscape was not uniform. Lika and a small part of Western Slavonija were the highest contaminated areas, while the Adriatic shore and Eastern Slavonija were significantly less contaminated. The ratio between the highest (near Gračac) and the least contaminated area of Croatia was about 50:1 with respect to Chernobyl-derived ^{137}Cs fallout [2].

A relatively short contamination period combined with great differences in the timing and amount of rain at the time of and immediately

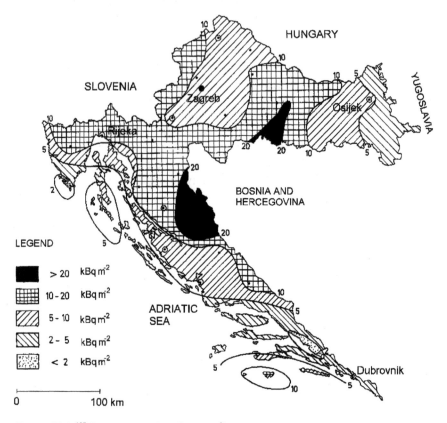

Figure 10.1 ¹³⁷Cs concentration (kBq m^{−2}) in the first 25 cm of vertical soil profiles in Croatia.

following the accident produced the significant variations in Chernobyl-derived ¹³⁷Cs contamination that were first observed. Variations in the soil concentrations of weapon-testing-derived ¹³⁷Cs contamination levels were considered to be the result of local meteorological conditions during each of the peak ¹³⁷Cs-fallout periods that took place over many years. As a consequence of deposition by numerous events over an extended period, which minimizes any local variation, total weapon-testing-derived cesium pollution can be regarded as generally uniform over the whole Croatian territory.

Deposited cesium penetrates slowly from the soil surface into deeper soil layers [3] depending strongly on the soil type [2]. Sorption processes can further retard the ¹³⁷Cs migration rate. The relative abundance of clay and mica minerals, particularly illite, results in the rapid and nearly irreversible cesium immobilization in the topsoil layer [4]. Meanwhile, cesium, as well as the other radionuclides that behave like cations, can be moved upward by plant uptake. This process depends on various factors: plant species, sorption and desorption processes in soil, mineral soil composition, grain size and soil types, lateral cesium migration, and so on.

Certain plant species are known as cesium pollution indicators, but the uptake by each individual plant can be very different. In the first place, it depends on the presence of free cesium in the species' root system zone and competitive effects of potassium [5–10]. Different soil types show differences in the ratio of sorbed to fixed cesium, in soil size fractions, in pH value, and content of organic matter, as well as in ^{137}Cs vertical distribution profiles and, consequently, in cesium transfer from soil to plants [11–18].

Even after relatively homogeneous contamination, all of these factors could introduce a wide range of contaminant variability in a local area. The representativity of any single-point taken sample could be questioned. Although additional samples could be taken near the sample in question during or shortly after the initial sampling, this option would be difficult to enact some years after the contamination. In Table 10.1, this is illustrated by the results of ^{137}Cs and ^{40}K determination in soil samples collected inside a circle of radius 150 m at Milanov vrh in the Gorski Kotar area. The terrain is a carbonate one, and the soil thickness ranges from a few centimeters up to several meters or more, with soil completely missing in significant areas where carbonate rocks are exposed. Moreover, the soil itself is not homogeneous. Soil horizons are differently developed, and the thicknesses of layers of organic matter are also very different. Gravel-size rock fragments are found on few sampling points. The activities of ^{137}Cs deposited in the first 15 cm of vertical soil profiles were found to vary throughout the circle of radius 150 m and across the 5-year period by almost an order of magnitude. Although higher ^{40}K concentrations were found in soils developed on limestone than in soils developed on dolomite, the activities of naturally occurring ^{40}K are very similar regardless of where the sample was collected or the year of collection.

Table 10.1 Activities of ^{137}Cs and ^{40}K (Bq kg^{-1}) in soils collected inside a circle of radius 150 m at Milanov vrh

Sampled	^{137}Cs	^{40}K	Soil taken
1994	51.2 ± 0.8*	507.0 ± 22.1	Beneath fir tree in forest
1994	208.0 ± 1.3	367.1 ± 22.9	Beneath fir tree in forest
1995	60.4 ± 0.8	497.3 ± 8.8	Beneath spruce tree in forest
1996	134.0 ± 1.2	349.2 ± 8.1	Beneath spruce tree in forest
1996	152.5 ± 1.3	282.0 ± 6.4	Beneath fir tree in forest
1997	157.0 ± 2.2	449.4 ± 14.8	From narrow forest meadow
1998	156.2 ± 2.0	405.7 ± 12.6	Beneath spruce tree beside forest road
1998	441.6 ± 3.7	378.2 ± 14.1	From the middle of wide forest meadow
1998	273.4 ± 2.7	433.9 ± 12.3	Beneath fir tree beside forest road
Range	51.2–441.6	282.0–507.0	
Mean ± σ	170.5 ± 120.6	407.6 ± 72.5	

Note
*Counting error.

Alternatively, plants and the bees that visit them can provide a means of detecting and monitoring radionuclide pollution over large areas. Depending on the honey bee pasture types and the plant uptake factors mentioned previously, ^{137}Cs appears in measurable levels in various types of honey [19–21]. Heather plants, *Calluna vulgaris* especially, are species well-known as indicators of cesium pollution [22, 23].

Honey bees and their products have been used as indicators and monitors of a variety of environmental pollution because of their ability to reflect the immediate environmental conditions [24–29]. In searching for and gathering food, honey bees set up flight patterns, which change as available sources or preferences change. The total potential foraging area of a honey bee colony can be presented as a large circle extending out from the hive. Honey bees readily fly up to 4 km in all directions from their hive and thus have access to an area of about 50 km^2 [30]. Because of diminishing returns with respect to the economics of the energy consumed by bees during very long foraging flights, a somewhat smaller area of some 15 to 20 km^2 can be treated as being well covered by honey bees in their nectar gathering. It is very important to note that over such a large area, all of the numerous different environmental factors are included in the samples produced by the bees.

While collecting nectar and honeydew, honey bees provide a composite sample from thousands of different points spread across a large area. On a typical day, a colony of honey bees will make several tens to hundreds of thousands of foraging flights [31]. Depending on the amount of nectar in each honey sac, between 100 000 and 150 000 foraging flights are needed to produce 1 kg of honey [32]. To fill its honey sac, on average a foraging bee needs to visit 80 to 150 individual flowers [33]. Thus, the honey inside each beehive represents a random average sample collected from several tens of millions of single points over a period of time. It seems that a sample of honey is probably the best composited random sample and, as such, provides the most representative values for the average concentrations of bioavailable elements in an area's environment.

Sampling and analysis

Since 1990, 12 stationary apiaries (five placed in the Gorski Kotar area), stocked with *Apis mellifera carnica*, have been used for environmental monitoring of radionuclides. In 1994, six colonies were placed in fir and spruce woods in the Gorski Kotar area and new measurements of radionuclides and selected elements were begun, increasing the network of hives being monitored by 1995.

Measurements of radionuclides and selected elements in soil, coniferous trees, and different honey types have been carried out to: (i) follow the behavior and the fate of cesium in the environment, (ii) determine the indicator capability of honey for cesium, (iii) examine selected cation concentrations in different honey types collected from meadow and forest areas far from any known source of pollution, and (iv) determine the indicator capability of honey for bioavailable elements in the environment.

The Gorski Kotar area is relatively far away from any significant source of environmental pollution. This part of Croatia is exposed only to pollutants that are deposited as fallout from global atmospheric contamination. Honey, soil, fir and spruce branch samples are collected regularly at six locations: Milanov vrh, Tršće, Lividraga, Suha rečina, Fužine, and Zalesina. The positions of the soil, honey, fir, and spruce sampling locations are shown in Figure 10.2.

Soil sampling and analysis

Soil samples were taken regularly at the six above-mentioned locations and at three additional locations in the Gorski Kotar area during the period from 1994 to 1998. Each sample was a composite taken from one area of approximately 500 cm², from the surface down to a depth of 15 cm. At each of the sampled locations, all samples were collected inside a circle

● SAMPLES OF SOIL, FIR, SPRUCE AND HONEY
○ SAMPLES OF SOIL AND HONEY
◆ SAMPLES OF HONEY

Figure 10.2 Sketch map of the Gorski Kotar area, Croatia, indicating soil, honey, fir, and spruce sampling locations.

of radius 150 m. Two main soil types were analyzed; soils developed on the Paleozoic bedrock and Quaternary lacustrine sediments (predominantly silicate soils), and soils developed on the Mesozoic limestones and dolomites (predominantly carbonate soils).

Air-dried soil was passed repeatedly through a 2-mm sieve and quartered to produce material with a grain size less than 0.5 mm. The sieved fraction was then dried at 105°C to a constant weight and stored in counting vessels of volume 125 cm^3 and known geometry for gamma-spectrometric analysis. Prior to X-ray fluorescence (XRF) analyses, sieved and dried soil samples were pressed into pellets.

Honey sampling, analysis, and results

A control series of nectar honey types was collected during the summer months from 1990 to 1996 from the whole Croatian territory. Since 1993, samples of honey have been collected regularly in the summer and early autumn from the Gorski Kotar area, Croatia. A long-time series of various types of honeydew honey and heather honey was collected in Austria, Germany, and Slovenia from 1952 through 1995.

Honey samples were collected mechanically, by extracting honey from combs. Honey types (nectar honey, mixed nectar and honeydew honey, and honeydew honey) were identified on the basis of pollen analyses [34] and electrical conductivity measurements [35] carried out by using a multi-range conductivity meter HI 8733 (Hanna Instruments). Radionuclide activity and selected element concentrations in honey were determined by gamma-ray spectrometry and the XRF method.

A standard sample of 300 pollen grains was used for pollen and honey type determination. Results of pollen determination in some typical Croatian bush-tree and meadow nectar honeys collected in the period 1991–1993 are presented in Table 10.2. A detailed pollen determination of honey samples collected during 1990 was not done. Nectar honey types (meadow, mixed, or bush-tree) were selected on the basis of the prevailing honey bee pasture on the respective locations. *Castanea sativa, Robinia pseudoacacia*, and *Tilia* sp. were dominant pollen types in nectar bush-tree honey collected between 1991 and 1993. *Tilia* sp. was not found in samples of honey from 1992. *Crataegus* sp. was identified only in honey collected in 1992, *Rubus* sp. only in honey collected in 1992 and 1995. Among meadow pollen types in honeys collected in 1991, *Leguminosae* and *Umbelliferae* were more prevalent than *Taraxacum officinale, Trifolium* sp., and *Onobrychis viciaefolia*. In nectar meadow honeys collected in 1992 and 1993, *Compositae* and *Brassicaceae* dominated over *Umbelliferae* and *Rosaceae*. Among pollen grains determined, in honey collected in the period 1994–1996, *Castanea sativa, Robinia pseudoacacia, Taraxacum officinale, Tilia* sp., *Centaurea* sp., *Trifolium* sp., *Leguminosae, Umbelliferae*, and *Brassicaceae* were more prevalent than *Lotus corniculatus, Onobrychis*

Table 10.2 Pollen determination results of typical Croatian mixed bush-tree and meadow nectar honey collected in early 1990s

Sample	Pollen type (%)								
	A	B	C	D	E	F	G	H	O
1	42		4		20	10	18	6	
2	67				9		24		
3	18			31	6	6			39
4	25			46	10	19			
5	39		8			20	13	10	10
6	24					67	9		
7			7	33		14		7	39
8	27		6	32					35
9	21	30		22		25	2		
10	9	3		45		36			7
11	40	10		29		14			7
12	6	30		27		24		6	7
13		20	10	29		16		5	20
14	44	9	4	28				3	12

A, *Castanea sativa*; B, *Tilia* sp.; C, *Robinia pseudoacacia*; D, *Leguminosae*; E, *Onobrychis viciaefolia*; F, *Umbelliferae*; G, *Trifolim* sp.; H, *Taraxacum officinale*; O, Others (*Achillea millefolium, Lotus corniculatus, Satureja montana, Prunus* sp., *Rubus* sp., *Crataegus* sp., *Sinapis* sp., *Brassicaceae, Compositae, Gramineae, Rosaceae*).

viciaefolia, Plantago sp., *Salvia* sp., *Campanula* sp., *Anthyllis* sp., *Alectorolopus* sp., and *Thymus* sp.

In contrast to nectar honey that would have been obtained primarily from blossoms, honeydew is a sugar solution yielded by the hindgut of homopteran insects. Honeydew appears on deciduous trees sporadically, but predominantly in coniferous woods. Conifers are inhabited by homopteran insects, among which leaf-lice and shield-shaped-lice prevail. Conifers are inhabited by green fir-lice (*Cinaria pectinatae*, mainly, but *C. pilicornis, C. viridescens, Lachnus grossus.* or *L. piceae* can be found frequently in fir and spruce forests in the Gorski Kotar area). Occasionally, shield-shaped-lice like *Physokermes piceae* or *P. hemycryphus* can also be found on the fir and spruce trees. These insects pierce the bark of the youngest branches and the needles of fir and spruce trees in search of food. Honeydew is the secretion of these insects.

The results of electrical conductivity measurements and pollen analyses were used to distinguish nectar honey, mixed nectar and honeydew honey, and honeydew types of honey. In the case when pollen grains were usually present and electrical conductivity was less than $0.7\,\mathrm{mS\,cm^{-1}}$, the sample was classified as nectar honey. If pollen grains were present and electrical conductivity was found in the range $0.7–1.0\,\mathrm{mS\,cm^{-1}}$, the sample was classified as mixed nectar and honeydew honey. In cases when pollen grains were absent or very rare and electrical conductivity exceeded $1.0\,\mathrm{mS\,cm^{-1}}$, the sample was classified as honeydew honey.

Coniferous tree sampling and analysis

The youngest segments of fir and spruce branches (including accompany-ing bark and needles) were taken as composites up to 6m above the ground. At each of the observed locations, 15 trees of both fir and spruce were marked inside a circle of less than 150m radius. Branches were cut annually in early autumn, at the end of September or early October. The tips of the fir branches grown that year were collected only in 1994. In each of following years, the tips of the branches and older branch seg-ments, including the segments grown in 1994, were taken for analyses. The tips of the spruce branches were collected for the first time in 1995. In each of following years, the tips of the branches and 1-year-old spruce branch segments were also taken for analyses. In order to check the possible changes in radionuclide activity during a year, the tips of the branches and 1-year-old branch segments were sampled monthly from a single fir tree during 1996 and 1997. Few fir and spruce trees were harvested in autumn 1994 and 1995. Only ^{137}Cs and ^{40}K activities were measured in tree rings as well as separately in needles and in the wooden parts of branches, includ-ing the bark. All samples were dried at 105°C to constant weight, homoge-nized, stored in counting vessels of volume 125 cm^3 and known geometry, and measured by the gamma-spectrometric method. Prior to XRF analy-ses, samples were pressed into pellets.

The gamma-spectrometric method

The activities of ^{137}Cs and ^{40}K were determined by gamma-ray spectrome-try, using a low-background hyper-pure germanium (HPGe) semiconduc-tor detector system coupled to a 4096 channel analyzer. Depending on sample mass and activity, spectra were recorded for times ranging from 80 000 to 150 000 seconds, and analyzed with a personal computer (PC) using GENIE PC Canberra software. The activities of ^{40}K and ^{137}Cs were calculated from the 661.6 and 1460.7 keV peaks, respectively. Double counting errors were taken as the detection limit. The activities of ^{137}Cs in samples were recalculated on July 1 of each year of the sample's collection.

The XRF method

Samples of soil, fir, and spruce material pressed into pellets or samples of honey in native form were placed in counting vessels. Specimens were excited by a ^{109}Cd annular source IPR, 25mCi. Emitted characteristic X-rays were detected by the system's Si-detector (resolution 165eV at 5.9 keV) and Canberra MCA S-100 software. Counting times were 10 000 to 50 000 seconds. United States National Bureau of Standards (NBS) Orchard Leaves SRM 1571 and Soil 5 were used for quantitative analysis.

X-ray spectra were evaluated by International Atomic Energy Agency (IAEA) software QXAS-AXIL, using the procedure "Simple Quantitative Analysis – Elemental Sensitivities" [36].

Statistical evaluation

The majority of sampled honey was collected from a mixture of silicate and carbonate terrains. Less than one-third of honey samples originated from well-known, strictly silicate or carbonate terrain. Because only a small number of well-defined samples of soil or honey were available, only the *t*-test was used in statistical evaluation of collected data. Statistical analyses of nectar honey compared to soil type were not done because of the small number of samples collected from the strictly silicate or carbonate terrains. However, no significant differences have been found for the mixed nectar and honeydew honey as well. Taking into account the aforementioned facts, the average element concentrations in all of the soil samples measured were taken for transfer factor calculations.

Radionuclide activities in honey during the 1990s in Croatia

The average and the range of ^{137}Cs and ^{40}K activities found in nectar honey (meadow nectar, bush-tree, and mixed honey) that was collected between 1990 and 1996 in Croatia are presented in Table 10.3. Previously documented trends showed year to year reductions in the activity levels of ^{137}Cs in bush-tree and meadow nectar honey types [19, 37]. This finding was confirmed by following the ^{137}Cs activity in nectar honey types up to 1996. Ten years after the serious cesium contamination event of the Chernobyl accident, ^{137}Cs activity in nectar honey types has become very low, frequently below the instrument detection limit. On the basis of data presented in Table 10.3, it is evident that for each successive year, ^{137}Cs activity in nectar

Table 10.3 Activities of ^{137}Cs and ^{40}K (Bq kg^{-1}) in Croatian nectar honey types (mixed, meadow, or bush-tree honey) collected between 1990 and 1996, Gorski Kotar area excluded

Year	Number of samples	^{137}Cs		^{40}K	
		Range	Mean ± σ	Range	Mean ± σ
1990	12	0.5–7.9	4.0 ± 2.4	18.8–30.1	24.9 ± 3.6
1991	16	0.4–3.9	1.9 ± 1.1	16.2–33.2	24.6 ± 4.6
1992	11	0.4–1.2	0.7 ± 0.3	18.7–29.5	24.5 ± 4.0
1993	17	0.2–0.9	0.5 ± 0.2	17.3–45.4	27.8 ± 9.0
1994	20	0.0–0.7	0.3 ± 0.2	13.8–41.9	27.4 ± 10.1
1995	10	0.0–0.4	0.1 ± 0.1	15.2–43.6	24.1 ± 7.9
1996	10	0.0–0.3	<0.1 ± nd	11.5–44.9	25.9 ± 12.5

Note
nd, not determined.

honey types decreased by approximately half the activity obtained in the previous year. Meanwhile, ^{40}K activities in nectar honey types have remained more or less the same during the whole time period studied.

The soil macroelement potassium is a mixture of radioactive ^{40}K and stable ^{39}K, and ^{40}K activity of approximately 309 Bq kg^{-1} corresponds to 1 percent of total potassium. The K$^+$ ion is the member of the same homologous series to which Cs$^+$ belongs. Both ions are taken up by plants and competitive effects of potassium on the cesium uptake cannot be excluded [6, 7, 9]. While the potassium content was constant with time, the ^{137}Cs content in nectar honey types decreased significantly with time. This fact cannot simply be a consequence of ^{137}Cs radioactive decay. This isotope has a relatively long half-life of 30.17 years. The ^{137}Cs content in plants is a consequence of cesium bioavailability in the plants' root system zone. Plants are able to take up only free cesium because sorbed cesium is not bioavailable [10, 12, 13, 16, 18]. It seems that about 10 years after initial deposition, the average cesium bioavailability for most of the meadow and bush-tree nectar plants became insignificantly low compared to its bioavailability immediately after the original ^{137}Cs deposition.

Significant differences in the activity levels and long-term behavior of ^{137}Cs were found between groups of nectar honey (Table 10.3) and honey groups (collected from the Gorski Kotar area) containing honeydew honey (Table 10.4). Honey groups collected from the Gorski Kotar area include nectar (predominantly meadow) honey, mixed nectar and honeydew honey and, more or less pure fir and/or spruce honeydew honey. The ^{137}Cs activities in honeydew honey are significantly higher, at level $P < 0.01$, than in nectar honey types. In numerous honeydew honey samples collected from 1993 to 1996 in the Gorski Kotar area, cesium was found in relatively high concentrations; more than 10 times (or even over 100 times in some cases) higher than in nectar honey from Croatia in the same respective year.

Table 10.4 Activities of ^{137}Cs and ^{40}K (Bq kg^{-1}) in honey from the Gorski Kotar area, collected between 1993 and 1999

Year	Number of samples	^{137}Cs		^{40}K	
		Range	Mean $\pm \sigma$	Range	Mean $\pm \sigma$
1993	16	0.9–46.5	21.0 $\pm$ 12.8	19.1–138.4	87.9 $\pm$ 27.1
1994	11	2.6–21.5	13.3 $\pm$ 6.6	46.5–143.3	105.0 $\pm$ 29.4
1995	9	0.8–19.7	8.0 $\pm$ 6.5	13.3–102.6	58.2 $\pm$ 34.1
1996	28	0.1–21.6	5.7 $\pm$ 5.1	14.4–135.6	46.0 $\pm$ 28.3
1997	12	0.0–27.8	8.5 $\pm$ nd	39.0–108.9	61.8 $\pm$ 20.2
1998	22	0.0–19.0	6.7 $\pm$ 4.9	8.0–137.3	66.9 $\pm$ 30.6
1999	7	0.5–16.1	5.9 $\pm$ nd	25.4–77.6	50.0 $\pm$ 20.5

Note
nd, not determined.

This finding suggests than honeydew honey could be used as the indicator of ^{137}Cs pollution years after a contamination event. Although the presented data clearly indicated a general trend toward a decreasing value of ^{137}Cs activity in honey collected in the Gorski Kotar area, these honey samples were composed of numerous nectar and mixed honeys (each collected in a single year). Few honeydew honey samples were collected in each year, as indicated by the significantly lower ^{40}K activity mean value. However, the average ^{137}Cs activity in more or less pure fir and/or spruce honeydew honey remained almost the same at $15.1\,Bq\,kg^{-1}$ during the whole research period. The activity of ^{137}Cs in honeydew honey decreased very slowly with time as a consequence of its radioactive decay and bioavailability from coniferous plants.

Honey as a long-term indicator of ^{137}Cs pollution in Central European countries

As was mentioned earlier, honeydew honey can be used as an indicator of ^{137}Cs pollution over a very long time. Honey is a stable product, lasting for decades or even hundreds of years, if properly stored. By measuring ^{137}Cs in archived samples of collected honey, it was thought it would be possible to retrospectively detect radioactive contamination events. Numerous honey samples that had been collected and kept by beekeepers over a period from 1952 to 1995 were found for locations in Central European countries located between the North sea and the Adriatic sea (Austria, Germany, and Slovenia). Analysis of these honey samples detected ^{137}Cs activities, which were corrected for radioactive decay and recalculated on July 1 of each year of the sample's collection. The ^{137}Cs determination in Austrian spruce honeydew honey samples collected in the Alps region from 1952 to 1994 (mainly in a circle of radius 30 km around Lunz am See)

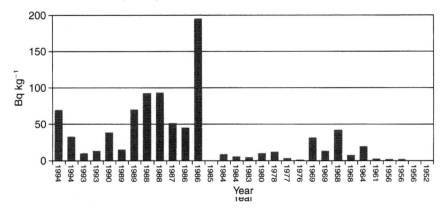

Figure 10.3 Activities of ^{137}Cs in Austrian dominantly spruce honeydew honey collected in the Alps region, mainly in the surroundings of Lunz am See, between 1952 and 1994.

Table 10.5 Activities of ^{137}Cs and ^{40}K (Bq kg^{-1}) in honey collected between 1965 and 1995 in Germany

Year	^{40}K	^{137}Cs	Honey type
1995	116.9 ± 13.8*	11.0 ± 0.3	honeydew/meadow
1995	141.8 ± 19.7	12.9 ± 0.4	honeydew/meadow
1993	109.3 ± 13.8	1.1 ± 0.1	meadow/honeydew
1992	100.0 ± 14.1	2.2 ± 0.2	meadow/honeydew
1991	128.6 ± 14.3	7.2 ± 0.3	meadow/honeydew
1991	97.2 ± 26.9	93.4 ± 1.3	heather (*Calluna vulgaris*)
1990	121.4 ± 13.7	3.9 ± 0.2	meadow/honeydew
1988	113.0 ± 21.9	112.8 ± 1.3	honeydew
1988	95.8 ± 14.1	24.7 ± 0.5	honeydew
1987	161.3 ± 14.1	10.5 ± 0.3	honeydew/meadow
1987	109.7 ± 18.6	180.1 ± 1.5	heather (*Calluna vulgaris*)
1987	135.0 ± 19.9	32.3 ± 0.5	honeydew
1987	152.4 ± 28.1	36.4 ± 0.8	honeydew
1986	209.9 ± 18.1	6.4 ± 0.3	honeydew
1977	76.6 ± 13.7	320.1 ± 1.9	heather (*Calluna vulgaris*)
1977	93.0 ± 13.8	125.4 ± 1.2	heather (*Calluna vulgaris*)
1975	149.5 ± 13.7	23.9 ± 0.5	honeydew
1969	82.2 ± 13.8	176.1 ± 1.6	heather (*Calluna vulgaris*)
1968	106.3 ± 14.0	238.1 ± 1.9	heather (*Calluna vulgaris*)
1966	106.8 ± 15.0	447.0 ± 2.7	heather (*Calluna vulgaris*)
1965	178.4 ± 13.9	19.9 ± 0.6	honeydew
1965	175.9 ± 14.0	14.0 ± 0.5	honeydew
1965	168.8 ± 13.8	0.7 ± 0.2	meadow/honeydew

Note
*Counting error; /first component prevails.

clearly indicates cesium fallout peaks (Figure 10.3). The expected steady decrease in cesium activity following the initial fallout peaks from weapon testing and the much larger fallout from the Chernobyl accident has not occurred. In both cases, there was a reduction in cesium levels following the initial exposure events, but the activity levels did not return to the pre-exposure conditions. Apparent reductions from 1989 through 1993, following the peak of the 1986 Chernobyl release, were contradicted by increasingly higher levels of ^{137}Cs in 1994.

Similar ^{137}Cs behavior was found in different (heather, honeydew, mixed meadow and honeydew) honey samples collected in Germany (Table 10.5), as well as in mixed meadow/chestnut and honeydew honey samples collected in Slovenia (Table 10.6). Different cesium activities have been found for the same year for the same or very similar types of honey in both countries. Such results could be explained by the fact that samples of honey were collected from different locations, locations that have been contaminated differently by cesium during the main fallout events in the past. The ^{137}Cs activities found in the honey of heather plants (*Calluna vulgaris*) are significantly higher than in honeydew honey. Heather is considered to be an excellent cesium pollution indicator, and so apparently is

Table 10.6 Activities of ^{137}Cs and ^{40}K (Bq kg^{-1}) in honey collected between 1986 and 1994 in Slovenia

Year	^{40}K	^{137}Cs	Honey type
1995	97.5 ± 4.0*	3.5 ± 0.2	meadow//spruce honeydew
1995	93.2 ± 3.9	20.6 ± 0.4	meadow/fir honeydew
1994	75.9 ± 3.7	35.5 ± 0.5	spruce honeydew/meadow
1993	121.5 ± 4.2	13.6 ± 0.3	chestnut/spruce honeydew
1992	58.3 ± 3.4	3.7 ± 0.2	chestnut/meadow
1991	116.1 ± 4.2	42.7 ± 0.6	fir honeydew/meadow
1991	95.9 ± 4.0	49.0 ± 0.6	spruce honeydew/meadow
1990	110.4 ± 4.4	50.6 ± 0.7	spruce honeydew/meadow
1989	76.0 ± 3.7	35.6 ± 0.5	meadow/spruce honeydew
1988	154.2 ± 4.7	41.8 ± 0.6	spruce honeydew/chestnut
1988	80.3 ± 3.7	10.4 ± 0.3	meadow//fir honeydew
1987	150.3 ± 4.8	37.4 ± 0.6	spruce honeydew/chestnut
1987	67.3 ± 3.6	8.3 ± 0.3	meadow//fir honeydew

Note
*Counting error; /first component prevails; //second component is traceable.

heather honey. Heather may be a better cesium monitor than coniferous plants. However, heather grows in open areas, rather than in thick fir or spruce forests. Because it is relatively easy to detect ^{137}Cs in both heather honey and honeydew honeys, the honeys from heather and honeydew from fir and spruce can be used as indicators of radioactive cesium pollution for long-term monitoring of contamination events.

Transfer of radionuclides and selected elements from soils through conifers and into honey

The results of these studies examine the difference in the uptake dynamics of bioavailable inorganic elements from soils into the nectar of flowers versus into the phloem of coniferous trees and passage through the hindgut of aphids into honeydew. The honey from meadow plants would have been obtained by the bees primarily from blossoms. Another insect feeding on coniferous trees produced the honeydew that was then stolen by the bees.

Radionuclides and selected elements in soils

The majority of the results presented here on radionuclides and selected element concentrations in soils have already been published [38]. All of the elemental concentrations measured, excluding Zn, were significantly higher in soils developed on carbonate bedrocks than in the predominantly silicate soils sampled in the Gorski Kotar area (Table 10.7). This was especially evident in the case of Rb, Sr, Mn, and Fe (at level $P < 0.01$) as well as in case of Ca and Ni at level $P < 0.05$. The highest concentrations

Table 10.7 Radionuclide activities (Bq kg⁻¹) and selected element concentrations (mg kg⁻¹) in Gorski Kotar soils

Element	All soils (n = 43[a] and 14)		Soils developed on carbonate bedrock (n = 25[a] and 7)		Soils developed on silicate bedrock (n = 18[a] and 7)	
	Range	Mean ± σ	Range	Mean ± σ	Range	Mean ± σ
^{40}K[a]	259–715	402 ± 78	282–507	410 ± 60	259–715	392 ± 102
^{137}Cs[a]	43.1–500.0	180.4 ± 112.6	51.2–500.0	192.4 ± 126.0	43.1–367.3	163.8 ± 94.7
^{40}K	259–715	411 ± 106	342–507	423 ± 50	259–715	398 ± 147
^{137}Cs	43.1–500.0	176.0 ± 126.6	51.2–500.0	199.5 ± 143.7	43.1–342.0	152.5 ± 113.0
Rb	76.5–177.3	123.9 ± 33.0	124.7–177.3	150.7 ± 18.6	76.5–109.8	97.1 ± 12.3
Ca	1400–50400	11600 ± nd	6550–50400	20550 ± 16650	1400–4100	2650 ± 900
Sr	35.6–115.3	76.7 ± 26.7	79.0–115.3	99.9 ± 12.6	35.6–69.8	53.5 ± 11.6
Ni	10.1–59.4	25.2 ± 14.8	20.7–59.4	37.0 ± 11.8	10.1–16.3	13.5 ± 2.0
Cu	19.9–61.3	36.8 ± 11.2	19.9–61.3	41.6 ± 13.1	21.5–42.7	32.1 ± 7.1
Zn	47.4–874.0	181.6 ± nd	108.0–255.6	177.9 ± 50.2	47.4–874.0	185.2 ± nd
Pb	31.6–97.9	64.1 ± 21.9	47.4–97.9	72.8 ± 19.2	31.6–94.2	55.3 ± 22.3
Co	4.6–20.7	13.3 ± 5.5	13.6–20.7	18.0 ± 2.6	4.6–11.6	8.5 ± 2.5
Hg	0.186–0.645	0.347 ± 0.144	0.239–0.645	0.397 ± 0.178	0.186–0.462	0.297 ± 0.086
Fe	23600–70300	45300 ± 15250	50800–70300	58650 ± 6950	23600–40600	31950 ± 6400
Mn	190–1420	775 ± 433	380–1420	1072 ± 397	190–790	478 ± 211
Cr	30.5–92.8	50.9 ± 18.4	32.2–92.8	56.0 ± 23.1	30.5–60.1	45.8 ± 11.7

Note
nd, not determined.

of these elements in predominantly silicate soils seldom exceeded the lowest concentrations found in soils developed from the carbonate bedrock. The activities of ^{40}K and ^{137}Cs in both soil types were fairly similar as well as concentrations of Cu, Zn, Pb, and Cr. Single element concentrations and radionuclide activities, excluding Ca, Zn, and ^{137}Cs, differed by less than an order of magnitude. Very wide ranges of calcium concentrations in the soil were a consequence of the soil type from which the sample was taken. Silicate soils are relatively pure calcium, while in soils developed on a carbonate bedrock, the calcium concentration depends on the type (limestone or dolomite) and abundance of the carbonate component. ^{137}Cs is a human-produced pollutant. In the case of a high zinc concentration found in one sample, local pollution caused by uncontrolled garbage disposal beside a forest road near where the sample was taken was suspected.

For soils the mean concentration of elements and the radionuclide activity were based on a limited number of samples ($n = 7$–14), each representing a single point at a specified depth. The soil samples may not accurately represent the true mean for soils across the entire area generally sampled by the bees. In order to check the representativeness of the mean ^{40}K and ^{137}Cs activity values, calculated on a limited sample number, an additional 29 soil samples were analyzed. However, no significant differences for the soil means as a consequence of the number of samples were detected at level $P < 0.05$ for either radionuclide.

Radionuclides and selected elements in distinctive fir and spruce parts

An overview of measured radionuclide activities and selected element concentrations obtained in the youngest terminal shoot of the fir and spruce branches sampled from the Gorski Kotar area is presented in Table 10.8. Only rubidium and cesium concentrations in the tips of the fir branches were significantly higher than in the spruce tips. The concentrations of all the other elements studied, excluding zinc, iron, and lead, were nearly equal or insignificantly lower in the spruce than in the fir tips of the branches.

Identical distributions of measured radionuclide activities and selected element concentrations were found in 1-year-old fir compared to 1-year-old spruce branch segments. The concentrations or activities of the majority of the elements studied were lower in the 1-year-old branch segments compared to the branch tops. In the case of Rb, ^{40}K, and ^{137}Cs they were lower (Rb and ^{137}Cs were significantly lower at $P < 0.05$) in 1-year-old branch segments than in the tops, while calcium, lead, iron, and manganese concentrations increased insignificantly.

Cation mobility in the youngest parts of fir branches was observed in up to 4-year-old fir branches. The ^{40}K activities decreased significantly in older branch segments, and in 6- to 8-year-old segments appeared to become more

Table 10.8 Radionuclide activities (Bq kg^{-1}) and selected element concentrations (mg kg^{-1}) in the top of fir and spruce branch shoots from the Gorski Kotar area

	Fir tops (n = 30)		Spruce tops (n = 24)	
Element	Range	Mean ± σ	Range	Mean ± σ
^{40}K	180.5–362.9	250.4 ± 85.7	139.0–318.9	213.3 ± 65.7
^{137}Cs	11.5–110.6	46.0 ± 32.7	4.9–26.1	14.0 ± 7.6
Rb	2.5–26.4	10.1 ± 6.7	1.1–12.2	5.3 ± 3.4
Ca	5610–20550	9890 ± 4290	5230–15530	8580 ± 3160
Sr	<0.3–42.1	<4.5 ± nd	<1.2–9.2	<3.4 ± nd
Ni	<0.4–1.2	<0.8 ± nd	<0.3–1.8	<0.8 ± nd
Cu	2.9–17.2	5.4 ± 2.8	2.1–14.9	5.2 ± 2.8
Zn	14.3–45.7	24.7 ± 9.4	17.6–101.0	34.7 ± 17.9
Pb	9.0–42.7	21.4 ± 10.4	9.5–69.1	23.2 ± 12.9
Fe	56.4–195.0	93.5 ± 36.3	45.2–250.9	102.0 ± 56.0
Mn	<113–1718	<469 ± nd	<81–1323	<352 ± nd
Cr	<0.8–2.6	<1.4 ± nd	<0.6–3.2	<1.5 ± nd

Note
nd, not determined.

or less constant [39]. Significant reductions in ^{137}Cs activity in the youngest fir branch segments could be followed in up to 2- or 3-year-old branch segments. In 5- to 6-year-old segments, ^{137}Cs activity is also more or less constant. Rubidium concentrations in up to 2-year-old branch segments decreased (in 1-year-old branches significantly at level $P < 0.05$) rapidly; in older segments concentrations were mainly constant. By contrast, the concentration of calcium, lead, iron, manganese, and chromium increased in older branch segments. Zinc, copper, nickel, and strontium concentrations were found to be almost constant in all fir branch segments, up to 4 years old.

The ^{137}Cs translocation in the tops of fir branch shoots was confirmed by data obtained from monthly sampling of tops and from 1-year-old branch segments from a single fir tree (Figure 10.4). Although the cesium content in older branch segments decreased, the question concerning the cesium source (translocations from older segments only and/or additional fir cesium uptake from soil) in the tops of the branches is still unresolved. In any case, ^{137}Cs activities are on average higher (excluding the tops and year-old segments) in fir needles than in the wooden parts and bark of branch segments [39]. On the other hand, cesium migrates into older tree growth rings also, as reported for the French white fir growth ring [40]. Cesium was not found in the growth rings of fir harvested at Crni lazi [39] that grew before approximately 1925. However, the similar radial ^{137}Cs distribution in both the French and the Crni lazi fir trees is opposite to the ^{137}Cs distribution in the sugi tree rings harvested from Japan [41]. In the sugi tree rings, the highest ^{137}Cs activities were found in the oldest tree rings. Additionally, a whole fir tree may act as a "cesium reservoir," but it

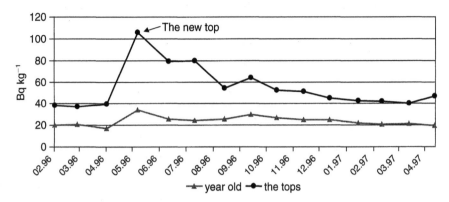

Figure 10.4 Activities of ^{137}Cs in monthly samples of the top of the branch shoots and 1-year-old branch segments, taken on a single fir tree from Gorski Kotar, Croatia.

loses older needles (more than 10-year-old needles are very rare) and the cesium contained in these needles, which may then be returned to the soil and later taken back up again by the tree. For long-term observations, even cesium radioactive decay must be taken into account. However, at present the samples collected and data analyzed are still insufficient to answer the question concerning where the whole cesium content present in fir tops arises from each new year.

Transfer of radionuclides and selected elements from soil into youngest fir and spruce branches

Transfer factors of radionuclides and selected elements from soils into the youngest parts of coniferous branches (Table 10.9) were expressed as a percentage, defined as the ratio between element concentration in coniferous material and in soil, and multiplied by 100. The uptake of the majority of the cations studied from soil into the youngest branches of fir was more or less similar. Fir trees showed a higher cesium, manganese, rubidium, calcium, and strontium uptake than spruce, but the difference is significant in the case of ^{137}Cs only. Among all the cations studied, calcium, manganese, potassium, and lead showed the highest bioavailability. The calcium concentration in fir material, especially in older branch segments, very often exceeded the calcium concentration in soils. High transfer factors for zinc, copper, and cesium confirmed the capability of coniferous plants as pollution indicators in the case of soils contaminated by ^{137}Cs, Cu or Pb. The transfer of all other elements studied from soil into young coniferous parts was below 10 percent, and in the case of iron, was only 0.24 percent.

It is interesting to note that the ^{137}Cs transfer factor from soil into spruce (6.5 percent) was almost double the rubidium transfer (3.3

Table 10.9 Transfer factor (%) of radionuclides and selected elements from soil
($n = 14$) into youngest segments of coniferous tree branches ($n = 132$),
fir (up to 4 years old; $n = 90$), and spruce (up to 1 year old; $n = 42$)
branches together with needles

Element	Coniferous tree (%) ($n = 132$)	Fir branches (%) ($n = 90$)	Spruce branches (%) ($n = 42$)
^{40}K	46.7 ± 14.4	47.7 ± 15.0	44.6 ± 13.1
^{137}Cs	12.6 ± 12.0	15.4 ± 13.4	6.5 ± 3.6
Rb	4.3 ± 3.7	4.8 ± 4.0	3.3 ± 2.5
Ca	96.0 ± 33.3	104 ± 33.7	78.1 ± 25.0
Sr	$<5.9 \pm nd$	$<6.6 \pm nd$	$<4.6 \pm nd$
Ni	$<3.1 \pm nd$	$<3.2 \pm nd$	$<2.8 \pm nd$
Cu	13.5 ± 5.9	13.7 ± 5.4	13.0 ± 7.0
Zn	15.5 ± 6.2	14.5 ± 4.6	17.6 ± 8.4
Pb	36.6 ± 17.7	36.5 ± 14.5	36.7 ± 23.4
Fe	0.24 ± 0.10	0.25 ± 0.08	0.23 ± 0.13
Mn	$<70.4 \pm nd$	$<81.4 \pm nd$	$<46.8 \pm nd$
Cr	$<2.9 \pm nd$	$<3.0 \pm nd$	$<2.7 \pm nd$

Note
nd, not determined.

percent), although transfer of rubidium from soils into spruce needles was
approximately one order of magnitude higher than the transfer of the
stable soil microelement cesium [42]. The bioavailability of stable cesium
is low because it is strongly adsorbed. The higher bioavailability of radioactive pollutant ^{137}Cs is a consequence of the fact that ^{137}Cs is not completely adsorbed yet, and penetrates very slowly into deeper soil layers [2].

Radionuclides and selected elements in various honey types

Concentrations of selected elements and radionuclide activities in different
types of honey (predominantly nectar honey from meadows, mixed meadow
nectar and honeydew honey, and honeydew honey from fir and spruce
forests) from the Gorski Kotar area are presented in Table 10.10. Among all
of the studied elements, potassium comprised about 90 percent or more of
the total cation amount in honey. The mean potassium content was the
lowest in meadow honey (≈ 0.09 percent), followed by mixed honeys (≈ 0.15
percent), and the highest in honeydew honey (≈ 0.28 percent). Strontium
and nickel concentrations in honey were the lowest in meadow honey and
the highest in honeydew honey. Compared to ^{40}K, the concentrations of Sr
and Ni were about three orders of magnitude lower, even hundred of thousands times lower in comparison to the total potassium content.

Without exception, the concentration of all measured elements was
significantly higher (at level $P < 0.01$, only for Zn at level $P < 0.05$) in
honeydew honey compared to meadow nectar honey or mixed honey. In
comparison with meadow nectar honeys, honeydew honey showed the

Table 10.10 Radionuclide activities ($Bq\,kg^{-1}$) and selected element concentrations ($mg\,kg^{-1}$) in different types of honey from the Gorski Kotar area

Element	Nectar honey (n = 18)		Mixed nectar and honeydew honey (n = 29)		Honeydew honey (n = 35)	
	Range	Mean ± σ	Range	Mean ± σ	Range	Mean ± σ
^{40}K	8.4–63.6	27.7 ± 15.2	15.9–108.9	46.7 ± 23.4	43.8 –143.3	86.6 ± 24.3
^{137}Cs	0.0–1.0	0.43 ± 0.36	0.9–6.8	3.25 ± 1.56	4.8–46.5	15.1 ± 9.3
Rb	0.001–0.350	0.071 ± nd	0.01–0.85	0.24 ± 0.21	0.52–4.24	1.86 ± 1.13
Ca	7–162	67.7 ± 41.4	20–254	94.0 ± 63.7	131–593	287.4 ± 131.0
Sr	<0.001–0.028	<0.007 ± nd	<0.002–0.034	<0.010 ± nd	<0.01–0.069	<0.032 ± nd
Ni	<0.005–0.188	<0.057 ± nd	<0.019–0.211	<0.079 ± nd	<0.105–0.472	<0.243 ± nd
Cu	0.04–0.78	0.32 ± 0.22	0.1–1.07	0.44 ± 0.32	0.64–2.40	1.42 ± 0.60
Zn	0.03–0.71	0.18 ± 0.15	0.06–0.94	0.26 ± 0.21	0.26–10.9	1.32 ± nd
Pb	0.19–2.77	0.82 ± 0.67	0.22–4.32	1.23 ± 0.90	0.84–7.08	3.47 ± 1.53
Fe	0.5–13.1	4.4 ± 3.5	1.1–20.2	5.7 ± 4.9	5.4–34.6	17.7 ± 9.1
Mn	<0.1–4.7	<1.1 ± nd	<0.2–6.7	<1.5 ± nd	<1.4–11.9	<4.84 ± nd
Cr	0.03–0.41	0.18 ± 0.12	0.03–0.83	0.24 ± 0.21	0.28–1.22	0.68 ± 0.31

Note
nd, not determined.

highest concentrations of Cs, Rb, Zn, and Mn. Potassium concentration increased by a factor of three, while the concentration of the majority of the other elements studied was approximately a factor of four higher in honeydew honey than in meadow honey. The greatest difference in concentrations was found in the case of ^{137}Cs and Rb. The mean concentrations of these two elements were found to be more than one order of magnitude higher in honeydew honey than in meadow honey.

Although copper concentrations in honey are very similar to those reported earlier [25, 28], zinc concentrations in honey from the Gorski Kotar area were several times higher than those found by Leita *et al.* [26], but one order of magnitude lower than was found in nectar and honeydew honey in Czechoslovakia [28]. Lead concentrations in honey from the Gorski Kotar area were one order of magnitude higher than those found by Jones [25], but several authors [26, 28] have reported very similar lead concentrations in honeys. For all of the element concentrations studied, no significant differences, at level $P < 0.05$, were found for honeydew honey, regarding the respective soil types where the honey was collected. No significant differences have been found in the case of mixed meadow nectar and honeydew honey as well.

Transfer of radionuclides and selected elements from soil in various honey types

Transfer factors from soils into the honey types studied (expressed as a percentage) were defined as the ratio between element concentration in honey and in soil, multiplied by 100. The transfer factors for each element and radionuclide from the soil into the honey types are presented in Table 10.11.

Table 10.11 Transfer factors (%) of radionuclides and selected elements from soil ($n = 14$) into predominantly nectar ($n = 18$) honey, mixed (nectar and honeydew; $n = 29$) honey, and honeydew ($n = 35$) honey

Element	Nectar honey (%)	Mixed honey (%)	Honeydew honey (%)
[40]K	6.73 ± 3.69	11.4 ± 5.68	21.1 ± 5.91
[137]Cs	0.246 ± 0.203	1.85 ± 0.88	8.56 ± 5.26
Rb	$0.057 \pm nd$	0.197 ± 0.168	1.50 ± 0.91
Ca	0.583 ± 0.357	0.810 ± 0.550	2.48 ± 1.13
Sr	$<0.009 \pm nd$	$<0.013 \pm nd$	$<0.042 \pm nd$
Ni	$<0.225 \pm nd$	$<0.316 \pm nd$	$<0.960 \pm nd$
Cu	0.859 ± 0.593	1.21 ± 0.88	3.85 ± 1.63
Zn	0.097 ± 0.085	0.145 ± 0.115	$0.72 \pm nd$
Pb	1.28 ± 1.04	1.92 ± 1.41	5.42 ± 2.39
Fe	0.010 ± 0.008	0.013 ± 0.011	0.039 ± 0.020
Mn	$<0.146 \pm nd$	$<0.188 \pm nd$	$<0.62 \pm nd$
Cr	0.357 ± 0.230	0.476 ± 0.414	1.34 ± 0.61

Note
nd, not determined.

Among the soil macronutrients, consisting of potassium, iron, and calcium, only potassium showed significant transfer from soil into all of the honey types studied. Calcium transfer from soil into the various honeys was approximately one order of magnitude lower than the transfer of potassium, while the transfer of iron was about three orders of magnitude lower than potassium. Zn and Mn transfer were very similar in all types of honey, demonstrating a low transfer from soils into honey. Transfer factors for Sr and Fe from soils into all of the honey samples generally were very low.

The lowest transfers were found into meadow honey. Without exception, transfer factors from soil into honey for all of the elements measured were significantly higher into mixed floral meadow and forest honeys (containing both nectar and honeydew) than into nectar meadow honey. Rb and ^{137}Cs showed significantly higher (at $P < 0.001$) transfers into mixed honeys compared to honey from meadows, although rubidium transfer into meadow honey was still very low (0.197 percent). Only ^{40}K (6.73 percent) and Pb (1.28 percent) showed transfers greater than 1 percent into meadow honey, followed by Cu (0.859 percent) and Ca (0.583 percent).

Mixed meadow and honeydew honey demonstrated an 11.4 percent transfer of ^{40}K from soil, 1.85 percent for ^{137}Cs, 1.92 percent for Pb, and 1.21 percent for Cu. All other elements were considerably lower than a 1 percent transfer. Only Ca with a value of 0.810 percent was even near the 1 percent level.

Among the honey types studied, honeydew honey exhibited the best indicator capabilities of all of the elements and radionuclides. By comparison with the aforementioned types of honey, the transfer of ^{137}Cs and Rb into honeydew honey was an order of magnitude higher, with transfers of 8.56 and 1.50 percent, respectively. Transfers for ^{40}K (21.1 percent), Pb (5.42 percent), Cu (3.85 percent), Ca (2.48 percent), Cr (1.34 percent), and Ni (<0.960) nearly equalled or exceeded 1 percent. On the other hand, the soil macroelement Fe showed a very low transfer (0.039 percent) from soils into honeydew honey. Strontium showed a similar behavior (<0.042 percent), so it seems that honey could not be used very successfully as an indicator of environmental pollution with strontium radioactive isotopes. The results show that samples of honey, especially honeydew honey, can indicate the consequences of global pollution events as well as events on a local scale.

Transfer of radionuclides and selected elements from the youngest fir and spruce branch parts in honeydew honey

Transfer factors (expressed as a percentage) from the youngest coniferous parts into honeydew honey are presented in Table 10.12. Only ^{137}Cs (68.2 percent) showed a transfer of over 50 percent into honeydew honey, followed by Cr (<45.8 percent), ^{40}K (45.2 percent), Rb (34.7 percent), Cu (28.6 percent), Fe (16 percent), and Pb (14.8 percent). Transfers of

Table 10.12 Transfer factors (%) of radionuclides and selected elements from youngest coniferous tree branch segments ($n = 132$) into honeydew ($n = 35$) honey

Element	Transfer factor (%)
^{40}K	45.2 ± 12.7
^{137}Cs	68.2 ± 41.9
Rb	34.7 ± 21.0
Ca	2.58 ± 1.18
Sr	$<0.70 \pm$ nd
Ni	$<31.5 \pm$ nd
Cu	28.6 ± 12.2
Zn	$4.68 \pm$ nd
Pb	14.8 ± 6.54
Fe	16.0 ± 8.3
Mn	$<0.88 \pm$ nd
Cr	$<45.8 \pm$ nd

Note
nd, not determined.

strontium and manganese are below 1 percent from fir and spruce youngest branch segments into honeydew honey. Among the soil macronutrients, including potassium, iron, and calcium, only the potassium showed almost identical transfer from soils into coniferous material as well as from conifers into honeydew honey. Very high calcium (as well as manganese) transfer from soil into conifers is followed by low transfer of both elements from conifers into honeydew honey. Iron shows the opposite behavior (chromium also), with very high transfer from coniferous material into honeydew honey. It seems that very high rubidium and cesium transfers from conifers into honeydew honey can be connected with the high mobility of both elements into the youngest fir and spruce branch parts.

Honeydew honey, among all of the honey types studied, best reflected persistent bioavailable inorganic contaminants in the environment. Because of the high transfer from conifers, honeydew honey could be used as an indicator for ^{137}Cs, Cr, Rb, Cu, Pb, and Ni pollution in areas covered with fir or spruce forests. In the case of ^{137}Cs, it seems that honeydew honey can be a very good indicator of pollution, even a long time after contamination. Honey bees collect honeydew from tens of millions of different single points covering forest areas of some tens of square kilometers. The consequence is that honey presents probably the best composited sample and the most representative values for the average concentrations of bioavailable elements in a given environment.

Concluding remarks

For the first time, concentrations of ^{137}Cs in various Croatian honeys collected during the 1990s are presented. Results found in Austria, Germany, and Slovenia from 1952 to 1995 are also discussed in order to reconstitute the scenario of ^{137}Cs pollution in Europe.

The concentrations of ^{137}Cs, ^{40}K, Ca, Fe, Rb, Sr, Cu, Zn, Pb, Ni, Mn, and Cr in Croatian soils, coniferous tree branches, and honeys have been measured. The analysis of the transfer of these radionuclides and selected elements from soils through conifers and into various types of honeys clearly reveals the relevance of the honeydew honey as an indicator of ^{137}Cs, Cr, Rb, Cu, Pb, and Ni pollution.

Acknowledgment

The authors thank Dr Helmut Horn, University of Hochenheim, Landesanstalt für Bienenkunde, Germany, Dr Herman Pechhacker, Institute for Beekeeping, Lunz am See, Austria, and Dr Janez Poklukar, Kmetijski Institute, Ljubljana, Slovenia, for the honeys collected from Germany, Austria, and Slovenia, respectively.

References

1 Faith, W.L. and Atkisson, A.A. (1972). *Air Pollution.* John Wiley and Sons, Inc., New York, USA, p. 393.
2 Barišić, D., Vertačnik, A. and Lulić, S., (1999). Caesium contamination and vertical distribution in undisturbed soils in Croatia. *J. Environ. Radioact.* **46**, 361–374.
3 Filipović-Vinceković, N., Barišić, D., Mašić, N. and Lulić, S. (1991). Distribution of fallout radionuclides through soil surface layer. *J. Rad. Nucl. Chem. Art.* **148**, 53–62.
4 Cremers, A., Elsen, A., Depreter, P. and Maes, A. (1988). Quantitative analysis of radiocaesium retention in soils. *Nature* **335**, 247–249.
5 Livens, F.R., Horrill, A.D. and Singleton, D.L. (1991). Distribution of radiocesium in the soil–plant systems of upland areas of Europe. *Health Phys.* **60**, 539–545.
6 Shaw, G. and Bell, J.N.B. (1991). Competitive effects of potassium and ammonium on caesium uptake kinetics in wheat. *J. Environ. Radioact.* **13**, 283–296.
7 Shaw, G., Hewamanna, R., Lillywhite, J. and Bell, J.N.B. (1992). Radiocaesium uptake and translocation in wheat with reference to the transfer factor concept and ion competition effects. *J. Environ. Radioact.* **16**, 167–180.
8 Bilo, M., Steffens, W., Fuhr, F. and Pfeffer, K.H. (1993). Uptake of $^{134/137}$Cs in soil by cereals as a function of several soil parameters of three soil types in Upper Swabia and North Rhine-Westphalia (FRG). *J. Environ. Radioact.* **19**, 25–39.
9 Robinson, W.L. and Stone, E.L. (1992). The effect of potassium on the uptake of ^{137}Cs in food crops grown on coral soils: Coconut at Bikini atoll. *Health. Phys.* **62**, 496–511.
10 Gerzabek, M.H., Mohamad, S.A. and Muck, K. (1992). Cesium-137 in soil texture fractions and its impact on cesium-137 soil-to-plant transfer. *Commun. Soil Sci. Plant Anal.* **23**, 321–330.

11 Shand, C.A., Cheshire, M.V., Smith, S., Vidal, M. and Rauret, G. (1994). Distribution of radiocaesium in organic soil. *J. Environ. Radioact.* **23**, 285–302.

12 Hird, A.B., Rimmer, D.L. and Livens, F.R. (1995). Total caesium-fixing potentials of acid organic soils. *J. Environ. Radioact.* **26**, 103–118.

13 Shenber, M.A. and Eriksson, Å. (1993). Sorption behaviour of caesium in various soils. *J. Environ. Radioact.* **19**, 41–51.

14 Bunzl, K., Förster, H., Kracke, W. and Schimmack, W. (1994). Residence times of fallout $^{239+240}$Pu, ^{238}Pu, ^{241}Am and ^{137}Cs in the upper horizons of an undisturbed grassland soil. *J. Environ. Radioact.* **22**, 11–27.

15 Barišić, D., Lulić, S., Kezić, N. and Vertačnik, A. (1992). ^{137}Cs in flowers, pollen and honey from the Republic of Croatia four years after the Chernobyl accident. *Apidologie* **23**, 71–78.

16 Sandalls, J. and Bennett, L. (1992). Radiocaesium in upland herbage in Cumbria, UK: A three year field study. *J. Environ. Radioact.* **16**, 147–165.

17 Zach, R., Hawkins, J.L. and Mayoh, K.R. (1989). Transfer of fallout cesium-137 and natural potassium-40 in boreal environment. *J. Environ. Radioact.* **10**, 19–45.

18 Cheshire, M.V. and Shand, C. (1991). Translocation and plant availability of radio caesium in an organic soil. *Plant & Soil* **134**, 287–296.

19 Barišić, D., Lulić, S., Vertačnik, A., Dražić, M. and Kezić, N. (1994). ^{40}K, ^{134}Cs and ^{137}Cs in pollen, honey and soil surface layer in Croatia. *Apidologie* **25**, 585–595.

20 Bunzl, K., Kracke, W. and Vorwohl, G. (1988). Transfer of Chernobyl-derived ^{134}Cs, ^{137}Cs, ^{131}J and ^{103}Ru from flowers to honey and pollen. *J. Environ. Radioact.* **6**, 261–269.

21 Molzahn, D. and Assmann-Werthmüller, U. (1993). Caesium radioactivity in several selected species of honey. *Sci. Total Environ.* **130/131**, 95–108.

22 Jackson, D. (1989). Chernobyl-derived ^{137}Cs and ^{134}Cs in heather plants in northwest England. *Health Phys.* **57**, 485–489.

23 Assmann-Werthmüller, U., Werthmüller, K. and Molzahn, D. (1991). Cesium contamination of heather honey. *J. Rad. Nucl. Chem. Art.* **149**, 123–129.

24 Tonelli, D., Gattavecchia, E., Ghini, S., Porrini, C., Celli, G. and Mercuri, A.M. (1990). Honey bees and their products as indicators of environmental radioactive pollution. *J. Rad. Nucl. Chem. Art.* **141**, 427–436.

25 Jones, K.C. (1987). Honey as an indicator of heavy metal contamination. *Water Air Soil Pollut.* **33**, 179–189.

26 Leita, L., Muhlbachova, G., Cesco, S., Barbattini, R. and Mondini, C. (1996). Investigation of the use of honey bees and honey bee products to assess heavy metals contamination. *Environ. Monit. Ass.* **43**, 1–9.

27 Bromenshenk, J.J., Carlson, S.R., Simpson, J.C. and Thomas, J.M. (1985). Pollution monitoring of Puget Sound with honey bees. *Science* **227**, 632–634.

28 Veleminsky, M., Laznička, P. and Stary, P. (1990). Honeybees (*Apis mellifera*) as environmental monitors of heavy metals in Czechoslovakia. *Acta Entomol. Bohemoslov.* **87**, 37–44.

29 Horn, U., Helbig, M., Molzahn, D. and Hentschel, E.J. (1996). The transfer of Ra-226 to honey and the possible use of the honey bee as a bioindicator in the uranium mining area of the Wismut region (German). *Apidologie* **27**, 211–217.

30 Hoopingarner, R.A. and Waller, G.D. (1993). Crop pollination. In: *The Hive*

and the Honey Bee (Graham, J.M., Ed.). Dadant & Sons, Hamilton, IL, pp. 1043–1082.

31 Bromenshenk, J.J., Smith, G.C., King, B.E., Seccomb, R.A., Alnasser, G., Loeser, M.R., Henderson, C.B. and Wrobel, C.L. (1998). New and improved methods for monitoring air quality and the terrestrial environment: Applications at Aberdeen proving ground – Edgwood area. *US Army Center for Environmental Health Research, Technical Report* **I.1–I.55**.

32 Tomašec, I. (1949). *Biologija pčela.* Nakladni zavod Hrvatske, Zagreb, Hrvatska, p. 94.

33 Free, J.B. (1993). *Insect Pollination of Crops.* Academic Press, London, UK, p. 684.

34 Louveaux, J., Mauruzio, A. and Vorwohl, G. (1978). Methods of melissopalynology. *Bee World* **59**, 139–157.

35 Vorwohl, G. (1964). Messung der elektrischen Leitfähigkeit des Honigs und der Verwendung der Messwerte zur Sortendiagnose und zum Nachweis von Verfälschungen mit Zuckerfütterungshonig. *Z. Bienenforsch* **7**, 37–47.

36 IAEA-QXAS Version 3.2 (1995). Manual, International Atomic Energy Agency, Vienna, Austria.

37 Barišić, D., Lulić, S., Vertačnik, A., Dražić, M. and Kezić, N. (1995). Long term behaviour of ^{137}Cs and ^{40}K in honey in Croatia. In: *Proceedings of the International Symposium on "Bee Breeding on the Islands"* (Kezić, N., Ed.), Island of Vis, Croatia, pp. 56–59.

38 Barišić, D., Vertačnik, A., Bromenshenk, J.J., Kezić, N., Lulić, S., Hus, M., Kraljević, P., Šimpraga, M. and Seletković, Z. (1999). Radionuclides and selected elements in soil and honey from Gorski Kotar, Croatia. *Apidologie* **30**, 277–287.

39 Kezić, N., Hus, M., Seletković, Z., Kraljević, P., Pechhacker, H., Barišić, D., Lulić, S. and Vertačnik, A. (1997). Honeydew honey as a long term indicator of ^{137}Cs pollution. *IAEA-TECDOC* **964**, 54–61.

40 Garrec, J.P., Suzuki, T., Mahara, Y., Santry, D.C., Miyahara, S., Sugahara, M., Zheng, J. and Kudo, A. (1995). Plutonium in tree rings from France and Japan. *Appl. Radiat. Isot.* **46**, 1271–1278.

41 Momoshima, N., Eto, I., Kofuji, H., Takashima, Y., Koike, M., Imaizumi, Y. and Harada, T. (1995). Distribution and chemical characteristics of cations in annual rings of Japanese cedar. *J. Environ. Qual.* **24**, 1141–1149.

42 Wyttenbach, A., Ferrer, V. and Tobler, R. (1995). The concentration ratios plant to soil for the stable elements Cs, Rb and K. *Sci. Total Environ.* **173/174**, 361–367.

11 Use of honey bees as bioindicators of environmental pollution in Italy

C. Porrini, S. Ghini, S. Girotti,
A.G. Sabatini, E. Gattavecchia, and
G. Celli

Summary

In Italy the use of honey bees in environmental monitoring goes back to the early 1980s with initiatives to control pesticides in agro-ecosystems.

In a pesticide monitoring station comprising two beehives fitted with underbaskets, a critical threshold of mortality was set of about 350 bees per week per station. Only when this threshold was exceeded were chemical analyses performed on the dead bees to determine the active ingredients responsible for their death. The two sets of data, mortality and residues, were processed using the Index of Environmental Hazard (IEH), which allows monthly assessments of the level of environmental contamination in the area investigated.

Heavy metal pollution (Pb, Ni, Cr) may be monitored with both bees and honey. It is in fact possible to integrate the data derived from these two matrixes to provide more complete information regarding the presence of contaminants in the environment.

Since the incident of 1986 involving the nuclear plant at Chernobyl, studies on the absorption and transfer of radionuclides to beehives have undergone a remarkable increase. The research still continues today with the monitoring of long-life radionuclides such as Cs-137 and Sr-90. All the studies have shown pollen and bees to be highly sensitive indicators of radioactive contamination.

Recently our research team demonstrated for the first time that honey bees can be used for the environmental detection of the phytopathogenic microorganism *Erwinia amylovora*, the causal agent of Fire Blight, the most destructive bacterial disease affecting Rosaceous plants. A new ultrasensitive and specific diagnostic technique (PCR-ELISA) was developed to permit automated detection of *E. amylovora* in pollen. It was shown that honey bees could reveal the presence of *E. amylovora* both in already disease-affected areas and in areas where no evident symptoms had yet appeared on the plants. Therefore, border areas likely to be hit by Fire Blight can be constantly monitored using bee-hives.

Introduction

The use of honey bees for environmental monitoring purpose dates back to 1935 when Svoboda [1] reported the negative repercussions of industrial pollutants on bees that foraged in densely populated, industrialized areas in Czechoslovakia. In subsequent years, numerous studies were undertaken to test the effectiveness of this hymenopteran as a biological indicator of the presence of contaminants in the environment.

A distinction should be made first of all between a biological indicator and a test organism. The former is an organism, or group of species, which indicates particular conditions in the environment through specific signals [2]. These signals may derive either from the presence of a single substance or the joint presence of several substances. A test organism, on the other hand, is used to assess the toxicity of a specific substance in laboratory tests and/or field trials. The distinction is clearly made for the sake of convenience, as the same organism, e.g. the honey bee, may be used as either an indicator or a test organism in different circumstances [3].

Honey bees are good biological indicators because they *reveal* the chemical impairment of the environment they live in through two signals: one is more evident, that is high mortality (in the case of pesticides), while the other is less so, consisting of residues present within their bodies or in beehive products (in the case of other contaminants like heavy metals and radionuclides) that may be detected by means of suitable laboratory analyses [4].

The effectiveness of honey bees as an ecological detector is founded upon several ethological and morphological characteristics, which are listed below [5]:

- an easy-to-breed, almost ubiquitous organism with modest food requirements;
- its body is covered with hairs, which makes it particularly liable to hold the materials and substances it comes into contact with;
- a high sensitivity to most phytopharmocological products, revealing when they are improperly spread through the environment (e.g. during flowering, in the presence of wind, etc.);
- a very high rate of reproduction and a relatively short average life-span, so that the colony undergoes rapid, continuous regeneration;
- great mobility and a flying range that allows a vast area to be monitored;
- high efficiency in ground surveys (numerous inspections per day);
- almost all environmental sectors (soil, vegetation, water, and air) are sampled;

- numerous indicators (through foraging) for each station (beehive) are provided;
- a variety of materials are brought into the hive (nectar, pollen, honey-dew, propolis, and water) and can be stored according to verifiable criteria.

Moreover, the management costs of monitoring with honey bees are extremely low, especially in proportion to the large number of samples that may be taken.

Bees interact strongly with the environment surrounding the hive. An average-sized population may consist of as many as 20 000 individuals and at certain times of the year their number can almost double. Approximately one-fourth of the population – the older worker bees that have already performed a whole series of other tasks inside the hive – is responsible for retrieving from the outside everything the colony needs to survive and develop. To this end, each bee makes on average about 10 trips a day, visiting a total of about 1000 flowers from which it draws nectar and pollen. Therefore, based on an empirical calculation, it can be estimated that a colony of bees gathers approximately 10 million micro-samples of nectar and pollen every day, as well as other substances such as honeydew, propolis, and water.

All these substances are conveyed into the hive, where they are stored. The incessant foraging normally occurs within a few hundred meters from the hives. However, should food supplies be scarce, the foragers are capable of venturing out much further, even kilometers away, in search of a sufficiently rich pasture, which they will then report to other members of the hive. The area normally reconnoitered by an average-sized family is acknowledged to be about 7 km² [6]. Within this range, the bees carefully survey the flowering of the species that are most appealing to them, send out explorers and constantly gather specimens from the various areas which are later to be stored in the hive.

The honey bees keep their territory constantly under control; thus any contaminants present therein are intercepted and carried into the hive, where they will become available for chemical analysis. It also follows that bees can quickly perceive any trends or changes occurring within the environments they inhabit and disclose them with equal promptness. The honey bee may therefore be considered a biological indicator.

However, there are several limitations to the use of bees for assessing the state of health of the environment [5]:

- the temperature must be at least 10°C in order for them to fly; consequently, at certain latitudes they cannot be used in wintertime;
- the forager bees may not all return to their hive; some may stray off-course (ending up in other hives), undergo a natural death, or be killed by pesticides or other xenobiotics;

- it is difficult to take a real-time census of an entire family in terms of its stage of development and the age of its members;
- there is an uncontrollable tendency of families to choose their food sources autonomously.

This latter limitation is viewed as the most serious obstacle to reliable environmental monitoring, but another characteristic behavior of bees must also be considered – loyalty to food sources (so that bees will visit a given botanical species as long as it continues to flower). The degree of loyalty varies according to subspecies: The German bee (*Apis mellifera mellifera*) appears to be the most loyal to a food source (0.8 percent deviation from the original source in every flight), whereas the Italian bee (*A. m. ligustica*) is the most fickle (14 percent) [7]. The type of co-evolutionary relationship established between these two subspecies and their respective environments of origin – continental and Mediterranean – may account for such behavioral differences. In any event, the relative disloyalty of Italian bees to a given pasture offsets the tendency to choose food sources autonomously, both during the same day and on subsequent days, and thus permits a more reliable monitoring of the territory.

Monitoring of pesticides

For about twenty years our research group at the "Guido Grandi" Department of Entomology of the University of Bologna has been studying the relationship between bees and the pesticides spread throughout the agro-ecosystem. These synthetic compounds have undergone extensive use, especially since the end of the Second World War, and have contributed significantly to the degradation of the rural ecosystem.

In itself, a cultivated field is a highly unstable ecosystem, and the advent of mechanical equipment, with the consequent introduction of single-crop farming, has further exacerbated the instability of the agro-ecosystem. The plants grown often show low resistance and are more vulnerable to the attacks of parasites, which farmers control with pesticides.

The growing use of these substances in farming has proved to be a double-edged sword. It is estimated that from the Second World War to 1974, although the use of insecticides underwent a tenfold increase, the damage caused by insects nearly doubled [8]. This also suggests that the number of pesticide-resistant species grew enormously.

Furthermore, the widespread use of pesticides has contributed to a further simplification of the agro-ecosystem, which has occurred above all at the expense of insects that are reputedly "man's allies," i.e. predators, parasitoids, and pollinators. The latter are often more likely to be harmed by pesticides than the target arthropod pests themselves and thus the exposure to potential infestation by phytophaga increases rather than decreases.

As noted previously, honey bees are extremely sensitive to pesticides. The number of dead bees in front of the hive is therefore the most important variable to be considered for these contaminants [9]. Not all pesticides have a lethal effect; in fact, many fungicides and other compounds used for different purposes often have a less severe impact on bees than the majority of insecticides. As a result, honey bees are a good direct indicator of insecticides and respond to their presence in the environment with a mortality that varies according to a number of factors: the toxicity (for bees) of the active ingredient used (LD_{50}) [10, 11], the presence and extension of flowering among cultivated or spontaneous plants, the presence of honey bees on the site and at the time of the chemical treatments, the means used to distribute the pesticide, and the presence of wind. Many bees struck directly by an insecticide will not have enough strength to return to their hive and will die on the field or during their return flight (high acute toxicity of the product). Usually, however, the treatment does not fully affect all the forager bees that are in the field at a given time; some will be only marginally hit and will eventually die in the hive (residue accumulation), sharing their fate with other bees that subsequently visit the flowers of the treated plants (whenever the active ingredients have no repellent effect) or gather nectar and pollen from spontaneous species growing either alongside treated crops or in nearby areas contaminated by drift. In the case of compounds that are not particularly dangerous, the insect acts as an indirect indicator, i.e. not sensitive but exposed, and will provide us with information in the form of residues.

Bees may effectively indicate the presence of pesticides in intensively cultivated areas with little spontaneous flora; in this case the insect is obliged to forage among cultivated species or in nearby areas and is thus likely to come into contact with any pesticides that have been sprayed. On the other hand, in areas abounding in wild plant species, bees have a broader choice of food sources and are therefore a less reliable indicator of the chemical impact on a cultivated field and its surroundings.

Nonetheless, other matrixes such as honey [12, 13] and pollen may also provide useful indications as to the diffusion of pesticides within the environment. In a study conducted in Castenaso, in the province of Bologna (Figure 11.1), between March and September 1995, pollen specimens were gathered weekly from two stations situated, respectively, in an urban area and in a rural area, both in the vicinity of a watercourse. Although it was forbidden to perform chemical treatments during the flowering season, traces of one or more pesticides were found in over one-third of the samples drawn from both stations, even in periods that were not normal for field crop growing. Moreover, given that all the specimens testing positive for pesticides contained pollen of both cultivated and spontaneously growing plants, it could be inferred that crops had also been treated in the presence of wind, causing pesticides to drift into surrounding areas. This mismanagement of pesticides, extremely harmful for the agro-

Figure 11.1 Italy: Geographic distribution of the areas where honey bees have been used for environmental monitoring (ER = Emilia-Romagna region, F = Friuli-Venezia Giulia region, T = Tuscany region).

ecosystem, could be traced both to the vegetable and flower gardens of amateur growers in the area covered by the urban station and to the cultivated farmland prevalent in the other station.

Capture of pesticides by bees

Figure 11.2 represents the transfer of polluting substances from the atmosphere to other environmental sectors. It clearly and succinctly illustrates how bees come into contact with the chemical compounds scattered throughout the environment.

Since bees move from flower to flower, alight on branches and leaves, ingest water from many sources (ditches, puddles, canals, brooks, fountains, dew, etc.) and intercept particles suspended in the atmosphere with their hairy bodies, they are capable of disclosing the presence of environmental contamination. In a small study conducted in collaboration with Sauro Tiraferri of the Regional Bureau for Prevention and the

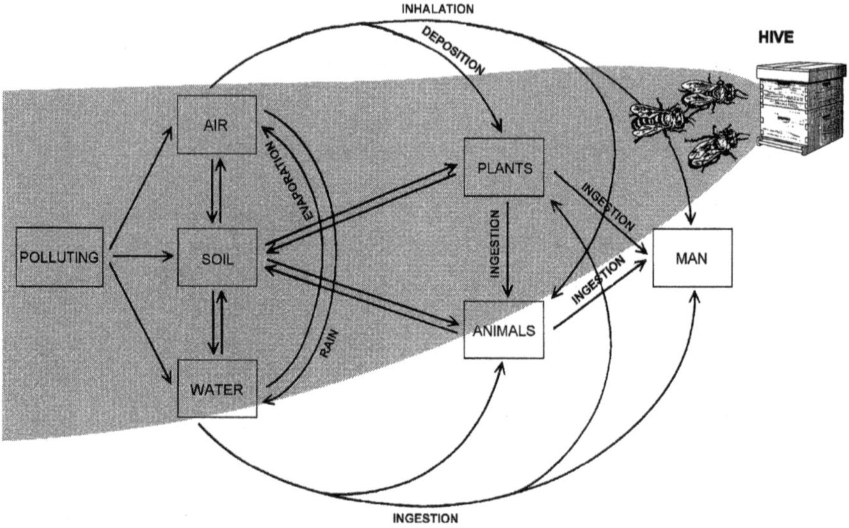

Figure 11.2 Chart of polluting substance diffusion in the environment. Honey bees may capture pollutants diffused in the air, deposited on plant surfaces and on the soil, and assimilate them from water, as shown in the gray area indicating the environmental sectors visited by the bee.

Environment of Rimini, it was shown how bees capture insecticidal compounds mainly by ingestion.

Nineteen specimens of bees that had died of poisoning were analyzed in the trial. The bees were first washed with acetone, to detect compounds deposited on the bees' surface, and then homogenized, to detect compounds accumulated inside the bees. Both the washing solutions and homogenized bees were examined to determine whether any pesticide residues were present. In the 16 samples testing positive, seven different types of insecticides were found.

Dimethoate and ethiofencarb, detected in eight and three samples respectively, were by far the most prevalent in the homogenized specimens, i.e. inside the bees' bodies. Parathion was detected predominantly and monocrotophos exclusively inside the bodies, whereas azinphos-methyl, omethoate, and methylparathion were present in varying percentages in the bees' bodies both internally and externally (Figure 11.3).

The study does not take into account a series of variables, such as the type of crop treated, the time lapse between the chemical treatment and the poisoning of the bees, and the botanical species visited. Nevertheless, it gives an idea of how the various pesticides may affect the insect and, above all, how bees capture them. The seven compounds detected achieve their effect through both contact (in particular dimethoate, parathion, and methyl-parathion) and ingestion; ethiofencarb and monocrotophos are also systemic.

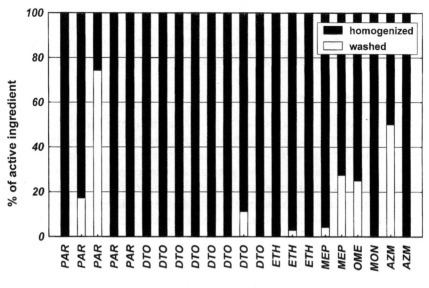

Figure 11.3 Active ingredients registered internally ("homogenized") and externally ("washed") in honey bees in various bee killings
(PAR = parathion, DTO = dimethoate, ETH = ethiofencarb,
MEP = methyl parathion, OME = omethoate, MON = monocrotophos, AZM = azinphos-methyl).

Monitoring stations

The design of a beehive monitoring station must take several factors into
consideration. Each colony making it up shows specific attitudes and territorial preferences; consequently the surveying of the territory may be
enhanced and diversified by increasing the number of families present –
that is, the more beehives there are, the more reliable the environmental
monitoring [14]. It is, however, necessary to consider the management
costs of a monitoring network. Increasing the number of colonies in a
monitoring station means raising the cost of inspections, including health
checks, as well as the cost of sample gathering and laboratory analyses.
Therefore, the choice of two beehives represents a compromise between
the financial resources available for the study and the goal of obtaining as
much information as possible from the bees in order to determine, on the
basis of suitable criteria, the state of health of the territory.

The families of bees making up the monitoring stations must be in a
good state of health, of average "strength" and, above all, similar in kind.
The "strength" of a colony is assessed by carrying out an accurate examination of the hives in front of which forager bees' flight activity is

observed, the age of the queen, the number of frames occupied by the bees, the number of frames for brood, its compactness or discontinuity, the percentage of new and old brood, the presence of drone cells and royal cells, the number of frames with food stores, the percentage of old and new honey, and the quantity of pollen present.

Location of monitoring stations

To ensure that the beehives will be positioned in appropriate locations for monitoring purposes, the area to undergo investigation must be chosen taking into account the use to which the territory is put, its orographic features, the composition of the vegetation, the presence or absence of "shelter" zones and natural areas, and the impact of human activity. In order to maximize the bees' intake of pollutants, reference may be made to a study by Colombo *et al.* [15] which takes into consideration two factors: the dispersal of the pollutant – assessed using the methods of atmospheric diffusion and the fallout of materials from an ideal emitting stack, taking into account wind conditions and the classes of atmospheric stability typical of the place to be monitored – and the manner in which contaminants are taken in by the colony, according to the ethology of foraging and the distribution of bees in the territory surrounding the hive. For this purpose, it is necessary to have detailed information about the botanical species inhabiting the area, how they succeed one another during a given period and their power to attract bees, as well as other more general information regarding weather conditions and physical traits of the territory.

Evaluation of bee mortality

An assessment of bee mortality is fundamental to ensure reliable monitoring of pesticides with this biological indicator.

Taking a census of dead honey bees presents several problems that many authors have attempted to overcome by means of various solutions. The cages devised by researchers can only give an incomplete picture of total mortality. These structures capture only the bees that have managed to return to the hive and are expelled, on dying, by their companions. The bees that have died on the field or on their way back to the hive cannot be counted. This fraction of mortality varies according to the cause of death: if the compound has a strong destructive power (highly toxic), the bees dying "away from home" will represent a large percentage. If, on the contrary, the compound produces a more gradual effect, a majority of bees will succeed in returning to the hive. Whenever a substance induces a slow poisoning, its lethal effect will result in a progressive, often unapparent, depopulation of the colony that the beekeeper cannot detect in time [16].

Over the years, various researchers [17–22] have employed different kinds of structures for collecting dead bees in field and semi-field experiments to assess the hazards deriving from the pesticides used in farming.

The cages must be efficient and meet several requisites:

- no interference with the flight and the normal activities of bees on the flight board;
- prevent access to possible predators, e.g. wasps;
- allow the bees to be easily counted;
- withstand different climatic conditions;
- easy to set up and remove;
- above all they must be inexpensive.

One of the biggest drawbacks of these structures is that the bees may become so accustomed to the presence of the cage that they begin to consider it part of the hive and therefore clear away the dead bees from it [23, 24].

In 1960 Gary [25] designed a rectangular cage made of wire, which researchers preferred to others because of its practicality. In Italy, a slightly modified version of Gary's cage [26] has been widely used both for environmental monitoring with bees and in field trials conducted to assess the hazards of pesticides [9, 12, 13, 27–31]. Although similar to the cage developed by Gary in 1960, it is simpler, consisting only of metal wire netting without film and lubricants, and the part serving as a cover has been reduced to a thin strip.

In 1987 Marchetti *et al.* [26] conducted tests on the effectiveness of the modified version of Gary's cage. Marked groups of dead bees were introduced into the hives to simulate five levels of mortality: 50, 100, 200, 400, and 800 bees. For the five levels of mortality, the following mean percentages of recovery were reported: 92.0, 93.2, 94.1, 87.8, and 87.2 percent, respectively. This type of cage thus appeared to be a highly effective means of collecting dead bees despite its simple structure. It was observed, however, that after a period of time the bees came to consider these cages as part of the hive itself and thus "sweeper" bees removed the dead bees collected in them. Consequently, a new cage was conceived, the so-called underbasket [23]. This basket is made up of a wooden frame and two metal wire nets, a bottom one with a fine mesh and a top one with a larger mesh. The structure is not integral with the beehive as it is positioned beneath the entrance. A special cover may be applied to prevent the dead bees collected in the baskets from rotting or being washed away by rain. Tests were conducted to assess the efficacy of the modified version of Gary's cage and the underbasket. The findings showed the underbasket to be more effective and reliable over time. In fact, with Gary's cages between 65.2 and 90.4 percent of the dead bees were recovered whereas the percentage achieved with the underbasket was between 71.4 and 96.4 percent [32].

Subsequently other trials were conducted to evaluate the efficacy of various types of cages taking into account the influence of variables such as time [33], whether death was natural or artificial [34], and the weekly holding capacity of dead bees, which, as mentioned previously, may be removed by different animals such as wasps and ants [35]. The latter study placed the holding capacity of the various kinds of receptacles in correlation with the environment (simplified or complex) and season (spring and summer).

The efficacy of Gary's cage was found to vary according to the above factors: 97.6 ± 3.2 percent in spring and 84.8 ± 14.9 percent in summer in the simplified environment; 66.1 ± 26.6 percent in spring and 85.2 ± 19.1 percent in summer in the complex environment. The underbasket, also tested in the four situations, showed much greater consistency: 99 ± 1.1, 97.3 ± 1.3, 98.9 ± 1.1, and 90.3 ± 7.4 percent, respectively. In all the experiments, therefore, the underbasket proved to be more reliable. It has thus been chosen for our own research projects. However, further efforts are being made to devise a trap that can completely prevent saprophages from removing dead bees.

Electronic bee counters

These devices should be able to count bees going in and out of the hive. It should thus also be possible to determine the number of bees dying out on the field.

In the past, various contrivances have been built. As far back as 1925 Lundie [36] sought to monitor beehive activity with a device consisting of thin metal strips that the bees were supposed to lift as they flew in or out, thereby making an electric contact. Various technological solutions were subsequently proposed: mechanical systems [37], hydraulic devices [38], a system with heat probes and oscillating cylinders [39], weighing procedures [40, 41], as well as a variety of other strategies such as the assessment of collected pollen [42] and a count of bees coming back after the beehive entrance had been closed [43, 44]. However, all these techniques had a drawback in that they were highly imprecise and required constant maintenance. The introduction of photoelectric technology led to the development of new systems [45–54]. Unfortunately, none of them has proved to be satisfactory for the purposes of environmental monitoring, although a recently introduced model promises to be much more reliable [55, 56]. Our own instrument comprises a transit detection module and a central data-gathering and storage unit, controlled by a program capable of following every single movement of the insect and signaling any irregular transit situations. Although in theory it should present no limits, in practice it has not yet proved to be wholly reliable for its intended purpose [57].

Critical threshold of mortality

The physiological mortality of a beehive is not easy to ascertain. It depends on numerous variables such as the season, the strength of the family, and the surrounding environment. Nonetheless, taking into consideration the number of eggs laid by the queen during the season, the number of brood cells occupied, and the number of adult bees, it may be hypothesized that about 1000 bees die naturally on any given day during the period of densest population, i.e. from May to July [58, 59]. Smaller values are recorded both before and after these months. The calculations made on the basis of the bees collected in cages will be an underestimate because, as noted previously, the efficacy of this gathering method varies according to the environment the hives are situated in and the season (see section on *"Evaluation of bee mortality"*).

Various experiments have been undertaken for the purpose of determining a critical threshold of mortality that may be assessed using underbaskets. Our study takes into account observations made in non-contaminated areas where this type of trap was shown to capture weekly up to 2.5 percent (175 dead bees per hive) of the maximum natural mortality.

Six monitoring stations, each comprising two hives, were posted in the township of Castenaso (province of Bologna). In the months of April and May 1997, the dead bees found in the baskets positioned in front of the hives were retrieved and analyzed weekly. The aim of the study was to identify the level of mortality at which pesticide residues began to appear; this level would thus be considered as the threshold marking the boundary between natural and induced mortality.

During the experiments, 48 samples of dead bees were collected and analyzed. Seventeen samples exceeded the critical threshold hypothesized of 350 dead bees per week per station (the sum of two hives). Fourteen (82.3 percent) were found positive, i.e. they contained at least one residue of a compound, whereas only 19.3 percent of the samples below the threshold of 350 dead bees were found to be positive. Of the latter samples, 66.6 percent fell within the bracket of mortality ranging from 300 to 350 dead bees (Figure 11.4) [60].

On the basis of these findings, therefore, it can be deduced that the critical threshold of mortality in a station comprising two hives is 300–350 dead bees per week. However, in some cases it may also be useful to analyze bee specimens that do not reach the critical threshold of mortality as they can provide evidence of active principles harmless for bees but dangerous for the environment. In addition, whenever the number of dead bees collected from the two hives in a station differs significantly, it is advisable to perform a separate palynological and chemical analysis on the two samples so that the crops treated may be more precisely identified.

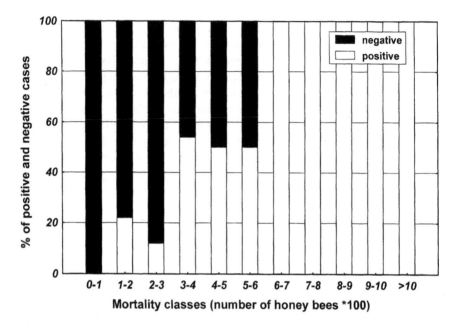

Figure 11.4 Relation between the number of dead bees and the presence of pesticides residues.

Chemical analysis of dead bee specimens

Different methods of analysis may be used for the qualitative and quantitative determination of carbamates and organophosphorus insecticides:

- gas liquid chromatography (GLC)
- high-performance liquid chromatography (HPLC)
- enzymatic methods based on the inhibition of acetylcholinesterase
- immunoenzymatic methods.

The extractive method chosen will depend on the type of analysis adopted and the matrix in which the analyzed substances are contained, which may or may not require the extract to be purified. The particular matrix analyzed in this study makes all the above-described methods more difficult to apply. As the bee's body is a complete organism, it contains a myriad of molecules that may interfere with the analytic processes in many different ways. Therefore, the extraction and purification of the compounds to be determined take on great importance.

At present the most widely used method of analysis is gas chromatography. Increasingly sophisticated instruments allow more reliable qualitative and quantitative determinations of these compounds to be carried out.

Combining gas chromatography and mass spectrometry (GLC-MS) is a particularly useful means for achieving a rapid identification: the components leaving the gas chromatography apparatus are injected directly into a mass spectrometer for the analysis of each chromatographic peak. The data provided with regard to the molecular structure of the compounds allow them to be clearly identified. In recent years a broad variety of analytic methods have been developed as alternatives or complements to gas chromatography. These methods involve the use, for example, of enzymatic reagents or biosensors which, in order to determine the pesticides being examined, exploit their inhibition of the enzyme's activity in the presence of an appropriate substrate and detection system. Various families of pesticides can be determined using enzymes: dithiocarbamate fungicides inhibit the aldehyde dehydrogenase, sulfonylurea herbicides inhibit acetolactate synthase, whereas organophosphorus insecticides and carbamates inhibit acetylcholinesterase [61].

Acetylcholinesterase is by far the most widely used enzyme in the preparation of biosensors for determining pesticides, both because organophosphorus insecticides and carbamates represent over half of the entire insecticide market and because the acetylcholinesterase commercially available has a high degree of purity and specificity of action and may be paired with many transducers (potentiometric, amperometric) in both flow and nonflow systems [62]. The specific tendency of organophosphorus pesticides and carbamates to inhibit acetylcholinesterase has been exploited for the purpose of determining these compounds, which are first separated by means of HPLC, then detected through a post-column reaction with immobilized acetylcholinesterase [63].

Various immunoenzymatic assays have been developed and several kits are now commercially available [64]. The aim of these techniques is to achieve a simpler, faster, and less costly determination of these compounds than is possible with GLC, while preserving the high degree of sensitivity and specificity that such analyses require. They are particularly useful for conducting a preliminary screening in laboratories where a larger number of samples must be analyzed [65].

The classic method for determining pesticide residues in bee samples is as follows [66]. Ten grams of bees are taken and mixed with 50 ml of water; then the following steps are performed:

- extraction with acetone
- separation in methylene chloride
- purification in the column
- gas chromatographic analysis with an NPD (nitrogen phosphorus detector) or ECD (electron capture detector).

Most of the chemical analyses for our experiments are conducted at the Technical Department of the Regional Bureau for Prevention and the

Environment (ARPA) of Emilia Romagna (provincial section of Rimini), the laboratory of the Centro Studi Ambientali (CSA – Environmental Studies Center) research institute of Rimini and the National Institute of Apiculture of Bologna (INA). For the analyses carried out at INA, at present we prefer to use a method in which the sample to be analyzed is purified by means of phosphatic precipitation so that interfering elements are eliminated before it undergoes separation with methylene chloride. This method ensures better chromatograms for a multiresidual technique as well as cleaner instruments [67].

Intense efforts are being undertaken both to overcome detection limits and to increase the number of determinable compounds, also for the purpose of developing alternative analytical methods that are simpler, cheaper, and require less solvent. Researchers are currently experimenting new methods for purifying samples and new extraction systems specially tailored to the bee's body that rely on techniques such as "gel permeation" [68] and solid phase extraction. A particular focus is placed on the study of enzymatic and immunoenzymatic assays for classes of compounds; these methods may allow samples to be passed through a fast preliminary qualitative screening and thus drastically reduce the number of analyses to be performed.

Palynological analysis

The identification of pollen may be of great importance in the field of monitoring with bees: a palynological analysis of bee products or of honey bees themselves provides precise information about the plants visited, and this information will be all the more precise if detailed knowledge is available regarding the vegetation present within the bees' flying range.

The most suitable matrix will be chosen according to the aim of the monitoring project: bees are used for monitoring pesticides, as they allow the determination of the crops that may have undergone treatment; when heavy metals are being monitored, honey is analyzed in addition to the bees themselves; bees and pollen are mainly used in the case of radionuclides, whereas pollen is the most suitable matrix for the detection of phytopathogenic microorganisms.

A sophisticated environment-monitoring technique, called the marker pollen technique, has been developed; it provides for the sowing, in specific areas, of botanical species that are particularly appealing to bees, and which are not present in the research area. This method is being tested for the monitoring of *Erwinia amylovora* and may be very useful in territories characterized by a homogeneity of crops (e.g. extensive single-crop farms).

As to sample-preparing techniques, the classic methods of melissopalynology are used for honey [69]. Some methods developed for bee samples provide for the bees' bodies to be first washed with acetone and then placed in a centrifuge so that the sediment can be collected. If the same sample is also to be used for determining the presence of pesticide residues, ether is added to the bees, which undergo a brief sonication; the bees are separated from the ether and the sediment is collected by means of centrifugation. In all cases, following extraction, the sediment – which contains pollen grains or pollen as such – is transferred onto a glass plate, englobed within a drop of glycerinated gelatine and examined under an optical microscope.

The terminology used to identify pollen types is based on a standard nomenclature [70, 71]. In some cases, it is possible to trace both the crop the pollen has originated from and its location by comparing the pollinic spectrum of each bee sample analyzed in the laboratory, the specially pre-pared crop-growing maps, and on-site observations [72, 73].

Data processing

Abundant information about the territory being surveyed may be derived by analyzing the data gathered with this monitoring method. It is possible not only to draw up a list of the compounds used in different areas in dif-ferent periods, but also to prepare detailed monthly maps showing the degree of chemical contamination within the territory itself, rated using a two-way Index of Environmental Hazards (IEH) [74, 75]. This index is obtained by crossing the mortality class of a station with the Index of Pes-ticide Toxicity (IPT) of the substances detected through the chemical analysis of dead bees from that station (Table 11.1). Table 11.1 has been compiled taking into account several factors. As noted previously, bees

Table 11.1 Index of Environmental Hazard (IEH*) using IPT (Index of Pesticide Toxicity) and mortality classes, as specified in the text

		Mortality classes (monthly mean of the number of dead bees per week)			
		0–200 D_4	*200–400* D_2	*400–800* C_3	*>800* C_1
Residue-free samples or mortality below critical threshold					
0	<IPT< 0.125	D_3	D_1	C_2	B_3
0.125	<IPT< 0.25	D_2	C_3	C_1	B_2
0.25	<IPT< 0.375	D_1	C_2	B_3	B_1
0.375	<IPT< 0.5	C_3	C_1	B_2	A_4
0.5	<IPT< 0.625	C_2	B_3	B_1	A_3
0.625	<IPT< 0.75	C_1	B_2	A_4	A_2
0.75	<IPT< 0.875	B_3	B_1	A_3	A_1
	IPT> 0.875	B_2	A_4	A_2	A_1

Note
*IEH: A_1, persistent; A_2, worrying; A_3, substantial; A_4, considerable; B_1, high elevated; B_2, important; B_3, widespread; C_1, medium average; C_2, medium-low; C_3, moderate; D_1, low; D_2, limited; D_3, minimal; D_4, absent.

dying on the field cannot be counted in the cages. Therefore, in some cases where a relatively small number of dead bees are found, subsequent chemical analysis may reveal them to contain an extremely toxic compound; on the contrary, a high rate of mortality may be associated with a compound that is only slightly toxic. In the former case, it may be plausibly assumed that the bees came only marginally into contact with the compound, or that the latter was powerful enough to kill many bees instantly. In the latter case, it is likely that another product was to blame for the bees' death but that it degraded rapidly, or that the slow-acting effect of the compound detected allowed the bees to return to the hive and subsequently die.

Mortality classes have been set on the basis of nearly 5000 weekly samplings carried out in recent years in monitoring stations in Italy (Table 11.2). The mortality class is derived from the monthly mean of weekly mortality values. The IPT can be calculated using the formula:

$$IPT = f_{corr} \sum_{c=1}^{N} \frac{(ct)_c(fp)_c}{N}$$

where $(ct)_c$ is the compound toxicity class with respect to bees, normalized to the highest value, $(fp)_c$ is the compound persistence factor, normalized to the highest value, N is the number of bee samples testing positive, and f_{corr} is the correction factor. This factor must be used only when, in the same month, some of the honey bee samples being subjected to chemical analysis (because they have exceeded the critical threshold of mortality) were positive while others were negative; the purpose of the correction factor is to give a "weight" to the latter samples in the formula. It is calculated as the ratio of the mean number of dead bees corresponding to negative samples and the overall mean number in the period taken into account. Only values greater than or equal to 1 are considered. When several pesticide residues are found in a single bee sample, the numerator in the formula for that sample can be obtained by means of a suitable averaging procedure. This index represents only a first step toward a more sophisticated processing of the data obtained by monitoring pesticides with bees. Every aspect of the model proposed is thus susceptible to improvement. First, it has been "calibrated" for a manual count of dead

Table 11.2 Samples percentage for each bee mortality class

Samples percentage	Classes
80	0–200
15	200–400
4	400–800
1	>800

bees and thus if electronic bee counters prove to be successful in the future, the model will have to be reformulated. Second, it may be integrated with other environmental or compound-linked parameters, in addition to persistence.

Pesticide monitoring levels

The systems for monitoring pesticides with bees may involve different levels of complexity and sensitivity, depending on the context and the objectives pursued (Table 11.3) [76]. Table 11.3 shows the six alternatives deemed most feasible; they differ in their objectives and the type of commitment required. Obviously as increasingly sensitive and complex methods are adopted, the human and financial costs will increase and the sphere of application will necessarily be restricted. If, for example, every Italian beekeeper installed cages for collecting dead bees (level I), it would be theoretically possible to set up at least 80 000 monitoring stations at a minimal expense; this would be unlikely at level VI, given the heavy financial burden.

For each technique it will be necessary to determine the level of monitoring best suited to the purposes being pursued, keeping in mind the type of environment concerned, its orography, how the soil is used, the vegetational composition, the impact of human activity, and the resources available. In every case, different methods may be used simultaneously.

Experiences in Italy

In Italy, many interesting results have been achieved using bees as bioindicators of pesticides in the agro-ecosystem. Since 1980 this strategy has been applied in 34 provinces, townships or inter-municipal territories through much of northern Italy. Overall, 400 monitoring stations have been installed to cover a total territory of 2800 km². Between 1983 and 1986, in particular, the analysis of 581 gathered samples of dead bees revealed which compounds were most widely used in that period in cultivated fields, above all in northern Italy [29, 30] (Table 11.4).

Dithiocarbamates, used as fungicides, are only slightly toxic for bees and may not be ascribed direct responsibility for their death. However, their vast presence in dead bees confirms them as being the most widespread compounds in cultivated fields. Although they are included in a low-toxicity class, they are believed to be potentially hazardous to human health as they contain a metabolite, ethylene thiourea, which, in large doses, causes damage to the thyroid. The dithiocarbamates were almost always found together with other products, insecticides for the most part, which were truly to blame for the bees' death. These include dimethoate, a compound serving a large variety of purposes but often misused. The use of parathion, a compound that has prevailed on the agricultural scene for

Table 11.3 Different levels (I–VI) of environmental monitoring with bees (modified)

	I	II	III	IV	V	VI
Context	Apiculture	Pollination	Pollination and monitoring	Monitoring	Monitoring	Monitoring in proportion to agent
Information obtainable	general data about damage	general data about damage-causal agent	general data about damage-causal agent	damage in proportion to agent	damage in proportion to agent	quantitative damage
Type of monitoring	general qualitative	specific qualitative	specific qualitative	qualitative quantitative	qualitative quantitative	qualitative quantitative
Count method	manual	manual	manual	manual	manual	automatic
Management	traditional	traditional	specific	specific	specific	specific
Equipment and techniques	traditional hive, cage for dead bees	traditional hive, cage for dead bees	traditional hive, cage for dead bees, trap for collecting pollen	traditional hive, cage for dead bees, trap for collecting pollen, collection of foragers	traditional hive, cage for dead bees, trap for collecting pollen, collection of foragers, evaluation of family strength	specific hive, electronic bee counter, cage for dead bees, trap for collecting pollen, collection of foragers, evaluation of family strength
Frequency of sample-gathering	1×7 days	1–2×7 days	1–3×7 days	1–5×7 days	3–5×7 days	continuous
Max. time of utilization	++++	+++	+++	++	+	+
Time commitment	+	+/++	+/++	+++	+++	+/++++
Professional qualification of operator	+	+	+	++	+++/++++	+++
Sensitivity	+	+/++	++/+++	++/+++	++++	+++
Costs	+	+	++	++	+++	++++
Field applicability	++++	+++	+++	++	+	+

Table 11.4 Pesticide monitoring with honey bees. Main active ingredients registered in the 4-year period 1983–1986 in Italy

Active ingredient	Positive samples (%)*
Dithiocarbamates (Mancozeb, Maneb, Metiram, Zineb, Ziram)	70.8
Dimethoate	15.3
Parathion	14.7
Azinphos-methyl	11.9
Carbaryl	11.0
Methyl parathion	10.4
Endosulfan	7.2
Omethoate	7.2
Methamidophos	2.4

Note
*Frequency of active ingredients found in 442 samples of dead bees being positive when subjected to chemical analysis (out of 581); the samples were taken from intensively farmed areas.

many years, is instead indicative of backward farming techniques and disregard for the environment. In fact, this compound is deadly for a large number of beneficial organisms.

In recent years more specific and more amply circumstantiated studies have been conducted in different areas such as the province of Forlì, the province of Ferrara, the territory of the Comunità Montana dell'Alto Tevere Umbro (Mountain Community of the Upper Umbrian Tiber Valley), the province of Pesaro, the City of Venice, the territory within the jurisdiction of the USL (local health authority) of Salò (in the province of Brescia, on the shore of Lake Garda), the town of Medole (province of Mantua), the town of Correggio (province of Reggio Emilia), the town of S. Martino in Rio (province of Reggio Emilia), the town of Guastalla (province of Reggio Emilia), the coast of Emilia-Romagna [77], and the town of Castenaso (province of Bologna) (Figure 11.1).

A 2-year study (1987–1988) conducted in the province of Ferrara showed that the compounds found in the highest percentages in dead bee samples were also the most widely sold in the province, confirming the bee's efficacy as a biological indicator [3, 78]. The pesticide monitoring strategy was first successfully implemented in the province of Forlì, an area of intensive orchard growing; with the backing of the Provincial Authority, it was possible to pursue the investigation uninterrupted from 1982 to 1993. The data gathered in this period brought to light a decided trend in improvement as far as the pesticide contamination of the agro-ecosystem of Forlì was concerned. This was ascribed above all to a new awareness among farmers, who were more careful about using pesticides properly and choosing products that did not threaten the environment and, in particular, were not harmful to useful insect species [79–81].

Monitoring case

In 1998 a monitoring project with bees was undertaken to identify any cases of pesticide abuse or misuse. The study was conducted in the townships of Castenaso (CAS), Granarolo Emilia (GRA), and Ozzano Emilia (OZZ) (province of Bologna), which together form a longitudinal territorial strip of 129 km^2 (Castenaso 33 km^2, Granarolo Emilia 34 km^2, and Ozzano Emilia 62 km^2) that runs from the flatlands north of the Via Emilia to the foothills of the Bolognese Apennines south of the same road (Figure 11.5).

A large variety of crops are grown within the territory in question, though cereals like wheat, corn and sorghum are by far the most common. Moreover, there are large plots of alfalfa and sugar beet as well as a myriad of smaller fields set aside for the cultivation of potatoes, various vegetables, fruit trees, grapevines (especially in hilly areas), and nurseries. Trees and wild herbaceous plants also inhabit many large areas, in particular along the Idice and Quaderna rivers.

In the 13 stations installed, the critical threshold of mortality was exceeded a total of 47 times (17 in Castenaso, 14 in Granarolo Emilia, and 16 in Ozzano Emilia): 10 in station OZZ 4, six in stations CAS 5, GRA 1, and GRA 3, five in station CAS 2, four in station OZZ 3, three in station CAS 4, two in station GRA 2, and one in stations CAS 1, CAS 3, CAS 6, OZZ 1, and OZZ 2. A total of 47 bee specimens were sent to the laboratory for analysis and 38 (80.9 percent) were found to contain at least one pesticide residue.

The data were analyzed on the basis of the IEH once a month for the purpose of delineating the trend in chemical contamination within the territory surveyed. The data gathered from station OZZ 3 are reported to serve as an example.

In the last two weeks of the month of May, when the survey began, a total of six bee killings were recorded. Subsequent chemical analyses showed all the dead bee samples to be positive. Judging from the overall data, the month of May (or rather the last 15 days of the month) could be considered moderately contaminated, with the exception of a peak at OZZ 4, where a very large number of dead bees was found (Figure 11.6).

A pollen analysis on the bees body and subsequent examination of crop-growing maps revealed that chemical treatments were most frequently undertaken to control grain aphids (*Rhopalosiphum padi, Sitobion avenae*, etc.). In the majority of cases the damage caused by these parasites – which, among other things, may be effectively controlled by ladybirds (*Coccinella 7-punctata, Adonia variegata*, etc.) – is less than the cost of the chemical treatment; nonetheless, farmers continue to perform such treatments, killing huge numbers of bees and ladybirds.

Aphids are insects that suck the sap of various plants through slender styles and then filter the sap to derive its nutritional elements. The left-

Stations map 1998

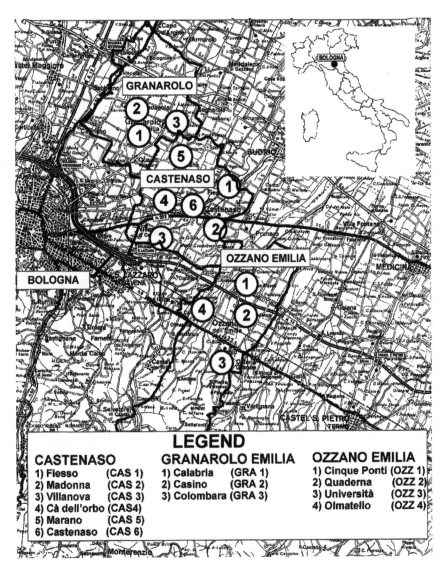

Figure 11.5 Monitoring stations in the areas of Castenaso, Granarolo Emilia, and Ozzano Emilia (province of Bologna) in 1998.

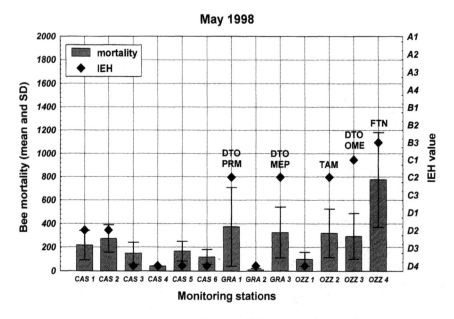

Figure 11.6 Mean honey bee mortality, pesticides, and IEH values registered in CAS, GRA, and OZZ monitoring stations in May, 1998 (DTO = dimethoate, PRM = pirimiphos-methyl, MEP = methyl-parathion, TAM = methamidophos, OME = omethoate, FTN = feni-trothion).

over, mainly sugary substances are deposited on leaves and stems, covering them in honeydew, which bees have a very strong liking for.

At station OZZ 3, only one mass death was discovered, on May 25. The pesticides found (dimethoate and omethoate) and the pollen present on the bees' bodies (above all *Papaver*, other wild species and ornamental species like *Magnolia*, *Pinus*, and *Gleditsia*) suggest that the bees gathered the honeydew secreted by aphids in the wheat fields around the monitoring station, especially in proximity to homes with gardens (Table 11.5 and Figure 11.7).

In June there were no fewer than 20 bee killings – almost half (42.5 percent) of the total recorded in the whole period under examination. Given the compounds found and the number of dead bees, rather high in some cases, the entire area could be considered in a state of medium to high contamination, which reached a peak at stations GRA 1, GRA 3, OZZ 3, and OZZ 4 (Figure 11.8). The majority of chemical treatments had been carried out to defend sugar beet and vegetable plots and above all vineyards against parasites. At station OZZ 3 the critical threshold was exceeded three times, on June 13, 20, and 27. In the first case, chemical analyses revealed the presence of fenitrothion, which – as could be

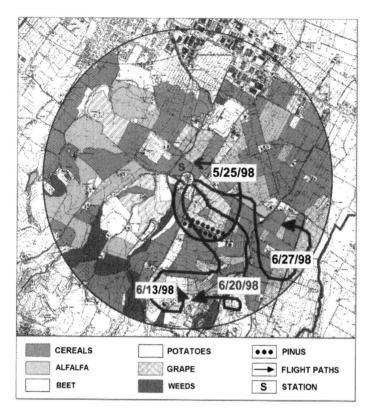

CEREALS   POTATOES   ••• PINUS

ALFALFA   GRAPE   → FLIGHT PATHS

BEET   WEEDS   S STATION

Figure 11.7 Crop-growing map of OZZ 3 station with the route of the honey
bees before bee killing.

inferred from the traces of pollen of *Pinus*, situated in an avenue southeast
of the station, and of a whole series of wild plants, mostly present in the
southern part of the monitoring area (Table 11.5 and Figure 11.7) – had
been used on vineyards situated about 800 meters away in that direction,
probably to control either grape thrips (*Drepanothrips reuteri*) or grape
tortrix (*Argyrotaenia pulchellana*).

In the second case, no pesticide residues were found in the bee sample
despite the high mortality (731 dead bees). This apparently contradictory
fact may have two plausible explanations: the compound either broke
down rapidly or was not among the substances sought. In any case, the
pollen traces indicated that the bees had foraged in the southernmost area,
characterized by uncultivated hillsides with an abundant flowering of
Hedysarum (Figure 11.7).

On June 27, the bees' death was ascribable to a treatment performed on
vineyards situated south of the station (prevalence of pollen of
Hedysarum) (Figure 11.7) to control grape moths (*Lobesia botrana*) and

Table 11.5 Pollen types detected on dead bees bodies in the OZZ 3 monitoring station

Pollen type	Date																							
	05/25/98	05/30/98	06/06/98	06/13/98	06/20/98	06/27/98	07/04/98	07/12/98	07/18/98	07/25/98	08/01/98	08/09/98	08/15/98	08/22/98	08/30/98	09/06/98	09/12/98	09/20/98	09/26/98	10/03/98	10/10/98	10/17/98	10/24/98	10/31/98
No. of dead bees	394	188	39	969	731	2352	274	268	63	93	80	27	30	120	7	29	29	0	44	46	4	20	0	25
Aesculus	0.5																							
Artemisia				0.5																				
Brassica form	0.5			0.5																				
Caryophyllaceae				4.5	1.5	1.0																		
Castanea	0.5			0.5		1.0																		
Chenopodiaceae				2.5	6.5	2.0																		
Clematis				24.7	3.0	4.0																		
Compositae A form				0.5																				
Compositae S form					3.0																			
Compositae T form				0.5																				
Convolvulus	2.5																							
Corylus	24.5			9.6	6.5	1.0																		
Cupressaceae	0.5					0.5																		
Dipsacaceae				1.0	1.0																			
Gleditsia	5.0																							
Graminaceae	5.0			0.5	2.5	2.5																		
Hedera	3.0																							
Hedysarum				9.1	31.5	62.0																		
Helianthus form					0.5	2.5																		
Juglans	0.5																							
Labiatae M form	6.0					0.5																		

Taxon				
Lotus corniculatus group	0.5	0.5	18.5	9.0
Magnolia	0.5	0.5		
Malva form			0.5	
Melilotus	1.5	1.5	1.0	
Moraceae/Urticaceae	1.5		0.5	
Oleaceae			1.0	
Papaver	2.5	10.6	7.5	3.0
Pinus	9.5	11.6	0.5	1.5
Polygonum lapathyfolium			0.5	0.5
Polygonum form	1.5		3.5	
Pyrus form			0.5	
Quercus	0.5			
Ranunculus form			1.5	
Rhamnus form	7.0		6.1	
Robinia	2.0	3.5	8.5	3.0
Rubus form	0.5	2.0	2.0	4.0
Rumex			0.5	
Sambucus nigra	1.0			
Trifolium repens group		4.5	1.5	1.5
Umbelliferae A form	23.0		1.0	
Zea				0.5
Total %	**100**	**100**	**100**	**100**

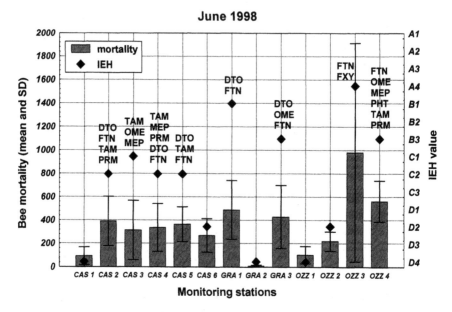

Figure 11.8 Mean honey bee mortality, pesticides, and IEH values registered in CAS, GRA, and OZZ monitoring stations in June, 1998 (DTO = dimethoate, FTN = fenitrothion, TAM = methamidophos, PRM = pirimiphos-methyl, OME = omethoate, MEP = methyl-parathion, FXY = fenoxycarb, PHT = phenthoate).

grape bud moths (*Eupoecilia ambiguella*). The product contained fenoxycarb, whose use is banned in Italy!

The long series of bee killings continued in the month of July. In fact, 23.4 percent of the cases occurred in this period. Thus from the standpoint of chemical contamination, this month did not differ greatly from June. The stations revealing the highest degree of contamination were CAS 2, CAS 5, and OZZ 4 (Figure 11.9). The treatments were aimed especially at vineyards, orchards, and vegetable fields. At station OZZ 3, the mortality never exceeded the critical threshold either in this month or in subsequent months.

In August, there was an abrupt drop in chemical treatments and consequently the number of mass bee killings declined drastically: only five out of the total of 47 cases, or 10.6 percent. From the pollen detected on the bodies of the dead bees, it could be inferred that treatments had been performed exclusively on vineyards and vegetable crops. Overall, the territory could be defined as moderately contaminated by pesticides. The stations with the highest IEH were CAS 5 and GRA 2 (Figure 11.10). In September and October, the last two months in which the study took place, the number of mass bee killings was lower than in the previous periods: two in September and three in October. Low levels of contamination were reported in all stations (Figures 11.11 and 11.12).

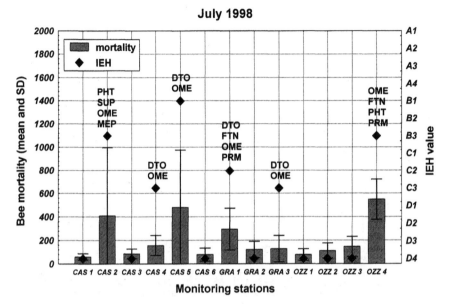

Figure 11.9 Mean honey bee mortality, pesticides, and IEH values registered in CAS, GRA, and OZZ monitoring stations in July, 1998 (PHT = phenthoate, SUP = methidathion, OME = omethoate, MEP = methyl parathion, DTO = dimethoate, FTN = fenitrothion, PRM = pirimiphos-methyl).

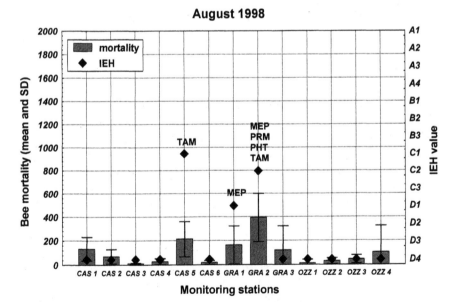

Figure 11.10 Mean honey bee mortality, pesticides, and IEH values registered in CAS, GRA, and OZZ monitoring stations in August, 1998 (TAM = methamidophos, MEP = methyl parathion, PRM = pirimiphos-methyl, PHT = phenthoate).

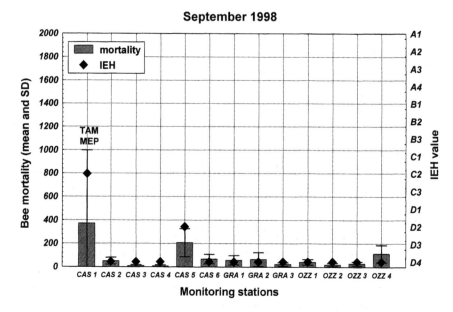

Figure 11.11 Mean honey bee mortality, pesticides, and IEH values registered in CAS, GRA, and OZZ monitoring stations in September, 1998 (TAM = methamidophos, MEP = methyl parathion).

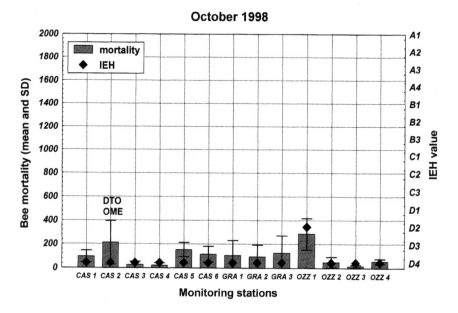

Figure 11.12 Mean honey bee mortality, pesticides, and IEH values registered in CAS, GRA, and OZZ monitoring stations in October, 1998 (DTO = dimethoate, OME = omethoate).

When the crop-growing maps are compared with the maps of pesticide contamination, the first impression is that the division of farmland into smaller fields results in a higher use of pesticides and consequent accentuation of drift phenomena. In fact, in stations OZZ 1 and OZZ 2, where the fields are larger than in other areas, the environment was generally less contaminated by pesticides. In other areas such as Castenaso, where it was possible to compare the data obtained in 1998 with historical data (monitoring started in 1991 in this area), the data gathered in the different stations showed variations in the use of pesticides and in mortality trends over the years that reflected a high fragmentation of agriculture.

In 1998, as in more recent years, the problem of dimethoate again came to the forefront. The widespread use of this compound (detected in 42.1 percent of the bee specimens analyzed), especially in the month of May, suggests that grain aphids are the target of the majority of the chemical treatments performed. This hypothesis has been further confirmed by palynological analyses. The year in question also saw an extensive use of fenitrothion, which in some cases was detected in dead bees even several weeks after spraying took place due to the slow release of the compound from the microcapsules containing it.

Methyl parathion, methamidophos, and methidathion, all too often found in dead bee samples, are hazardous both for the environment and for man and reflect a non-professional approach to farming and above all a lack of respect for the territory. There is also the issue of fenoxycarb. The sale and use of this compound is banned throughout Italy (Ministerial Decree 8.8.1995; G.U. n. 189 of 14.8.1995, an exception being made only for the province of Bolzano), in view of its toxic effects on beneficial entomofauna. Nonetheless, residues of this compound were found in a bee sample taken from station OZZ 3 in June.

By comparing the palynological data and pesticide residues found in dead bees with crop-growing maps it was possible to determine episodes of inappropriate pesticide use, such as treatments performed in windy weather conditions (detectable, for instance, through a high quantity of pollen from spontaneous plants found on the bee's body) or when the crops treated were flowering.

Using the IEH as our reference, it may be affirmed, in conclusion, that the most alarming situations were reported in stations CAS 2, CAS 5, GRA 1, GRA 3, OZZ 3, and OZZ 4.

Monitoring of heavy metals

In Italy, researchers began to investigate the relationship between these contaminants and bees at the beginning of the 1980s, focusing on two areas. The first involved an assessment of the quantities of heavy metals contained in beehive products, honey in particular, whereas the second area of research centered on the use of bees for monitoring these contaminants in urban or industrial areas.

The presence of metals in honey is tied to the latter's botanical origins but also depends on the type of soil it is produced in and the human activities taking place there. In order to carry out a thorough investigation, a honey-based reference material for the detection of trace elements must be produced and this was the task undertaken by our research team in collaboration with the Higher Institute of Health [82].

In 1986, the first tentative attempts were made to use bees for monitoring heavy metals in the cities of Rome [83], Florence, Arezzo, and Pisa [84]. The research was subsequently extended to Modena [72], Reggio Emilia, Bologna, Forlì, Ravenna, Rimini [85], and Cesena. Other studies were carried out in industrial areas or areas crossed by major roadways: Bologna Apennines, Valleys of Terni, and Valleys of Brescia (Figure 11.1).

Although some interesting data were obtained, the researchers realized that they had to focus their attention on the mechanisms by which honey bees capture heavy metals and the choice of a suitable matrix. One of the fundamental aspects that differentiate heavy metals from other pollutants such as pesticides is the method of their introduction into the territory and their environmental fate. Pesticides are scattered both in time and space and, depending on the type of chemical compound, its stability, and affinity with the target organism and the surrounding environment, they are degraded by various environmental factors over a greater or lesser period of time. Heavy metals, on the other hand, are emitted in a continuous manner by various natural and anthropical sources and, since they are not degraded, they are continuously kept "in play," thus entering the physical and biological cycles. Heavy metals present in the atmosphere can deposit on the hairy bodies of bees and be brought back to the hive with pollen, or they may be absorbed together with the nectar of the flowers, or through the water in puddles, ditches, fountains, and streams, or through the honeydew produced by aphids.

A number of variables have to be considered when using bees, or beehive products such as honey, to monitor heavy metals in the environment: the weather (rain and wind can clean the atmosphere or transfer heavy metals to other environmental sectors), the season (the nectar flow, which is usually greater in spring than in summer and autumn, could, emissions being equal, affect the pollutant by diluting it), and the botanical origin of the honey (the honeydew produced by aphids, like the nectar of flowers with an open morphology, is much more exposed to pollutants than the nectar of flowers with a closed morphology).

Capture of heavy metals by bees

To understand how bees are affected by different types of heavy metals and, in particular, how they capture them, a study was carried out analyzing 178 samples of forager bees caught on their return to hives in three different areas: an urban environment, an industrial area, and a rural location.

As can be seen from Figure 11.13, the lead in the urban and industrial areas is found in higher quantities in the "mineralized" material (accumulated inside the bees) than in the "washed" one (deposited on the bee's surface), in agreement with the findings of other authors [86], to a highly significant degree ($P < 0.0001$), while the ratio was inverted in the rural areas, that is, the metal in question was found at a higher rate in the "washed" material than in the "mineralized" sample ($P < 0.0005$). As regards nickel a significant difference was found only in the rural area ($P < 0.05$), where the amount was again higher in the "washed" material. This was also the case for chromium, which, however, exhibited a statistically significant difference in all three environments (urban, $P < 0.05$; industrial, $P < 0.005$; rural, $P < 0.005$).

As noted above, the "mineralized" material from the urban and industrial areas contained more lead than the "washed" samples, while the opposite was true for samples from the rural area; this could indicate that persistent contamination induces higher absorption of pollutants, by inhalation or ingestion, into bees' bodies during foraging. On the other hand, the fact that higher levels of all three metals were found in the "washed" material than in the "mineralized" material in the rural environment could suggest that the pollutants are in a transitory condition, and

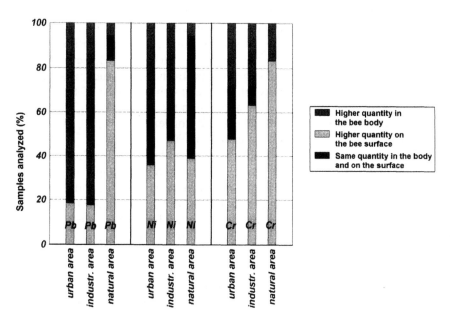

Figure 11.13 Quantity of heavy metals detected internally and externally in honey bees in relation to three different environments.

are more scattered throughout the atmosphere as they do not impregnate or deposit on the different environmental components visited by bees. Nickel and chromium do not behave like lead in the two more highly contaminated areas; this discrepancy is probably attributable to their different environmental fate but also reflects the high number of cases in which both the values ("washed" and "mineralized") were equal because they were below the limit detectable by the instrument [87].

Studies on the bee matrixes to be used

In a study conducted in the ceramics-manufacturing district of Sassuolo (province of Modena) and Scandiano (province of Reggio Emilia) (Figure 11.1), Cavalchi and Fornaciari [88] affirmed that the bee matrix was difficult to interpret due to the complexity of the factors influencing the assimilation of pollutants, but that honey and propolis appeared to be good biological markers of fluoride contamination.

In the city of Modena, the levels of lead, chromium, nickel, and cadmium in the air (as measured by automatic detectors) were compared with those contained in honey, larvae, and pollen samples taken monthly from beehives situated in the vicinity of the detectors themselves. The findings cannot be considered conclusive, as the average monthly data recorded by the automatic detectors referred to a single point in the atmosphere whereas the beehive data were referable to the area around the hive visited by bees in a given month. Nonetheless, they showed that the "fresh" honey recently imported into the honey chamber was the matrix that best reflected the trend in lead contamination of the atmosphere, as recorded by the detectors [89]. It was also observed, again with regard to lead in the honey matrix, that the values provided by the detectors were a reliable anticipation of the biological data. In the same study, the authors also tried to estimate the ratio of the mean concentration of the various contaminants in honey (in $\mu g/kg$) and that in the air (in $\mu g/m^3$), which may be estimated as approximately 1000–2000 for lead and nickel, 2000–4000 for chromium, and 3000–5000 for cadmium [90].

Bees and beehive products have been compared with other environmental markers [91]. The authors compared the percentages of lead and cadmium detected in bees, "fresh" honey, pollen, propolis, and royal jelly with the results derived from the analysis of clover (*Trifolium pratense* L.) and rain. Their findings showed no correlation between the heavy metals found on bees' bodies and in beehive matrixes and those detected with other "environmental markers."

Specific experiments have been undertaken to determine which of the various bee matrixes best represented the territory being monitored. To assess the various metals (chromium, nickel, copper, zinc, cadmium, lead, potassium, calcium, and magnesium) found in honey, pollen, and beeswax, several authors [14, 92, 93] have used not only standard statistical tests but also multivariate statistical analysis, which allows the most information to be obtained with a limited input of data. The authors concluded that the matrix which best illustrates the state of the environment is pollen, followed by wax and honey, respectively.

All these studies are no doubt of great interest and the results obtained can help us to understand how bees may be most effectively used to monitor environmental pollution. However, it is our opinion that when seeking to establish whether the bee may be considered a good environmental marker, it is not so important to find a good correlation with other environmental markers or monitored sites. Unlike other biological indicators, the bee is a traveling sensor that visits various environmental sectors (air, water, vegetation, soil); moreover, heavy metals are by now ubiquitous pollutants, each with their own environmental fate. It may be much more important to assess that each matrix is able to give repeatable data. For this purpose the two matrixes, bees and honey, were examined under the same conditions and using the same procedures. In the course of several monitoring sessions carried out over the past few years in different areas, large numbers of the same type of bees or honey samples were taken from the same hives: the bees were all foragers on return to the colony, while the honey was "fresh," recently imported into the honey chamber. The individual samples (43 in the 16 samplings of bees and 74 in the 29 on honey) were analyzed separately.

For the purposes of our statistical analysis, two types of non-parametric tests were used to compare the variability coefficients of the values obtained in the two matrixes: the Kolmogorov–Smirnov two-sample test and the Mann–Whitney U-test. The former is more accurate for small samples, while the latter is preferable for larger samples [94]. The statistical analyses thus conducted showed a significant difference for chromium (U-test $P < 0.05$) only. As regards the other metals, a slightly higher degree of reliability was observed for honey (Figure 11.14). Therefore, "fresh" honey seemed to be the best matrix to use in environmental monitoring programs, not only because sampling is easy, but also and above all because of the broad nature of the information it is able to supply, as it comes from the nectar collected for several days in large areas. Nonetheless, bees may successfully complement honey in order to provide more complete environmental information.

Chemical analysis of heavy metals

The analyses were first conducted at ARPA of Emilia Romagna (provincial section of Rimini) and subsequently in the laboratories of the CSA research institute in Rimini.

The bees were washed with a solution of 5 percent HNO_3 (Suprapur Merck 67 percent) in distilled water (Milli-Q Millipore) to determine the amounts of heavy metals deposited on the body. Subsequently they

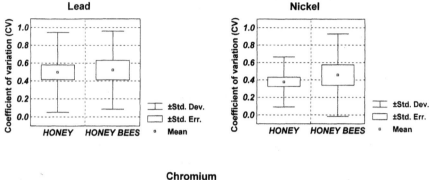

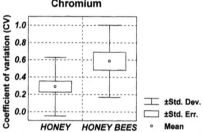

Figure 11.14 Comparison between honey bee and honey matrixes referred to data repeatability for lead, nickel, and chromium.

underwent a mineralization procedure: the bees were mixed with a solution of 1.5 ml HNO_3 concentrate (Suprapur Merck 67 percent) and 0.4 ml H_2O_2 (Aristair – BDH 30 percent) and placed in a microwave (MLS-1200 MEGA, Milestone) device at 500 W for 5 minutes. The mineralized material was mixed with distilled water to obtain 25 ml and analyzed, like the wash solution, by atomic absorption spectrometry (SpectrAA 220 G Varian) with a graphic heater and Zeeman effect corrector (GTA 110 Varian) and autosampler (PSD 110 Varian). For the analysis of honey, a direct dilution (1 to 20) with Triton X 100 (Merck) at a strength of 0.2 percent in distilled water was prepared and analyzed by standard addition method.

Reference values for heavy metals

Environmental pollution monitoring initiatives seek to establish a threshold of risk, making reference to limits which, if broken, should trigger an alarm. However, especially as far as heavy metals are concerned, it is always advisable to make reference to control areas that are sufficiently similar to those being investigated and are proven to be relatively free of contamination. Such controls are important because they allow the natural factors in play in the area under investigation to be taken into consideration.

The reference values used for our studies were drawn in part on the basis of data available in the literature and in part from our own experimental data. The literature does not provide homogeneous data, given the variety of research aims and methodological approaches. However, precisely for this reason, we can consider them because of the large number of environmental variables (Table 11.6).

On the basis of the data obtained during studies conducted over the last few years in many areas, our research team has been able to make a fairly reliable data analysis for the purpose of determining reference levels for "fresh" honey and bees. The minimum and maximum thresholds have been defined by calculating a quartile so as to derive two median values for a group of data: the low quartile and the high quartile (Table 11.7). From the two sets of data – drawn from the literature and experimentally – it was possible to derive the approximate reference values shown in Table 11.8.

The above reference values are subject to variation over time as they must be constantly updated on the basis of new data. Hence they will be automatically adapted to changing environmental situations. For example,

Table 11.6 Heavy metal concentration in honey (bibliographical data)

Heavy metal	Range (mg/kg)	Refs	Type of honey
Pb	0.016–0.8	[95]	Floral
Pb	0.02–0.37	[96]	Floral
Pb	0.02–0.52	[96]	Honeydew
Pb	0.01–1.10	[97]	Floral
Pb	<0.02–0.33	[98]	Forest
Pb	<0.02–0.06	[98]	Floral
Pb	0.001–0.289	[99]	
Pb	0.01–1.10	[100]	
Pb	0.024–1.667	[101]	
Pb	0.00–0.28	[102]	
Pb	0.00–0.94	[103]	Floral
Pb	0.0032–0.186	[82]	Sunflower
Pb	<0.06–1.31	[104]	Floral
Pb	<0.06–0.561	[104]	Eucalyptus
Pb	<0.06–0.204	[104]	Chestnut
Pb	<0.06–0.125	[104]	Acacia
Pb	<0.06–0.093	[104]	Citrus
Pb	0.054–0.075	[106]	
Cr	0.001–0.0039	[82]	Sunflower
Cr	0.016–0.197	[105]	Wild floral
Cr	0.012–0.205	[105]	Sunflower
Cr	0.017–0.579	[105]	Floral
Cr	0.037–0.618	[105]	Acacia
Cr	<0.08–0.11	[104]	Eucalyptus
Cr	<0.08–0.541	[104]	Floral
Ni	0.01–1.93	[102]	
Ni	0.017–0.049	[82]	Sunflower

Table 11.7 Heavy metals concentration in honey and in bees in all samples analyzed during our research

Matrix	Heavy metal	Sample number	Range (mg/kg)	Low quartile	High quartile
Honey	Pb	962	0.0013–1.74	0.02	0.111
Honey	Ni	972	0.004–3.23	0.02	0.208
Honey	Cr	963	0.0008–0.78	0.005	0.013
Bees	Pb	392	0.02–25	0.439	2.744
Bees	Ni	393	0.025–8.064	0.105	0.427
Bees	Cr	393	0.005–6.902	0.038	0.275

Table 11.8 Reference thresholds for heavy metal monitoring with honey and bees

Matrix	Heavy metal	Reference threshold	
		Low	High
Honey	Pb	0.02	0.115
Honey	Ni	0.02	0.2
Honey	Cr	0.005	0.015
Bees	Pb	0.40	2.0
Bees	Ni	0.10	0.40
Bees	Cr	0.04	0.25

the reference values with regard to lead were higher until quite recently, but the drastic decline in the use of lead in gasoline has resulted in falling rates which are reflected in the honey bee matrixes analyzed.

Method used for monitoring heavy metals

All the operations of sample gathering and preparation must be conducted so as to avoid every risk of contamination. The two beehives used for each station were prepared according to the procedure described in the section on "*Monitoring stations*" during the experimental period, which in Italy goes from April to September. Feeding must be completely avoided in order to stimulate the bees to forage naturally. Therefore, the hives must be prepared in advance so as to obtain strong, healthy colonies. The smoker, which is used to inspect the hives, must be used sparingly even though, according to the findings obtained in preliminary experiments, it should not release heavy metals (Table 11.9). The samples of "fresh" honey are drawn from the honeycomb or, in the latter's absence, from the nest. A portable refractometer is used to facilitate its identification. In uncapped honey, the humidity level must be higher than 19 percent. The sample is obtained either by aspirating the honey from the cells with a syringe or removing a portion of honeycomb, which is then squeezed in the laboratory. The bees to be gathered are the foragers on return to the

Table 11.9 Absorption by honey of heavy metals contained in smoke*

Matrix	Aliquot number	Pb (mg/kg) mean ± SD	Ni (mg/kg) mean ± SD	Cr (mg/kg) mean ± SD
Honey treated with smoke	5	0.011 ± 0.002	0.141 ± 0.006	0.006 ± 0.0002
Honey not treated	5	0.013 ± 0.0007	0.135 ± 0.009	0.006 ± 0.0008

Note
*The smoke was produced from cardboard and sprayed with an apicultural smoker directly onto the honey. All aliquots came from the same honey sample.

hive, but without their load of pollen. They may be captured when they settle on the flight board by means of an aspirator or other similar means.

Monitoring case

In 1999, an experiment was conducted in the city of Gravellona Toce (province of Verbania), situated near Lake Maggiore, to assess the presence of lead, nickel, and chromium emitted by vehicles and various industrial plants located in the area undergoing investigation. Three monitoring stations, each comprising two hives, were placed in strategic points of the city (GR); another was set up in the National Park of Val Grande, also near Lake Maggiore, to serve as a nearby control station (T1), and the last was placed in Chiusi della Verna (province of Arezzo), in the Apennines between Tuscany and Emilia, to be used as a remote control station (T2) (Figure 11.1).

The study involved various aspects, including a morphological analysis of the particles found in the samples, in order to determine the origin (traffic, urban waste incinerators, or industrial plants) of the contaminants investigated and the analysis of the heavy metal content inside the body and on the surface of bees. As the findings have not yet been published, just a few data are reported here. A sample of "fresh" honey was drawn from the hives every month and a sample of foragers every 2 weeks. Unfortunately, in the second control station (T2) it was not possible to obtain bee samples.

The data obtained were subjected to the Kruskal–Wallis test. The test revealed significant "between-month" differences ($P < 0.05$) in the three metals contained both in the honey and bee samples, while as regards the "between-station" differences, the results were statistically significant only for lead in bees. The comparison with the control stations (T1 and T2) revealed a significant difference only for chromium in honey and lead in bees.

The mean levels of lead fell between the minimum and maximum reference values. The overall mean recorded for honey was 1.2 and 2.7 times higher than in control stations T1 and T2, respectively, and 1.9 times higher in the case of bees (T1). Different trends were observed for the two matrixes: in honey there was an uptrend until August, followed by a downturn in September (Figure 11.15), whereas in bees the trend see-sawed (Figure 11.16).

The mean concentration of nickel in honey fell between the minimum and maximum reference thresholds until July only, and was 3.6 and 1.6 times higher than in the two control stations (T1 and T2); instead, the mean found in bee samples drawn from the three stations in Gravellona was 1.2 times lower than the mean in station T1. Both matrixes underwent see-saw trends with a very high peak detected in honey in August (Figures 11.17 and 11.18).

There was an uptrend in chromium levels, which peaked in August for honey and in September for bees, and subsequently a slight downtrend for both matrixes. With regard to the bees from control station T1, it is worth pointing out the high mean value of July: 3.66 mg/kg! The overall mean for honey was found to be 4 and 1.3 times higher than in T1 and T2, while as was the case with nickel, the mean values in bees from T1 were twice as high as the means obtained for the city stations (Figures 11.19 and 11.20).

Some conclusions may be drawn on the basis of the above findings. The relatively insignificant differences found between the quantities of heavy metals in the area under examination and in the control areas and the similar trends observed in the two areas with regard to lead in bees and nickel in honey demonstrate that these types of pollutants have by now become ubiquitous and it is thus difficult to find uncontaminated areas.

The two matrixes undergoing comparison, honey and bees, may complement each other in providing information about heavy metal contamination in the environment. Whereas honey may be used to obtain average data regarding a vast area, as it derives from nectar that has been collected in many places on different days, bees, given their behavioral characteristics, can provide us with more detailed information than honey, because the pollution detected on them may be ascribable only to the five or six days preceding their capture [107].

Monitoring of radionuclides

Small amounts of radionuclides are dispersed in the environment from the chimneys of nuclear power plants, factories, or health establishments. Alternatively, they may be discharged accidentally (unlike several decades ago when atomic experiments were performed deliberately in the atmosphere) and dispersed in the environment at a greater or lesser rate and over more or less extensive areas, remaining in certain matrices for periods of time ranging from several hours to hundreds of years, depending on their nature (half-life), the severity of the incident (which could range from an explosion in a nuclear power plant reactor to a small leak), and the medium into which they are discharged (air, water, or underground).

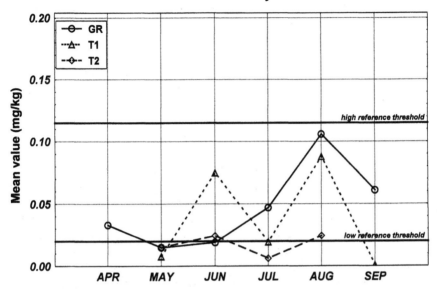

Lead in honey

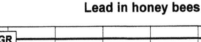

Figure 11.15 Lead level registered in honey in the monitoring stations of Gravellona (GR), Val Grande (T1), and Chiusi della Verna (T2) referred to the low and high thresholds (see Table 11.8).

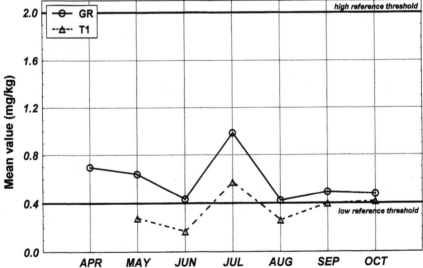

Lead in honey bees

Figure 11.16 Lead level registered in honey bees in the monitoring stations of Gravellona (GR) and Val Grande (T1) referred to the low and high thresholds (see Table 11.8).

Nickel in honey

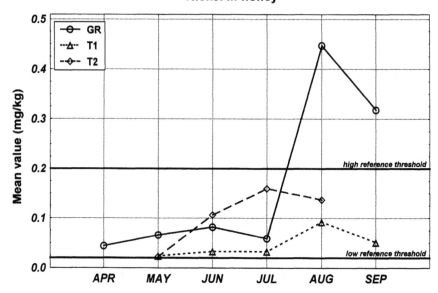

Figure 11.17 Nickel level registered in honey in the monitoring stations of Gravellona (GR), Val Grande (T1), and Chiusi della Verna (T2) referred to
the low and high thresholds (see Table 11.8).

Nickel in honey bees

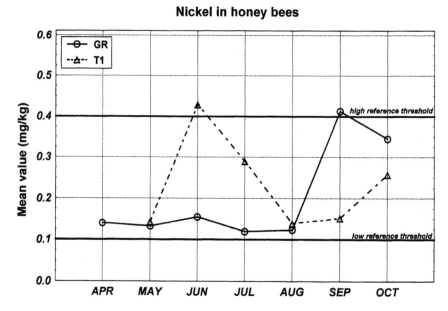

Figure 11.18 Nickel level registered in honey bees in the monitoring stations of
Gravellona (GR) and Val Grande (T1) referred to the low and high
thresholds (see Table 11.8).

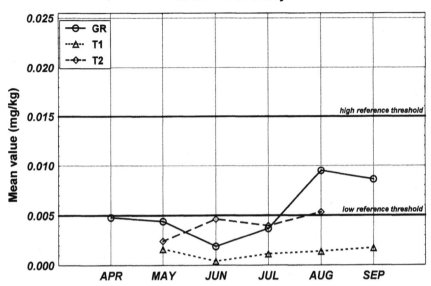

Figure 11.19 Chromium level registered in honey in the monitoring stations of Gravellona (GR), Val Grande (T1), and Chiusi della Verna (T2) referred to the low and high thresholds (see Table 11.8).

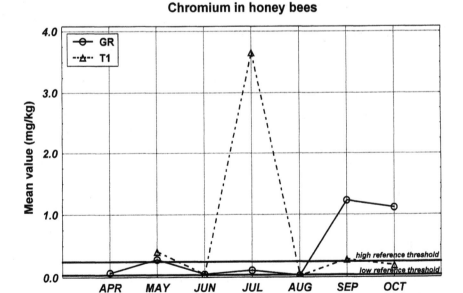

Figure 11.20 Chromium level registered in honey bees in the monitoring stations of Gravellona (GR) and Val Grande (T1) referred to the low and high thresholds (see Table 11.8).

The use of honey bees and hive products to monitor radioactivity dates back to the end of the 1950s and the beginning of the 1960s when Svoboda correlated a rise in the level of ^{90}Sr in honey with the nuclear experiments being performed in the atmosphere at the time [108]. Several years later, Racoveanu and co-workers came to a similar conclusion [109]. Honey bees were used to monitor radioactive emissions from nuclear power plants. The measurements of beta and gamma activity in honey coming from bee-hives in the area gave negative results [110]. However, other authors noted ^{7}Be, ^{137}Cs, ^{3}H, and ^{22}Na (although at infinitesimal levels) in honey produced near the Los Alamos National Laboratory in New Mexico [111]. For some years now (since before the Chernobyl incident), our research team, in collaboration with local beekeepers, has also been monitoring for possible radioactive waste in the area surrounding the Trino Vercellese and Caorso nuclear power plants (Figure 11.1). The radiometric measurements made by Dr Antonio Rossi of the Servizio di Fisica Sanitaria at the Ospedale Maggiore, Bologna, on samples of honey, wax, larvae, honeycomb, and bees have never revealed significant levels of activity as compared to the background level (unpublished data).

It was not until the Chernobyl state of emergency (April–May 1986) that the excellent efficacy of bees in detecting radioisotopes was unequivocally demonstrated. Since the Chernobyl incident, numerous experiments have been conducted involving bees, both with a view to evaluating the radioactive elements contained in the hive products and their transfer dynamics [112–123] and to assessing the efficacy of using bee colonies as biological indicators. In this context honey from various botanical origins was analyzed in an attempt to understand the dynamics of radionuclide fallout in the Tuscany region (Figure 11.1) [124]. Honeydew honeys proved to be the most contaminated, followed by nectar honeys. The authors correlated the different levels of radioactivity measured not only with the botanical origin but also with the presence of pollen grains in the honey, as well as with atmospheric events. Giovani *et al.* [125] measured radioactivity in honey in the Friuli-Venezia Giulia region (Figure 11.1) and used rainfall data for May 1986 to standardize the concentrations of ^{137}Cs and ^{134}Cs detected with ground deposition. They concluded that floral honey, together with acacia and honeydew honey, can be used to trace maps of radioactive contamination in a given area. In another research project carried out by our team, again in the context of Chernobyl [126], numerous samples of honey, bees, wax, and pollen were analyzed. The findings demonstrated that pollen was the most efficient indicator of atmospheric radionuclide contamination as it reflects that of the air. Honey bees can also be used profitably for monitoring, while there are numerous uncertainties to the use of honey.

Pollen accurately reflects levels of air contamination, thus proving to be an excellent indicator of radioactive pollution (Figures 11.21 to 11.23). Of all hive products, pollen is the best matrix for detecting radionuclides, due

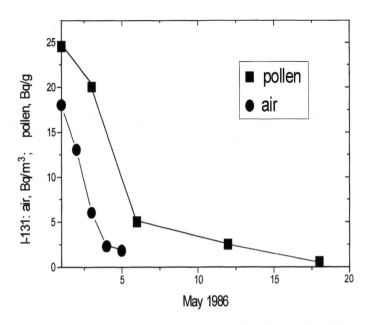

Figure 11.21 Levels of I-131 in air and in pollen during May 1986.

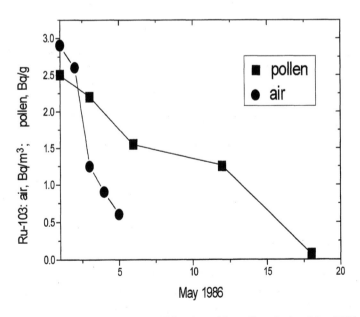

Figure 11.22 Levels of Ru-103 in air and in pollen during May 1986.

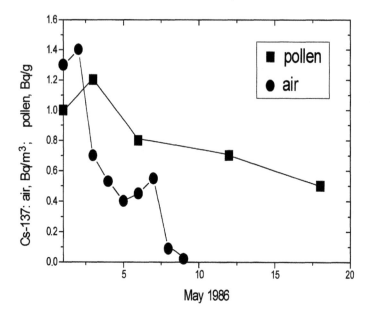

Figure 11.23 Levels of Cs-137 in air and in pollen during May 1986.

to its greater exposed surface area in comparison to other hive products and closer contact with the environment both before and during transport. In addition, pollen is easy to collect and has a greater importance than honey as the spectrum of flowers used as a source by bees is wider than for nectar during peak blossoming periods.

Bees themselves have also proved to be efficient indicators of environmental contamination; gamma spectrometry measurements show a higher correlation, compared to honey, between ^{137}Cs radioactivity in the bee matrix and ground deposition ($R = 0.95$) (Figure 11.24).

As far as honey is concerned, measurements made of the ^{137}Cs content in samples of different botanical origin collected from six Italian regions during the period May–July 1986 identified values of between about 3 and 360 Bq/kg (30–360 Bq/kg for honeydew honey, 20–180 Bq/kg for chestnut honey, and 5–60 Bq/kg for acacia honey). Subsequent research carried out during the period 1994–1996 on samples of honey coming from the same zone (Turin) (Figure 11.1) and nominally classified as lime honey revealed ^{137}Cs levels of between 1 and 5 Bq/kg, with the exception of two samples with ^{137}Cs of more than 30 Bq/kg. Subsequent analysis of pollens showed that the two honeys with the highest ^{137}Cs content contained chestnut pollen in amounts more than 50 percent, demonstrating that the radioactivity of honey is greatly influenced by the pollen content, in agreement with the findings of other authors [124].

One disturbing case emerged from bee and hive product research

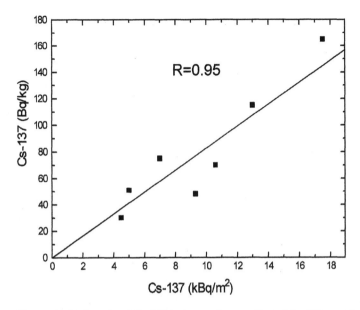

Figure 11.24 Levels of Cs-137 in honey bees collected in different locations on May 9, 1986, versus Cs-137 ground deposition.

during the post-Chernobyl period. In a sample of wax taken from a hive in the Emilia-Romagna region and in one of bees taken in the Friuli-Venezia Giulia region (Figure 11.1), both in May 1986, the gamma spectrum showed characteristic peaks of the ^{95}Zr–^{95}Nb pair in a percentage unlike that of any of the other matrices analyzed, i.e. considerably higher than the levels attributable to Chernobyl fall-out. Thus the radioactive contamination detected in these two samples was difficult to attribute to the incident in the power plant in the former Soviet Union. The fact was intriguing, but given that no plausible explanation could be found, the data were set aside. In July of 1987, however, a news report from the UK caused the "case" to be reopened. The English newspaper *The Independent* published an article on the English army's use of rapid-decay radioactive dust in simulations of nuclear incidents. In the article, army spokesmen confirmed that use of this radioactive material was indispensable to make the exercises realistic. This aroused the suspicion that it might be a common practice in NATO circles to perform simulations of this kind. Moreover, considering that there are NATO bases located in the two aforesaid Italian regions (and 14 years ago, before the fall of the Berlin Wall, there were certainly more than now) and the fact that the two elements ^{95}Zr and ^{95}Nb have a limited half-life (64 and 35 days, respectively), the report may plausibly explain the data collected by our researchers.

Bees have proven to be highly effective in detecting even very low levels of environmental radioactivity, as was demonstrated on the occasion of the leakage of ^{137}Cs at Algeciras (Spain) in the spring of 1998. In May 1998, our radiochemical laboratory detected an anomalous presence of cesium 137 in honey bee samples taken from environmental monitoring stations in the Bologna province (Castenaso, Granarolo Emilia, and Ozzano Emilia) (Figure 11.5) during the weeks May 17–24, May 30–June 6 and June 6–13. During the week May 24–30, the level of this radionuclide was instead below the limit of detectability at all stations (Figure 11.25).

Cesium 137 is an artificial radionuclide used in clinical, industrial, and research applications and one of the main radioactive products of fission reactions taking place in nuclear reactors. The possibility that the anomalous radioactivity derived from active nuclear plants can be rule out as the ^{137}Cs was not accompanied by the other radionuclides produced during fission. Towards the end of April 1998, an incident occurred at the Algeciras steel works in southern Spain with emissions of ^{137}Cs coming from a radioactive source no longer in use. This source ended up in the foundry. The fact that the presence of ^{137}Cs was interrupted for a week and then resumed is not unusual as the transport and soil deposition of air-dispersed pollutants is strictly linked to wind and precipitation. The levels of radioactivity were negligible and many times below every alarm threshold, but the bee matrix promptly revealed the presence, albeit minimal, of ^{137}Cs

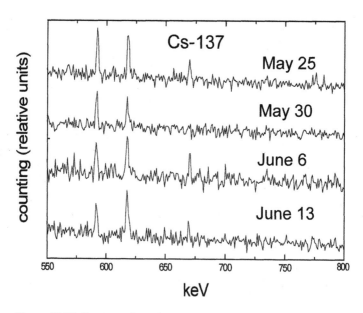

Figure 11.25 Comparative gamma spectrum of honey bee samples collected weekly from May 17 to June 13, 1998, at monitoring stations in the local authority areas of Castenaso, Granarolo Emilia, and Ozzano Emilia (province of Bologna).

in the atmosphere with an efficiency above that of traditional monitoring techniques.

Monitoring of phytopathogenic microorganisms

Recently our research team, which includes, besides the authors, the "La Carlina" cooperative, the Department of Plant Pathology of the University of Bologna, and the Plant Protection Service of the Region Emilia-Romagna, demonstrated for the first time how honey bees could be used as a bioindicator to detect the presence of phytopathogenic microorganisms in the environment, both for the purpose of prevention and for epidemiological studies. The study was conducted on *Erwinia amylovora* – the causal agent of a severe disease among *Rosaceae* known as Fire Blight – as part of a campaign against an epidemic which started in 1994 in Emilia-Romagna [127, 128].

Bacterial Fire Blight of Rosaceae

Fire Blight is caused by *E. amylovora*, a gram-negative bacterium belonging to the family of the *Enterobacteriaceae*. It is one of the most destructive diseases that can affect important plants like pear and apple trees and is also capable of causing damage to over 200 species belonging to the *Rosaceae* family [129, 130]. This disease is highly dangerous mainly for two reasons: first, it is almost impossible to prevent it from spreading, especially as it is carried by numerous vectors including, as reported in the literature, the honey bee [131–133]; and second, there are no reliable means for fighting it, although the disease has been known since 1780 and is the subject of intense research throughout the world. The lack of satisfactory disease control criteria, combined with the virulence of the pathogen itself, has led pear-growing to be totally abandoned in some areas particularly suited to the cultivation of fruit trees; this has occurred, for example, in some areas of the eastern United States [134, 135]. The only effective way to beat the disease is thus prevention, which means direct, early detection of disease symptoms [136].

The disease was described for the first time in the American State of New York in 1780. It gradually spread throughout the world, affecting New Zealand, Europe, and Asia. In Italy, the first sites of bacterial infection were observed in 1990 [137, 138]. In the region of Emilia-Romagna (Figure 11.1), *E. amylovora* was reported for the first time in 1994, when five foci of the disease were identified in pear orchards [139]. The provinces of Bologna, Ferrara, Modena, Reggio Emilia, Ravenna, and Forlì (Figure 11.1) together represent the major pear-growing area in Italy and one of the largest in Europe. Consequently, the appearance of Fire Blight stirred up great alarm and provided a large stimulus to research on this topic.

The idea of using bees to monitor *E. amylovora* is founded on the assumption that if bees can spread the disease by carrying the bacteria within the hive's range of action, it should also be possible to detect the presence of the bacteria itself in this area by looking for it on bees or in other materials carried by bees to their hive, as in the case of environmental pollutants. Like the other monitoring schemes using bees, the project first of all required the identification of the most suitable matrix. Among the beehive matrixes that could be used for this type of monitoring, the choice fell on pollen, deemed best suited to the purpose as samples are relatively easy to collect, little time passes between the bees' gathering of pollen and researchers' collection of samples to be analyzed, the bacteria survives longest in this matrix [131], and finally, it is possible to ascertain precisely which species the bees have visited.

Second, it was necessary to develop an original analytic method based on PCR-ELISA with chemiluminescent detection techniques. This required, among other things, the synthesis of a oligonucleotide hybridization probe for the plasmidic DNA of *E. amylovora*, which brought a significant improvement in terms of specificity and detection limits compared to the methods described previously in the literature [127, 128] (Figure 11.26). This ultra-sensitive, specific method was thus applied in order to detect the bacteria in samples of pollen gathered – using special traps – from beehives within whose range of action there were sites affected by Fire Blight and, for comparative purposes, in samples collected from what were considered disease-free areas.

In 1998 five stations, each comprising three hives, were installed in areas where the presence of Fire Blight had been ascertained and another station in an area considered to be disease-free. Samples were gathered twice weekly from the beginning of April until the middle of May, which corresponded to the main flowering period of the species of interest (pears, apples, and hawthorn). From May to July, samples were taken weekly according to whether there were any second flowerings or other possible sources of inoculation. The pollen trap was left on the hives for at least 2 hours. The pollen gathered from the traps was immediately frozen with dry-ice and stored at −20°C until the time of analysis. Each sampling was conducted so as to avoid possible cross-contamination.

At the end of the 1998 sampling campaign, the presence of *E. amylovora* had been detected in at least one sample taken from each of the stations situated in the infected area, except for one (station B, Figure 11.27), where the diseased pear trees had been completely eliminated the previous year. In the station located in the disease-free area (station A, Figure 11.27), on the other hand, no samples tested positive. All the positive samples contained pear pollen, while not all samples containing pear pollen tested positive.

The findings demonstrated the possibility of detecting the presence of *E. amylovora* in the environment by using bees to keep vast areas under

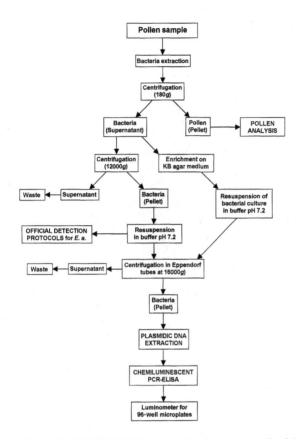

Figure 11.26 PCR-ELISA chemiluminescent method for detection of *Erwinia amylovora*.

continuous surveillance for long periods. The number of positive samples is not very high in proportion to the total samples gathered, but if monitoring is limited to the flowering period of the most important species (April–May), the percentage rises considerably. In any case, at least one sample containing *E. amylovora* was found in every station where foci of the disease had already been identified.

To confirm these initial findings, the experiment was repeated in 1999 with four stations: two of them were in areas considered disease-free (stations A and B, Figure 11.28) and two in areas severely affected by the disease (stations C and D, Figure 11.28). The experimental monitoring campaign of 1999 not only confirmed the 1998 findings for the stations situated in areas where the presence of Fire Blight had already been verified, but also revealed the presence of *E. amylovora* in the month of April in an area considered unaffected; several months later the disease devastatingly manifested itself in an orchard of young pear trees (station A, Figure 11.28).

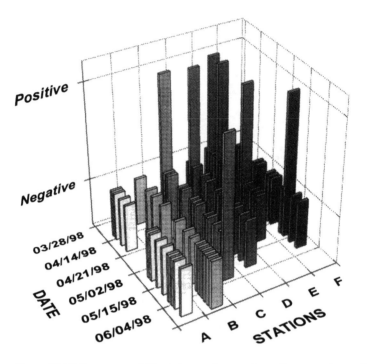

Figure 11.27 Presence of *Erwinia amylovora* in pollen samples collected in 1998 at six monitoring stations. Higher histograms indicate positive samples. Stations C, D, E, F were placed in infected areas, station B in an area where infected pear orchards were eliminated, and station A in an unaffected area.

The study thus provided experimental proof that bees can detect the presence of the bacterium before it manifests itself in visible symptoms on affected plants.

Following these encouraging results, the 2000 campaign was designed with the aim of assessing the operational efficacy of using honey bees for the early detection of Fire Blight. It would also seek, where possible, to predict the risk of the disease spreading to areas that were not yet affected but adjacent to identified disease-harboring sites. Moreover, marker pollens are being used in the hope of obtaining more precise information on contaminated plots frequented by bees.

Seven of the nine stations installed in 2000 were in a line perpendicular to the southeast front of expansion of the epidemic in the province of Forlì-Cesena (Figure 11.1); the first station was located near the most recently ascertained focus of the disease and the last station at a distance of about 28 km away, in the direction of a disease-free area. The other two stations were set up respectively in a heavily infected area and in a transition area containing both healthy orchards and blighted orchards; in the

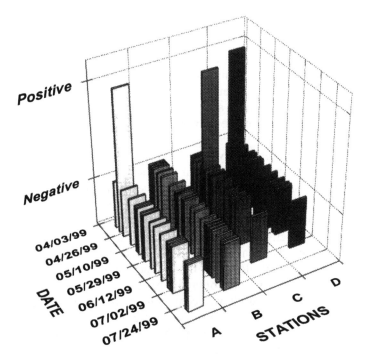

Figure 11.28 Presence of *Erwinia amylovora* in pollen samples collected in 1999 at four monitoring stations. Higher histograms indicate positive samples. Stations C and D were placed in infected areas, stations A an B in areas not affected by Fire Blight.

latter the researchers plan to test the so-called marker pollen method. If, as there is reason to believe, this experimental monitoring campaign shows a positive outcome, it will be possible to start using bee data to better and more promptly orient disease-prevention teams and keep hard-to-inspect areas under surveillance. It will also be possible to monitor the spreading of the disease and to try to predict where it will hit. Thus it may be hoped that this useful insect will be transformed from an alleged "plague-spreader" into a precious ally in the battle against Fire Blight.

These experiences have opened up new horizons for the use of honey bees as bioindicators by demonstrating that they may also be used to monitor phytopathogenic microorganisms present in the environment. The possible applications are many.

General conclusions

Using honey bees for environmental monitoring has produced interesting results in all the fields investigated by our research team. After twenty years of experience, we may venture to affirm that only bees can provide continuous, real-time detection of pesticide contamination at low operating costs. Only these insects, in fact, are capable of immediately and unequivocally revealing the improper use of pesticides and, in many cases, the location where the chemical treatment was performed.

The level of pesticide contamination can be defined on the basis of an index that takes different parameters into account. In the future, this index may be improved by including, above all, such factors as environmental fate and other factors ascribable to the various pesticides detected with bees.

Compared to pesticides, heavy metals are undoubtedly more complex to analyze, but the kind of information that honey bees can provide, by combining the factors of time and space, serves as a stimulus to further investigation. The heavy metals considered in our studies showed an upward trend both in honey and in bees; this trend needs to be examined and interpreted correctly in order to "calibrate" the beehive instrument with reference to the chemical and physical detection methods, and to use this calibrated instrument alongside the classic comparison with control areas presumed relatively free of contamination.

Another future area of study will center on analyzing the morphology of the particles transported by bees (already applied in other research sectors). This will aid researchers in identifying the origin (vehicles or industry) of heavy metal contamination.

Radionuclide monitoring, based on the methods developed during the Chernobyl emergency, is being continuously adopted in various areas. Different aspects merit attention, such as, for example, the resuspension of radioactive particles that occurs when fields are ploughed and their impact on forager bees.

The arrival in Italy of the phytopathogenic bacteria *Erwinia amylovora* has shown how honey bees, branded as dangerous carriers of the microorganism, may, if wisely exploited, prove to be an excellent tool for detecting the presence of the bacteria in a certain area before the disease has a chance to manifest itself, so that prompt action may be taken to halt its progress.

Despite the strides forward taken thus far, in the future broader research will have to focus on various aspects such as setting up colonies suited to environmental monitoring purposes, by selecting, for example, bees with a tendency to explore a vaster territory around the hive, and more continuously. Moreover, it will be necessary to develop more reliable, but inexpensive, facilities and techniques for monitoring the trends in the main beehive parameters (mortality, flight activity, expansion or decline in brood and food stores, etc.). Keeping expenses down is essential for promoting bee monitoring projects, at least in Italy. In fact, to gain approval these projects must compete with other similar ones, not only in terms of the information they provide – which may be original or serve to complement other types of surveys – but also in terms of operating costs.

Acknowledgments

The authors thank Drs Piotr Medrzycki, Lorenzo Monaco, and Grazia Lanzoni for their significant contributions.

References

1 Svoboda, J. (1961). Prumyslové otravy vcel arsenem (Industrial poisoning of bees by arsenic). *Ved. Pr. Vyzk. Ustavu Vcelarskeho CSAZV* **2**, 55–60.

2 Roembke, J. and Moltmann, J.F. (1996). *Applied Ecotoxicology.* CRC Press Inc., Boca Raton, FL, USA, p. 282.

3 Celli, G. and Porrini, C. (1991). L'ape, un efficace bioindicatore dei pesticidi. *Le Scienze* **274**, 42–54.

4 Celli, G. (1994). L'ape come indicatore biologico dei pesticidi. In: *Atti del Convegno: "L'ape come Insetto Test dell'Inquinamento Agricolo" P.F "Lotta Biologica e Integrata per la Difesa delle Colture Agrarie e delle Piante Fore-stali", March 28, 1992, Florence, Italy* (D'Ambrosio, M.T. and Accorti, M., Eds). Ministero Agricoltura e Foreste, Rome, Italy, pp. 15–20.

5 Accorti, M., Guarcini, R. and Persano Oddo, L. (1991). L'ape: Indicatore biologico e insetto test. *Redia,* **74**, n. 1 (appendix), 1–15.

6 Crane, E. (1984). Bees, honey and pollen as indicators of metals in the environment. *Bee World* **55**, 47–49.

7 Gasanov, S.O. (1967). Flower migration and specialization of honeybee of various races. In: *XXI International Congress of Apiculture* (Hambleton, J.I., Townsend, G.F. and Harnaj, V.D., Eds). Apimondia Ed., Bucharest, Romania, pp. 281–284.

8 Pimentel, D., Krummel, J., Gallahan, D., Houg, J., Merrill, A., Schreiner, I., Vittum, P., Koziol, F., Back, E., Yen, D. and Fiance, S. (1978). Benefits and costs of pesticide use in U.S. food production. *BioScience* **28/772**, 778–784.

9 Celli, G., Porrini, C., Radeghieri, P., Sabatini, A.G., Marcazzan, G.L., Colombo, R., Barbattini, R., Greatti, M. and D'Agaro, M. (1996). Honeybees (*Apis mellifera* L.) as bioindicators for the presence of pesticide in the agro-ecosystem. Field test. *Insect. Soc. Life* **1**, 207–212.

10 Atkins, E.L., Kellum, D. and Atkins, K.W. (1981). Reducing pesticide hazards to honey bees: Mortality prediction techniques and integrated management strategies. *Univ. Calif., Div. Agric. Sci. Leaf.* **2883**, 1–23.

11 Johansen, C.A. and Mayer, D.F. (1990). *Pollinator Protection. A Bee and Pes-ticide Handbook.* Wicwas Press, Cheshire, CT, USA, p. 212.

12 Celli, G. (1983). L'ape come insetto test della salute di un territorio. In: *Atti XIII Congresso Nazionale Italiano di Entomologia, Sestriere, Turin, Italy* (Arzone, A., Conti, M., Currado, I., Marletto, F., Pagiano, G., Ugolini, A. and Vidano, C., Eds), pp. 637–644.

13 Celli, G. and Porrini, C. (1987). Apicidi e residui di pesticidi nelle api e nell'alveare in Italia (1983–1986). *Boll. Ist. Ent. "Guido Grandi" Univ. Bologna* **32**, 75–86.

14 Di Giacomo, F., Franco, M.A., Giaccio, M., Prota, R., Floris, I., Chessa, M. and Sferlazzo, G. (1996). Valutazioni statistiche multivariate sul contenuto di metalli in mieli prodotti in prossimità di fonti di inquinamento. *Riv. Merceol.* **35**, 296–310.

15 Colombo, V., Lavagno, E. and Ravetto, P. (1990). Optimal configuration of a bee hive environmental monitoring network and significance of the retrievable information. In: *Interdisciplinary Conference on The Environment: Global Problems – Local Solutions, Hofstra University, Hempstead, NY, USA, 1990* (Hickey, J.E. and Longmire, L.A, Eds). Greenwood, Westport, CT, USA, pp. 397–414.

16 Barbattini, R. and Greatti, M. (1995). La mortalità delle api e il monitoraggio dell'inquinamento agricolo. *Inftore Fitopatol.* **45**, 13–17.

17 McIndoo, N.E. and Demuth, G.S. (1926). Effects on honeybees of spraying fruit trees with arsenicals. *US Dept. Agric. Bull.* **1364**, 1–32.

18 Rennie, J. (1927). Acarine disease in hive bees: Its cause, nature and control. *N. Scotland Coll. Agric. Bull.* **33**, 22–24.

19 Herman, F.A. and Brittain, W.H. (1933). Apple pollination studies in the Annapolis Valley, N.S., Canada. New series. *Canada Dept. Agric. Bull.* **162**, 158–193.

20 Johansen, C.A., Coffey, M.D. and Quist, J.A. (1957). Effect of insecticide treatments to alfalfa on honey bees, including insecticidal residue and honey flavor analyses. *J. Econ. Entomol.* **50**, 721–723.

21 Anderson, L.D. and Atkins, A.L., Jr. (1958). Effects of pesticides on bees. *Calif. Agric.* **12**, 3–4.

22 Anderson, L.D. and Atkins, A.L., Jr. (1958). Toxicity of pesticides to honeybee in laboratory and field test in Southern California, 1955–1956. *J. Econ. Entomol.* **51**, 103–108.

23 Accorti, M., Luti, F. and Tarducci, F. (1991). Methods for collecting data on natural mortality in bee. *Ethol. Ecol. Evol.* special Issue **1**, 123–126.

24 Illies, I., Mühlen, W., Dücker, G. and Sachser, N. (2000). A study of undertaking behaviour of honeybees (*Apis mellifera* L.) by use of different bee traps. In: *Hazards of Pesticides to Bees* (Pélissier, C. and Belzunces, L.P., Eds), *IOBC wprs Bull.*, **23**, p. 24.

25 Gary, N.E. (1960). A trap to quantitatively recover dead and abnormal bees from the hive. *J. Econ. Entomol.* **53**, 782–785.

26 Marchetti, S., Chiesa, F. and D'Agaro, M. (1987). Bee mortality following treatment with perizin in colonies of *Apis mellifera carnica* × *A. m. ligustica*. *Apicoltura* **3**, 157–172.

27 Celli, G. (1974). Condizioni di sopravvivenza dell'ape nei sistemi agricoli attuali. *Ann. Acc. Naz. Agr.* **4**, 395–411.

28 Celli, G., Porrini, C. and Tiraferri, S. (1984). Rapporti tra apicoltura e ambiente. L'ape come indicatore biologico dei pesticidi (con particolare riferimento alla provincia di Forlì) (Nota preventiva). *Boll. Ist. Ent. "G. Grandi" Univ. Bologna* **39**, 231–241.

29 Celli, G., Porrini, C. and Raboni, F. (1988). Monitoraggio con api della presenza dei Ditiocarbammati nell'ambiente (1983–1986). *Boll. Ist. Ent. "G. Grandi" Univ. Bologna* **43**, 195–205.

30 Celli, G., Porrini, C. and Tiraferri, S. (1988). Il problema degli apicidi in rapporto ai principi attivi responsabili (1983–1986). In: *Atti Giornate Fitopatologiche, Lecce, Italy (vol. 1)* (Brunelli, A. and Foschi, S., Eds), pp. 257–268.

31 Porrini, C. (1991). Rapporti tra api (*Apis mellifera* L.) e pesticidi. Valutazione degli effetti del quinalphos irrorato su erba medica. In: *Atti XVI Congresso Nazionale Italiano di Entomologia, September 23–28, 1991, Bari-Martina*

Franca (Taranto), Italy (De Lillo, E., Di Girolamo, E., Nuzzaci, G., Porcelli, F. and Todisco, F., Eds), pp. 545–553.

32 Chiesa, F., Barbattini, R., Greatti, M., D'Agaro, M. and Porrini, C. (1994). Confronto fra l'efficacia di raccolta di api morte in gabbie di tipo diverso. In: *Atti del Convegno del Gruppo di Ricerca: "L'Ape come Insetto Test dell'Inquinamento Agricolo" P.F "Lotta Biologica e Integrata per la Difesa delle Colture Agrarie e delle Piante Forestali", March 28, 1992, Florence, Italy* (D'Ambrosio, M.T. and Accorti, M., Eds). Ministero Agricoltura e Foreste, Rome, Italy, pp. 101–110.

33 Greatti, M., Barbattini, R., D'Agaro, M. and Nazzi, F. (1994). Effect of time on the efficiency of different traps for collecting dead honey bees. *Apicoltura* **9**, 67–72.

34 Greatti, M., Barbattini, R. and D'Agaro, M. (1994). Efficacia di diversi modelli di gabbie per la raccolta di api morte: Verifica mediante l'utilizzo di api marcate allo sfarfallamento. In: *Atti XVII Congresso Nazionale Italiano di Entomologia, June 13–18, 1994, Udine, Italy* (Frilli, F., Governatori, G. and Milani, N., Eds), pp. 839–842.

35 Brokmeier, K. (1996). *Valutazione della mortalità delle api (Apis mellifera L.) (Hymenoptera, Apidae) nel monitoraggio ambientale. Strutture e metodologie a confronto.* Degree Thesis, Biology, University of Bologna, Italy, p. 81.

36 Lundie, A.E. (1925). The flight activities of the honeybee. *USDA Bull.* **1328**, 1–37.

37 Fabergé, A.C. (1943). An apparatus for recording the bees living and entering a hive. *J. Sci. Instrum.* **20**, 28–31.

38 Chauvin, R. (1952). Nouvelle technique d'enregistrement de l'activité de la ruche. *Apiculteur* **96**, 9–14.

39 Chauvin, R. (1963). Essais d'enregistrement simultané des principaux phénomènes de la vie d'une ruche. *Ann. Abeille* **6**, 167–183.

40 Koulichkov, N.N. (1971). Method for determining number of foraging bees in the field. *Am. Bee J.* **11**, 268–269.

41 Chauvin, R. (1976). Sur la mesure de l'activité des abeilles au trou de vol d'une ruche à dix cadres. *Insectes Soc.* **35**, 75–82.

42 Free, J.B. and Preece, P.A. (1969). The effect of the size of honeybee colony on its foraging activity. *Insectes Soc.* **16**, 73–78.

43 Robinson, G.E. and Morse, R.A. (1982). Numbers of honey bees that remain out all night. *Bee World* **63**, 117–118.

44 Danka, R.G. and Gary, N.E. (1987). Estimating foraging population of honey bees from individual colonies. *J. Econ. Entomol.* **80**, 544–547.

45 Burrill, R.M. and Dietz, A. (1973). An automatic honeybee counting and recording device (Apicard) for possible system analysis of a standard colony. *Am. Bee J.* **113**, 216–218.

46 Erickson, E.H., Whitefoot, L.O. and Kissinger, W.A. (1973). Honeybee: A method of delimiting the complete profile of foraging from colony. *Environ. Entomol.* **2**, 531–535.

47 Spangler, H.J. (1969). Photoelectrical counting of outgoing and incoming honeybees. *J. Econ. Entomol.* **62**, 1183–1184.

48 Marletto, F. and Piton, P. (1983). Conta-api elettronico per la verifica dell'attività degli alveari. In: *Atti XIII Congresso Nazionale Italiano di Entomologia, Sestriere, Turin, Italy* (Arzone, A., Conti, M., Currado, I., Marletto, F., Pagiano, G., Ugolini, A. and Vidano, C., Eds), pp. 707–712.

49 Bühlmann, G. (1987). Messungen am Flugloch: März und April. *Schweiz. Bienenztg.* **110**, 132–140.

50 Bühlmann, G., Wille, M. and Imdorf, A. (1987). Messungen am Flugloch: Mai und Juni. *Schweiz. Bienenztg.* **110**, 245–258.

51 Bühlmann, G., Wille, M. and Imdorf, A. (1987). Messungen am Flugloch: Juli und August. *Schweiz. Bienenztg.* **110**, 293–303.

52 Bühlmann, G., Wille, M. and Imdorf, A. (1987). Messungen am Flugloch: September–Dezember. *Schweiz. Bienenztg.* **110**, 491–500.

53 Rickli, M., Bühlmann, G., Gerig, L., Herren, H., Schürch, H.S., Zeier, W. and Imdorf, A. (1989). Zur Anwendung eines elektronischen Bienenzählgerätes am Flugloch eines Bienenvolkes. *Apidologie* **20**, 305–315.

54 Cazzola, A. (1988). *Contatore elettronico per statistiche bioecologiche.* Degree Thesis, University of Bologna, Italy, p. 92.

55 Struye, M.H., Mortier, H.J., Arnold, G., Miniggio, C. and Borneck, R. (1994). Microprocessor-controlled monitoring of honeybee light activity at the hive entrance. *Apidologie* **25**, 384–395.

56 Struye, M.H. (2000). Possibilities and limitations of monitoring the flight activity of honeybees by means of BeeSCAN bee counters. In: *Hazards of Pesticides to Bees* (Pélissier, C. and Belzunces, L.P., Eds), *IOBC wprs Bull.* **23**, p. 27.

57 Porrini, C., Cazzola, A. and Menozzi, R. (1994). Prospettive per la realizzazione di un contatore elettronico per la valutazione della dinamica di popolazione delle api. In: *Atti del Convegno: "L'Ape come Insetto Test dell'Inquinamento Agricolo" P.F "Lotta Biologica e Integrata per la Difesa delle Colture Agrarie e delle Piante Forestali," March 28 1992, Florence, Italy* (D'Ambrosio, M.T. and Accorti, M., Eds). Ministero Agricoltura e Foreste, Rome, Italy, pp. 117–126.

58 Chauvin, R. (1968). *Traité de Biologie de l'Abeille.* Masson et Cie, Paris, France, 5 volumes, p. 2158.

59 Capelo, A., Casalone, P. and Ferrari, G. (1983). Un modello matematico per la valutazione del numero di api presenti in un alveare. *Apicolt. Mod.* **74**, 239–245.

60 Dalpero, A.P. (1998). *Estrazione e determinazione quali-quantitativa di pesticidi sulla matrice ape.* Degree Thesis, Pharmaceutical Chemistry and Technology, University of Bologna, Italy, p. 83.

61 Marty, J.L., Mionetto, N., Norguer, T., Ortega, F. and Roux, C. (1993). Enzime sensors for the detection of pesticides. *Biosens. Bioelectron.* **8**, 273–280.

62 Jeanty, G. and Marty, J.L. (1998). Detection of paroxon by continuous flow system based enzyme sensor. *Biosens. Bioelectron.* **13**, 213–218.

63 Gonzales, M.E.L. and Townshend, A. (1990). Flow-injection determination of paraoxon by inhibition of immobilized acetylcholinesterase. *Anal. Chim. Acta* **236**, 267–272.

64 Hock, B., Dankwardt, A., Kramer, A. and Marx, A. (1995). Immunochemical techniques: Antibody production for pesticide analysis. A review. *Anal. Chim. Acta* **311**, 393–405.

65 Roda, A., Rauch, P., Ferri, E., Girotti, S., Ghini, S., Carrea, G. and Bovara, R. (1994). Chemiluminescent flow sensor for the determination of paraoxon and aldicarb pesticides. *Anal. Chim. Acta* **294**, 35–42.

66 Ambrus, A., Lantos, J., Visi, E., Csatlos, I. and Sarvari, L. (1981). General method for determination of pesticide residues in samples of plant origin, soil and water. I. Extraction and cleanup. *J. Assoc. Off. Anal. Chem.* **64**, 733–769.

67 Sasaki, K., Suzuki, T. and Saito, Y. (1987). Simplified cleanup and chromatographic determination of organophosphorus pesticides in crops. *J. Assoc. Off. Anal. Chem.* **70**, 460–464.

68 Rossi, S., Dalpero, A.P., Ghini, S., Colombo, R., Sabatini, A.G. and Girotti, S. (2001). Multi residual method for gas chromatography analysis of pesticides in honeybees cleaned by gel permeation chromatography. *J. Chromatogr. A* **905**, 223–232.

69 Louveaux, J., Maurizio, A. and Vorwhol, G. (1978). Methods of melissopalynology. *Bee World* **59**, 139–157.

70 Zander, E. (1935). *Beiträge zur Herkunftsbestimmung bei Honig. I. Pollengestaltung und Herkunftsbestimmung bei Blütenhonig.* Reichsfachgruppe Imker, Berlin, Germany, p. 349.

71 Persano Oddo, L. and Ricciardelli D'Albore, G. (1989). Nomenclatura melissopalinologica. *Apicoltura* **5**, 63–72.

72 Mercuri, A.M. and Porrini, C. (1991). Melissopalinological analysis applied to air pollution studies in urban areas of Modena and Reggio Emilia. *Aerobiologia* **7**, 38–48.

73 Stefano, M.A. (1996). *Impiego del polline come marker nel monitoraggio dell'inquinamento da pesticidi, tramite api.* Degree Thesis, Biology, University of Bologna, Italy, p.171.

74 Porrini, C., Colombo, V. and Celli, G. (1996). The honey bee (*Apis mellifera* L.) as pesticide bioindicator. Evaluation of the degree of pollution by means of environmental hazard indexes. In: *Proceedings XX International Congress of Entomology, August 25–31, 1996, Florence, Italy*, p. 444.

75 Porrini, C. (1999). Metodologia impiegata nei programmi di monitoraggio dei pesticidi con api. In: *Atti del Workshop "Biomonitoraggio della Qualità dell'Aria sul Territorio Nazionale," November 26–27, 1998, Rome, Italy* (Piccini, C. and Salvati, S., Eds). ANPA, Rome, Italy, Series 2/1999, pp. 311–317.

76 Accorti, M. (1994). Influenza dell'ambiente sul comportamento e sulla biologia delle api nel monitoraggio ambientale. In: *Atti del Convegno: "L'Ape come Insetto Test dell'Inquinamento Agricolo' P.F "Lotta Biologica e Integrata per la Difesa delle Colture Agrarie e delle Piante Forestali", March 28, 1992, Florence, Italy* (D'Ambrosio, M.T. and Accorti, M., Eds). Ministero Agricoltura e Foreste, Rome, Italy, pp. 45–57.

77 Porrini, C., Celli, G. and Radeghieri, P. (1998). Monitoring of pesticides through the use of honeybees as bioindicators of the Emilia-Romagna coastline (1995–1996). *Ann. Chim.* **88**, 243–252.

78 Celli, G., Porrini, C., Baldi, M. and Ghigli, E. (1991). Pesticides in Ferrara province: Two years monitoring with honey bees (1987–1988). *Ethol. Ecol. Evol.* Special Issue **1**, 111–115.

79 Celli, G., Porrini, C. and Tiraferri, S. (1987). L'ape come insetto-test dei pesticidi. *Inftore Agr.* **38**, 59–63.

80 Porrini, C. (1991). I pesticidi in provincia di Forlì: Otto anni di monitoraggio con le api. *Vita della Provincia*, Special Edition April 11 1991, pp. 5–11.

81 Porrini, C. (1996). Le api in provincia di Forlì: Dodici anni di monitoraggio

ambientale. In: *La Biodiversità alle Soglie del 2000*. ES Ed., Bologna, Italy, pp. 27–30, 40–44.

82 Caroli, S., Forte, G., Iamiceli, A.L. and Galoppi, B. (1999). Determination of essential and potentially toxic trace elements in honey by inductively coupled plasma-based techniques. *Talanta* **50**, 327–336.

83 Accorti, M. and Persano Oddo, L. (1986). Un servizio di monitoraggio ambientale urbano: "Apincittà" *Inftore Agr.* **42**, 39–41.

84 Celli, G., Porrini, C., Balestra, V. and Menozzi, R. (1988). Monitoraggio di inquinanti atmosferici urbani mediante api. In: *Atti del Convegno "Salute e Ambiente," May 27–28, 1988, Cagliari, Italy* (Puggioni, G., Busonera, E. and Gatti, A.M., Eds), pp. 73–85.

85 Celli, G., Porrini, C., Radeghieri, P., Amati, S., Santi, F. and Gasparo, D. (1996). Monitoraggio degli inquinanti ambientali tramite bioindicatori. *Acer* **4**, 4–6.

86 Leita, L., Muhlbachova, G., Cesco, S., Barbattini, R. and Mondini, C. (1996). Investigation of the use of honey bees products to assess heavy metal contamination. *Environ. Monit. Assess.* **43**, 1–9.

87 Porrini, C., Celli, G., Radeghieri, P., Marini, S. and Maccagnani, B. (2000). Studies on the use of honeybees (*Apis mellifera* L.) as bioindicators of metals in the environment. *Insect Soc. Life* **3**, 153–159.

88 Cavalchi, B. and Fornaciari, S. (1983). Api, miele, polline e propoli come possibili indicatori di un inquinamento da piombo e fluoro – Una esperienza di monitoraggio biologico nel comprensorio ceramico di Sassuolo-Scandiano. In: *Atti del Seminario di Studi "I Biologi e l'Ambiente. Nuove Esperienze per la Sorveglianza Ecologica," February 17–18, 1983, Reggio Emilia, Italy* (Manzini, P. and Spaggiari, R., Eds), pp. 275–300.

89 Celli, G., Porrini, C., Siligardi, G. and Mazzali, P. (1988). Le calibrage de l'instrument abeille par rapport au plomb. In: *Proceedings XVIII International Congress of Entomology, August 3–9, 1988, Vancouver, Canada*, p. 467.

90 Balestra, V., Celli, G. and Porrini, C. (1992). Bees, honey, larvae and pollen in biomonitoring of atmospheric pollution. *Aerobiologia* **8**, 122–126.

91 Cesco, S., Barbattini, R. and Agabiti, M.F. (1994). Honey bees and bee products as possible indicators of cadmium and lead environmental pollution: An experience of biological monitoring in Portogruaro city (Venice, Italy). *Apicoltura* **9**, 103–118.

92 Franco, M.A., Chessa, M., Giaccio, M., Di Giacomo, F., Prota, R. and Sferlazzo, G. (1997). Bee pollen as indicator of environmental pollution by heavy metals. *Riv. Merceol.* **35**, 295–309.

93 Franco, M.A., Chessa, M., Sferlazzo, G., Giaccio, M., Di Giacomo, F., Prota, R. and Manca, G. (1998). Beeswax as an indicator of environmental pollution by heavy metals. *Riv. Merceol.* **37**, 3–11.

94 Siegel, S. (1956). *Non Parametric Statistics for the Behavioral Sciences.* McGraw-Hill Book Company, Inc., New York, USA, p. 269.

95 Altman, G. (1985). Belastung von Blütenhonig mit Schwermetallen und ihre Herkunft. *Apidologie* **16**, 197–198.

96 Bogdanov, S., Zimmerli, B. and Erard, M. (1985), Schwermetalle in Honig. *Mitt. Geb. Lebensmittelunters. Hyg.* **77**, 153–158.

97 Censi, A. and Cremasco, S. (1989). L'alimento miele in vista del mercato unico europeo: alcuni aspetti sanitari. In: *Atti del V Convegno Internazionale*

dell'*Apicoltura in Agricoltura, October 6–8, 1989, Lazise, Verona, Italy* (Cirone, R. and Manoli, P., Eds), pp. 3–11.

98 Erard, M., Miserez, A. and Zimmerli, B. (1982). Exposition des nourrissons au plomb, cadmium, zinc et sélénium de provenance alimentaire. *Mitt. Geb. Lebensmittelunters. Hyg.* **73**, 394–411.

99 Höffel, I. (1985). Schwermetalle in Bienen und Bienenprodukten. *Apidologie* **16**, 196–197.

100 Oddi, P. and Bertani, P. (1987). Contaminanti nel miele: Nota 1. Residui di Pb e Cd. In: *Atti S.I.S. Vet.* XLI, pp. 998–1000.

101 Otto, K. and Jekat, F. (1977). Experimentelle Untersuchungen über die Belastung eines Nahrungsmittels mit Rückständen von Blei, Zink, Cadmium. *Ernähr. Umsch.* **24**, 107–109.

102 Zalewski, W., Syrocka, K., Porzadek, K. and Lipinska, J. (1989). Analysis of heavy elements in bee products collected in Poland. In: *Proceedings of the XXXI International Congress of Apiculture, August 19–25, 1987, Warsaw, Poland* (Borneck, R., Ostach, H., Bornus, L. and Kuzba, A., Eds). Apimondia Ed., Bucharest, Romania, pp. 509–511.

103 Zitti, G. and Angelozzi, G. (1990). La concentrazione di piombo nei mieli abruzzesi. In: *Studio Analitico dei Mieli Abruzzesi (Vol. 3)* (Calvarese, S., Ed.). Regione Abruzzo, Italy, pp. 55–66.

104 Conti, M.E., Saccares, S., Cubadda, F., Cavallina, R., Tenoglio C.A. and Ciprotti, L. (1998). Il miele nel lazio: Indagine sul contenuto in metalli tracce e radionuclidi. *Riv. Sci. Aliment.* **2**, 107–119.

105 Trstenjak Petrovic, Z., Mandic, M.L., Grgic, J. and Grgic Z. (1994). Ash and chromium levels of some types of honey. *Z. Lebensm. Unters. Forsch.* **198**, 36–39.

106 Pinzauti, M., Frediani, D., Biondi, C., Belli, R., Panizzi, C., Cosimi, C. and Zummo, V. (1991). Impiego delle api nel rilevamento dell'inquinamento ambientale. *Analysis* **8**, 355–407.

107 Raes, H., Cornelis, R. and Rzenik, U. (1992). Distribution, accumulation and depuration of administered lead in adult honeybees. *Sci. Total Environt.* **113**, 269–279.

108 Svoboda, J. (1962). Teneur en strontium 90 dans les abeilles et dans leurs produits. *Bull. Apicole* **5**, 101–103.

109 Racoveanu, N., Popa, A. and Tarcitu, E. (1965). La radioactivité du miel d'abeilles dans la région de Bucarest. *Bull. Apic. Doc. Sci. Tech. Inf.* **2**, 147.

110 Gilbert, M.D. and Lisk, D. (1978). Honey as an environment indicator of radionuclide contamination. *Bull. Environ. Contam. Toxicol.* **18**, 241–251.

111 Wallwork-Barber, M.K., Ferenbaugh, R.W. and Gladney, E.S. (1982). The use of honey bees as monitors of environmental pollution. *Am. Bee J.* **122**, 770–772.

112 Accorti, M., Belli, M., Marchetti, A., Nicolai, P. and Persano Oddo, L. (1987). Osservazioni preliminari sulla distribuzione e la dinamica del ^{137}Cs e ^{134}Cs nelle componenti biologiche dell'alveare. *Boll. Soc. Ital. Ecol.* **8**, 93.

113 Albertazzi, S., Alessio, G. and Dangeri, P. (1987). Indagini sulla radioattività e sullo spettro pollinico di mieli dell'Italia settentrionale in relazione alle aree di produzione. *Boll. Soc. Ital. Ecol.* **8**, 96.

114 Bettoli, M.G., Sabatini, A.G. and Vecchi, M.A. (1987). Il miele prodotto in Italia dopo l'incidente di Chernobyl. *Apitalia* **14**, 5–7.

115 Bunzl, K., Kracke, W. and Vorwohl, G. (1988). Transfer of Chernobyl-derived ^{134}Cs, ^{137}Cs, ^{131}I and ^{103}Ru from flowers to honey and pollen. *J. Environ. Radioact.* **6**, 261–269.

116 Canteneur, R. (1987). Tchernobyl. Produits de la ruche et radioactivité. *Rev. Fr. of Apic* **2**, 86–88.

117 Gattavecchia, E., Ghini, S., Tonelli, D. and Porrini, C. (1987). Il miele italiano dopo Chernobyl. *L'Ape nostra Amica* **9**, 27.

118 Giovani, C., Padovani, R., Godeassi, M., Frilli, F., Barbattini, R. and Greatti, M. (1995). Radiocesio nei mieli millefiori e di melata del Friuli-Venezia Giulia. *Apicolt. Mod.* **86**, 59–66.

119 Klepsch, A. and Molzahn, D. (1987). Die radioaktive Belastung des Honigs in Hessen nach dem Reaktorunglück in Tschernobyl. *Die Biene* **3**, 109–114.

120 Ropolo, R., Patetta, A. and Manino, A. (1987). Osservazioni su radioattività e miele nel 1987 in Piemonte. *Apicolt. Mod.* **78**, 11–15.

121 Ropolo, R., Manino, A. and Patetta, A. (1988). Radioattività e miele in Piemonte un anno dopo. *Apicolt. Mod.* **79**, 147–151.

122 Fresquez, P.R., Armstrong, D.R. and Pratt, L.H. (1997). Radionuclides in bees and honey within and around Los-Alamos National Laboratory. *J. Environ. Sci. Health* **32**, 1309–1323.

123 Barisic, D., Vertacnik, A., Bromenshenk, J.J., Kezic, N., Lulic, S., Hus, M., Kraljevic, P., Simpraga, M. and Seletkovic, Z. (1999). Radionuclides and selected elements in soil and honey from Gorski Kotar, Croatia. *Apidologie* **30**, 277–287.

124 Pinzauti, M., Lazzeri, M. and Melosi, G. (1987). Dopo Chernobyl. L'ape insetto test nel rilevamento della radioattività ambientale. *Quad. Toscana Notizie* **2**, 1–50.

125 Giovani, C., Padovani, R., Godeassi, M., Frilli, F., Barbattini, R. and Greatti, M. (1995). Radiocesio nei mieli millefiori e di melata del Friuli-Venezia Giulia. *Apicolt. Mod.* **86**, 59–66.

126 Tonelli, D., Gattavecchia, E., Ghini, S., Porrini, C., Celli, G. and Mercuri, A.M. (1990). Honey bees and their products as indicators of environmental radioactive pollution. *J. Radioanal. Nucl. Chem., Art.* **141**, 427–436.

127 Merighi, M., Malaguti, S., Bazzi, C., Sandrini, A., Landini, S., Ghini, S. and Girotti, S. (1999). Specific detection of *Erwinia amylovora* by immunoenzymatic determination of PCR products. *Acta Horticult.* **489**, 39–42.

128 Merighi, M., Sandrini S., Landini, S., Ghini, S., Girotti, S., Malaguti, S. and Bazzi, C. (2000). Chemiluminescent and colorimetric detection of *Erwinia amylovora* by immunoenzymatic determination of PCR amplicons (PCR-ELISA) from plasmid pEA29. *Plant Disease* **84**, 49–54.

129 Mazzucchi, U. (1992). Sintomatologia. In: *Atti delle Giornate di Studio sul Colpo di Fuoco da Erwinia amylovora, Bologna, 1–2 Aprile 1991* (Mazzucchi, U., Ed.). Tecnoprint s.n.c., Bologna, Italy, pp. 7–17.

130 Mazzucchi, U. (1992). L'agente causale. In: *Atti delle Giornate di Studio sul Colpo di Fuoco da Erwinia amylovora, Bologna, 1–2 Aprile 1991* (Mazzucchi, U., Ed.). Tecnoprint s.n.c., Bologna, Italy, pp. 19–27.

131 Bazzi, C., Tagliati, M.E., Spina, F. and Bendini, L. (1992). Disseminazione di *Erwinia amylovora* a breve e a grande distanza. In: *Atti delle Giornate di Studio sul Colpo di Fuoco da Erwinia amylovora, Bologna, 1–2 Aprile 1991* (Mazzucchi, U., Ed.). Tecnoprint s.n.c., Bologna, Italy, pp. 29–40.

132 Stefani, E. (1992). Mezzi di lotta agronomici e biologici. In: *Atti delle Giornate di Studio sul Colpo di Fuoco da Erwinia amylovora, Bologna, 1–2 Aprile 1991* (Mazzucchi, U., Ed.). Tecnoprint s.n.c., Bologna, Italy, pp. 41–50.

133 Calzolari, A. and De Giovanni, G. (1992). La rete di monitoraggio in Emilia Romagna. In: *Atti delle Giornate di Studio sul Colpo di Fuoco da Erwinia amylovora, Bologna, 1–2 Aprile 19911* (Mazzucchi, U., Ed.). Tecnoprint s.n.c., Bologna, Italy, pp. 129–135.

134 Van der Zwet, T. and Beer, S.V. (1995). Fire blight: its nature, prevention and control. A practical guide to integrated disease management. *Agric. Inf. Bull.* **631**, 91.

135 Van der Zwet, T. (1996). Present worldwide distribution of fire blight. *Acta Horticult.* **411**, 7–8.

136 Missere, D., Carli, G., Spada, G., Mascanzoni, G. and Marangoni, B. (1999). Strategie di difesa. *Il Divulgatore* **1**, 35–50.

137 Cariddi, C. (1990). Colpo di fuoco sul pero. *Terra e Vita* **34**, 67–69.

138 D'Anna, R., Sesto, F., Arreddia, R., Armato, P., Cirrito, V., Albanese, G. and Granata, G. (1994). La rete di monitoraggio per il colpo di fuoco batterico delle pomacee in Sicilia. *Inftore Agr.* **50**, 57–59.

139 Calzolari, A., Finelli, F., Ponti, I. and Mazzoli, G.L. (1999). L'esperienza dell'Emilia Romagna nella lotta al colpo di fuoco. *Inftore Agr.* **14**, 65–70.

12 Typology of French acacia honeys based on their concentrations in metallic and nonmetallic elements

J. Devillers, J.C. Doré, C. Viel,
M. Marenco, F. Poirier-Duchêne,
N. Galand, and M. Subirana

Summary

The elemental analysis of 150 French acacia honeys (*Robinia pseudoacacia* L.) collected by beekeepers in apparently polluted and nonpolluted environments was performed by using inductively coupled plasma atomic emission spectrometry (ICP-AES) to measure significant concentrations of Ag, Ca, Cr, Co, Cu, Fe, Li, Mg, Mn, Mo, P, S, Zn, Al, Cd, Hg, Ni, and Pb. Fortunately, Cd, Hg, Ni, and Pb were not detected in the analyzed samples. Conversely, Ag, Cu, Al, Zn, and S were found in some samples located near industrial areas. Because a high variability was found in the concentration profiles, correspondence factor analysis was used to rationalize the data and provide a typology of the honeys based on the concentration of these different elements in the honeys. The results were confirmed by means of principal component analysis and hierarchical cluster analysis. Finally, the usefulness of the acacia honey as a bioindicator of heavy metal contamination is discussed.

Introduction

The continued expansion of industrial production and the growing use of chemicals in agriculture have led to an increase in the number and quantities of xenobiotics released into the different compartments of the biosphere [1]. The health risks to human and nonhuman biota associated with these chemicals are evaluated on the basis of critical and reliable information on exposures and on related adverse health effects [2]. In this process, the estimation of the environmental concentrations of the hazardous chemicals plays a key role. A number of precise technical sampling methods are available for monitoring pollutants in the environment. However, due to their high technicality and cost, they are generally not used routinely [2]. Conversely, bioindicators are now widely employed for estimating, at low cost, the level of contamination of organic and inorganic chemicals in aquatic and terrestrial ecosystems [e.g. 3–5].

Thus, honey bees commonly forage within 1.5 km of their hive and exceptionally as far as 10 to 12 km, depending on their need for food and its availability [6]. During their foraging flights, they visit numerous plants to gather nectar, pollen, honeydew, sap, and water. Honey bees also visit puddles, ponds, and other aquatic resources to collect the 10 to 40 liters of water which are necessary annually for the colony [7]. When honey bees settle on leaves, penetrate in the corolla of flowers to gather nutritive substances, and collect water in aquatic resources, they provide composite samples from thousands of different visited points spread across a broad area. Consequently, these insects and their products such as honey, wax, or royal jelly can provide a good idea of the level of contamination which can be found in air, soil, vegetation, and water in a radius of a few kilometers from their hive [8, 9].

Heavy metals, which are ubiquitous environmental pollutants, are found in all the compartments of the biosphere and in living species [e.g. 10–13], including honey bees and their products [14–24]. In this context, samples of French acacia (*Robinia pseudoacacia* L.) honeys, directly collected by beekeepers in hives located in media presenting different degrees of pollution, were analyzed for their concentrations of heavy metals and some other metallic and nonmetallic elements in order to see whether it was possible to find a relationship between industrialization and the levels of honey contamination by heavy metals and related compounds. An attempt was also made to provide a typology of the honey samples from the multivariate analysis of their concentrations of metallic and nonmetallic elements in relation to environmental variables.

Materials and methods

Sampling

Under the authority of the CNDA (National Center for the Development of Apiculture), beekeepers of various French departments were first contacted by letter to determine their interest in being involved in a study dealing with the elemental analysis of acacia honeys and their typology on the basis of environmental variables. A sampling protocol and material to collect and store the honey were then sent only to those beekeepers interested in the project and who agreed to provide all the necessary information to interpret the analytical results found with their honey(s). In the protocol, beekeepers were required to select one hive located in an unpolluted area and another near a source of pollution such as an industry, mine, highway, urban area, and so on. It was necessary to manually collect the honey samples by slow extraction from the combs. Beekeepers had to use the material provided for the study to avoid problems of external contamination by trace elements. The use of bee smokers was prohibited, and it was also forbidden to smoke during the sampling process. Honey samples had to be stored in small hermetically sealed containers which were certified as free of trace elements, and were sent out to the bee-keepers.

The environmental conditions around the hives had to be clearly described. It was also required to give some climatic information, such as the main direction of the winds, and so on. If the two hives selected by a beekeeper were located in the same department, the kilometric distance between them had to be provided. Finally, any unusual event (e.g. fire) also had to be mentioned.

A total of 150 different acacia honeys were obtained from various French departments (Figure 12.1). All samples were collected in May–June 1999. Honeys were sent by post to the analytical laboratory for determination of their metallic and nonmetallic element content.

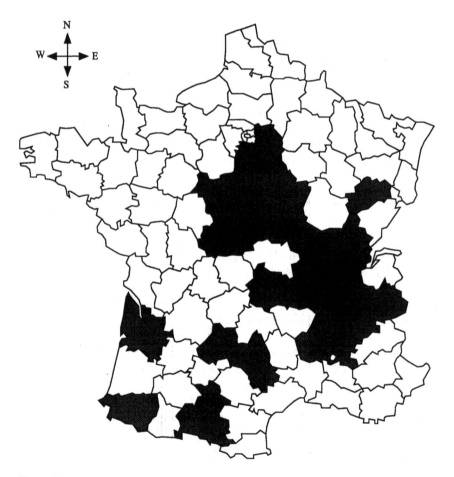

Figure 12.1 Honey sampling regions in France (in dark).

Analytical method

Prior to the preparation and chemical analysis of the honeys, the samples were coded and randomized to avoid identification of their location and characteristics by the chemists. The mineralization of the honey samples was performed in polypropylene-stoppered vials of volume 10 ml [Plastiques Gosselin, ref. TR 95 PPN 10TT (vials) and ref. B135 (stoppers)] by dissolution in HNO_3 at 69.5 percent (63.01 g/mol; $d = 1.409$) (Carlo Erba, ref. 408071). The nitric acid was diluted in a 2/3 ratio with water previously purified according to the guidelines of the French Pharmacopoeia (10th edition). For each honey sample, amounts of 1 g and 2 g, exactly weighed, were digested with 5 ml of the above acidic solution. Stoppered vials were placed in a bain-marie and warmed up to the temperature of mineralization of 60°C. After 3 to 4 hours under these experimental conditions, the volume of each vial was exactly adjusted to 10 ml with HNO_3 (2/3) and the mineralization at 60°C was continued as described above. The time required to obtain complete mineralization of a sample ranged from 6 to 7 hours and the product was analyzed after keeping it for 15 hours at room temperature. A solution of 5 ml was injected into an inductively coupled plasma atomic emission spectrometer (Panorama, Jobin & Yvon) previously calibrated for the 18 metallic and nonmetallic elements studied. The zero point was obtained from the acidic solution used to mineralize the honey and which corresponded with a blank. The wavelengths (nm) of the emission peaks of the 18 elements studied were the following: aluminum (Al), 396.152; cadmium (Cd), 226.502; calcium (Ca), 317.933; chromium (Cr), 267.716; cobalt (Co), 228.616; copper (Cu), 324.754; iron (Fe), 259.940; lead (Pb), 220.353; lithium (Li), 670.776; magnesium (Mg), 279.553; manganese (Mn), 257.610; mercury (Hg), 184.887; molybdenum (Mo), 202.032; nickel (Ni), 231.604; phosphorus (P), 178.225; silver (Ag), 328.068; sulfur (S), 180.672; zinc (Zn), 213.856. All samples were analyzed automatically in triplicate by using the spectrometer. In addition, for each sample, both quantities (i.e. 1 and 2 g) were analyzed. The standard deviations were always less than 5 percent. The limit of the detection of S, Al, Ni, Ca, Mg, P, and Pb in the honey samples was 1 ng/g. That for Hg was 0.5 ng/g while Ag, Cr, Fe, Li, and Mn were not detected at a concentration less than 0.2 ng/g. The limit of detection of Co, Cu, Mo, Cd, and Zn was 0.1 ng/g.

Data analysis

Statistical analyses were performed with ADE-4 [25], a powerful statistical software program designed specifically for the analysis of environmental data. ADE-4 includes the main linear multivariate analyses and numerous graphical tools for optimal data display.

Analytical results

The elemental analyses obtained from 1 or 2 g of honey yielded similar results, and hence were averaged. The number of positive responses (i.e. concentrations greater than the different limits of detection) for each metallic or nonmetallic element in the 150 honeys analyzed and their corresponding average, smallest, and highest concentrations (in mg/kg to raw (wet) weight) are given in Table 12.1. Detailed analytical results are listed in Table 12.2, except for elements with a frequency of positive responses less than 5 percent.

Table 12.1 shows that calcium (Ca), magnesium (Mg), and phosphorus (P) were detected in all the samples analyzed. The concentrations of these three elements show Gaussian distributions (graphs not given). The results obtained are not surprising because of the nature, role, and ubiquity of these fundamental elements. Manganese (Mn), is also significantly present in most of the honey samples. Aluminum (Al), molybdenum (Mo), and sulfur (S) have been detected in more than 50 percent of the samples, and to a lesser extent, copper (Cu) and zinc (Zn). About 30 percent of the analyzed samples include measurable concentrations of cobalt (Co) while about 20 percent of the honeys are contaminated with quantifiable concentrations of chromium (Cr). Table 12.1 shows that silver (Ag) has been detected in 10 samples with concentrations ranging from 0.08 to 2.16 ppm. Lithium was only measured in samples 6, 43, 44, 133, and 149 (Table 12.2)

Table 12.1 Number of positive responses (Nb/150) for the 18 elements studied with their corresponding mean, lowest, and highest concentrations (in ppm)

Element	Nb/150	Mean	Range
Ag	10	0.596	0.08–2.16
Ca	150	22.86	2.98–108.50
Cr	33	0.187	0.05–0.52
Co	46	0.091	0.03–0.25
Cu	72	0.163	0.03–2.30
Fe	107	1.167	0.13–10
Mg	150	8.708	1.43–109.50
Mn	141	0.777	0.06–10.34
Mo	86	0.441	0.07–0.81
P	150	73.45	32.12–397.5
S	84	15.39	1.60–67.66
Zn	67	0.746	0.04–5.96
Al	99	0.374	0.05–1.44
Li	5	0.07	0.02–0.24
Ni	0	na*	na
Hg	0	na	na
Cd	0	na	na
Pb	0	na	na

Note
*na, not applicable.

Table 12.2 Element concentrations (ppm) in acacia honeys collected in France

No.	Ag	Ca	Cr	Co	Cu	Fe	Mg	Mn	Mo	P	S	Zn	Al
1	<ld*	14.77	<ld	0.03	<ld	1.76	5.45	0.29	0.45	53.61	<ld	<ld	0.30
2	<ld	18.18	<ld	<ld	<ld	0.81	5.50	0.33	<ld	47.48	<ld	0.40	<ld
3	<ld	7.82	<ld	<ld	<ld	<ld	3.77	0.09	0.49	47.38	<ld	<ld	<ld
4	<ld	12.95	<ld	0.04	<ld	<ld	6.91	0.21	<ld	56.87	<ld	0.42	0.27
5	<ld	7.61	<ld	0.07	<ld	0.17	5.81	0.20	0.44	54.78	2.86	0.11	<ld
6	<ld	5.48	0.11	<ld	<ld	0.37	3.03	0.10	0.43	42.88	5.11	0.27	<ld
7	<ld	4.68	0.09	0.04	<ld	0.66	2.11	<ld	0.48	40.78	<ld	<ld	<ld
8	<ld	7.30	<ld	0.03	<ld	0.13	4.17	0.28	0.48	49.70	<ld	<ld	<ld
9	<ld	11.40	<ld	<ld	<ld	10.00	16.65	0.52	0.58	125	9.11	0.74	0.43
10	<ld	10.95	0.16	0.10	<ld	4.76	9.97	0.28	0.81	98.36	<ld	0.97	0.10
11	<ld	18.86	<ld	0.11	<ld	1.03	7.27	1.39	<ld	61.11	<ld	0.70	0.39
12	<ld	10.25	<ld	<ld	<ld	0.44	4.17	0.22	0.61	49.25	<ld	0.34	0.31
13	<ld	13.48	<ld	0.10	<ld	<ld	5.53	0.41	0.53	57.28	<ld	<ld	0.25
14	<ld	9.84	0.15	0.11	<ld	0.47	3.46	0.22	0.71	51.29	<ld	0.32	<ld
15	<ld	5.62	0.13	0.10	<ld	0.33	2.73	0.14	0.77	53.72	<ld	0.20	<ld
16	<ld	23.95	<ld	<ld	<ld	0.63	16.27	1.73	0.42	81.51	16.89	<ld	0.62
17	<ld	10.13	<ld	<ld	<ld	1.57	5.03	0.28	0.79	59.34	<ld	0.29	0.10
18	<ld	19.39	<ld	<ld	<ld	<ld	7.37	0.17	<ld	71.70	10.30	<ld	0.48
19	<ld	29.83	0.15	0.11	0.22	0.79	18.34	3.05	0.72	96.55	<ld	0.45	<ld
20	<ld	108.5	<ld	0.12	0.57	1.78	46.83	2.64	<ld	149.3	35.90	0.65	0.63
21	<ld	13.06	<ld	<ld	<ld	0.79	3.47	0.18	0.56	48.02	<ld	<ld	0.31
22	<ld	16.87	<ld	<ld	<ld	0.61	9.39	0.93	0.62	55.74	<ld	<ld	<ld
23	<ld	34.99	<ld	<ld	<ld	1.59	4.77	1.42	0.59	57.70	8.18	0.52	0.50
24	<ld	32.96	<ld	<ld	<ld	1.00	7.49	3.11	0.60	71.22	<ld	0.76	0.13
25	<ld	15.45	<ld	<ld	<ld	0.39	4.00	0.35	0.68	46.24	5.73	<ld	<ld
26	<ld	7.34	0.12	0.11	<ld	0.62	2.82	0.19	0.63	46.44	<ld	<ld	<ld
27	<ld	47.34	0.08	0.13	1.68	2.23	102.6	10.34	0.68	350.2	60.11	0.95	1.10
28	<ld	67.01	<ld	0.13	2.30	2.94	109.5	9.65	0.58	397.5	67.66	1.26	1.01
29	<ld	55.20	<ld	<ld	<ld	1.82	12.97	0.58	0.60	73.71	17.09	<ld	<ld
30	<ld	23.48	0.13	0.11	<ld	0.80	7.32	1.73	0.67	62.66	<ld	<ld	<ld
31	<ld	20.40	0.16	0.13	<ld	0.76	7.06	1.26	0.73	53.95	<ld	0.24	<ld
32	<ld	23.15	0.16	0.13	<ld	0.82	8.20	1.22	0.76	61.15	<ld	1.88	<ld
33	<ld	19.82	<ld	<ld	<ld	0.70	6.14	0.19	0.63	57.58	8.20	<ld	0.28
34	<ld	15.24	<ld	<ld	<ld	1.38	6.24	0.22	0.62	70.78	5.96	0.79	0.43
35	<ld	11.12	<ld	<ld	<ld	0.47	4.76	0.10	0.68	56.77	11.47	<ld	0.27
36	<ld	18.81	<ld	<ld	<ld	0.64	7.85	0.42	<ld	55.88	<ld	0.55	0.30
37	<ld	14.35	<ld	<ld	<ld	<ld	5.16	0.48	<ld	44.58	<ld	0.72	0.28
38	<ld	33.86	<ld	<ld	<ld	1.18	23.35	2.89	<ld	105.8	15.20	1.28	0.43
39	<ld	18.25	<ld	0.11	<ld	1.06	6.98	1.16	0.65	57.80	<ld	1.49	0.46
40	<ld	20.37	<ld	<ld	<ld	<ld	6.82	1.15	<ld	58.75	4.12	1.79	0.49
41	<ld	27.47	<ld	<ld	<ld	3.35	6.41	1.28	<ld	52.32	<ld	5.96	0.98
42	<ld	21.46	<ld	<ld	<ld	1.56	9.06	2.79	0.62	59.60	7.49	1.30	1.17
43	<ld	34.36	<ld	<ld	<ld	0.41	12.00	<ld	<ld	65.73	16.89	<ld	0.44
44	<ld	14.55	<ld	<ld	<ld	0.58	5.23	0.13	<ld	62.30	10.24	<ld	0.56
45	<ld	15.30	<ld	<ld	<ld	0.69	5.32	0.17	0.42	52.50	9.06	<ld	0.63
46	<ld	21.63	0.14	0.08	<ld	1.62	5.43	0.13	0.37	53.28	<ld	0.47	0.74
47	<ld	15.86	<ld	<ld	<ld	<ld	4.02	0.09	0.53	54.33	<ld	<ld	0.25
48	<ld	15.94	<ld	0.08	<ld	0.54	7.74	0.37	0.54	43.32	<ld	0.49	<ld
49	<ld	18.15	<ld	0.07	0.27	1.23	5.47	0.12	0.53	54.83	11.27	<ld	0.40
50	<ld	34.54	<ld	<ld	<ld	1.21	19.35	0.19	<ld	90.48	17.34	<ld	<ld

Table 12.2 Continued

No.	Ag	Ca	Cr	Co	Cu	Fe	Mg	Mn	Mo	P	S	Zn	Al
51	<ld	15.19	<ld	<ld	<ld	0.27	6.13	0.08	0.37	56.14	<ld	<ld	<ld
52	0.15	34.71	0.15	0.08	<ld	<ld	18.98	0.60	0.43	83.97	<ld	<ld	0.66
53	<ld	13.73	<ld	<ld	<ld	0.55	4.74	0.16	0.46	62.35	12.92	<ld	<ld
54	<ld	13.69	<ld	<ld	<ld	0.49	4.61	0.15	<ld	50.83	<ld	<ld	0.36
55	<ld	25.85	<ld	<ld	<ld	<ld	14.74	0.27	0.50	118.6	11.50	<ld	0.46
56	<ld	13.54	<ld	<ld	<ld	0.37	3.78	0.10	<ld	56.45	<ld	<ld	0.25
57	<ld	35.10	<ld	<ld	<ld	0.38	3.72	0.22	<ld	50.09	<ld	<ld	0.25
58	<ld	17.54	<ld	0.07	<ld	0.62	9.92	0.82	0.44	51.43	<ld	0.38	0.29
59	<ld	12.86	<ld	<ld	<ld	1.17	4.13	0.13	<ld	41.86	<ld	0.46	0.79
60	<ld	15.95	<ld	<ld	<ld	<ld	4.54	0.11	0.32	43.54	<ld	<ld	0.40
61	<ld	13.67	0.11	<ld	<ld	0.68	6.06	0.12	<ld	55.39	9.49	0.45	0.75
62	<ld	9.08	<ld	<ld	<ld	0.86	4.38	0.16	<ld	44.17	6.90	<ld	0.26
63	<ld	26.91	<ld	<ld	<ld	1.77	9.07	0.84	<ld	43.48	16.96	1.11	0.93
64	<ld	24.47	<ld	<ld	<ld	1.30	8.39	0.63	<ld	46.60	7.89	<ld	<ld
65	<ld	57.96	<ld	<ld	<ld	<ld	36.28	1.37	0.53	91.85	<ld	2.00	1.00
66	<ld	12.24	<ld	0.08	<ld	0.88	4.17	0.14	0.34	48.17	13.69	<ld	<ld
67	<ld	13.74	<ld	0.03	<ld	0.82	5.32	0.23	0.11	37.63	11.77	<ld	0.35
68	<ld	16.00	<ld	<ld	<ld	0.32	5.26	0.17	0.44	50.03	1.60	<ld	0.20
69	<ld	27.33	<ld	<ld	<ld	1.35	6.94	0.30	0.12	63.17	17.75	<ld	<ld
70	<ld	28.66	<ld	<ld	0.06	0.97	8.87	<ld	0.10	61.98	17.58	0.91	0.16
71	0.08	6.93	0.05	<ld	0.04	<ld	2.55	0.09	0.26	57.72	5.24	<ld	0.05
72	<ld	21.13	0.10	<ld	0.04	<ld	2.08	0.06	0.12	37.83	8.36	<ld	0.14
73	<ld	61.77	<ld	0.04	0.24	0.97	19.08	0.54	0.16	77.71	30.12	0.81	0.63
74	<ld	12.26	0.06	<ld	0.04	<ld	1.99	0.14	0.14	34.78	8.00	<ld	<ld
75	<ld	24.51	<ld	0.03	0.16	0.67	6.49	0.18	0.15	70.86	<ld	0.56	0.30
76	<ld	90.12	0.24	<ld	0.10	1.03	18.71	5.99	<ld	75.11	28.81	0.58	<ld
77	<ld	17.27	<ld	0.03	0.06	0.50	5.50	0.18	0.19	66.27	16.79	0.27	0.23
78	<ld	27.53	0.09	0.04	0.07	0.71	9.11	0.52	<ld	73.21	21.38	<ld	0.23
79	<ld	14.05	0.20	<ld	<ld	2.13	4.05	0.30	0.14	47.88	<ld	<ld	1.02
80	<ld	80.13	<ld	0.04	0.20	2.63	63.13	<ld	<ld	223.4	37.62	1.48	1.44
81	<ld	30.97	<ld	<ld	0.10	4.24	8.66	0.44	0.13	86.81	21.93	<ld	0.59
82	<ld	12.27	0.11	0.09	0.09	<ld	4.10	0.47	0.24	70.23	13.72	<ld	0.21
83	<ld	46.26	0.14	<ld	0.06	0.64	9.71	1.35	<ld	66.27	16.53	<ld	0.30
84	<ld	27.05	<ld	<ld	0.04	0.62	5.35	<ld	<ld	71.72	9.50	<ld	<ld
85	<ld	13.64	0.25	<ld	0.05	0.81	4.95	0.28	<ld	61.35	18.66	0.16	0.21
86	<ld	70.40	<ld	<ld	<ld	0.75	4.90	0.30	<ld	59.48	26.05	0.23	<ld
87	<ld	10.11	<ld	0.04	0.05	<ld	3.13	0.10	0.20	54.50	13.16	<ld	0.16
88	<ld	11.40	<ld	<ld	0.26	0.36	4.42	0.11	<ld	46.24	<ld	<ld	<ld
89	<ld	8.62	0.36	<ld	0.06	<ld	2.99	0.09	<ld	67.95	12.08	0.29	0.17
90	<ld	7.08	0.09	<ld	0.05	<ld	2.14	0.13	<ld	43.79	8.46	<ld	0.18
91	<ld	21.93	<ld	<ld	0.30	0.74	6.06	0.20	0.21	46.49	13.68	<ld	<ld
92	<ld	13.31	<ld	<ld	<ld	0.69	2.98	<ld	<ld	32.12	6.74	0.41	<ld
93	<ld	9.32	<ld	<ld	0.06	0.25	3.38	0.10	<ld	56.09	<ld	<ld	0.23
94	<ld	7.85	0.13	<ld	0.06	0.57	4.08	0.12	<ld	49.15	10.53	0.23	0.36
95	<ld	36.22	<ld	<ld	0.06	0.41	7.48	0.19	<ld	65.94	11.03	<ld	<ld
96	0.13	13.26	0.09	<ld	0.06	<ld	2.51	2.86	0.12	56.63	6.41	0.55	<ld
97	<ld	18.05	<ld	<ld	<ld	<ld	5.58	0.17	0.07	44.68	12.60	<ld	<ld
98	<ld	16.85	<ld	<ld	<ld	0.52	5.66	0.39	<ld	53.61	12.78	0.89	<ld
99	0.17	26.14	<ld	<ld	0.14	0.50	7.16	2.94	0.25	78.63	13.86	<ld	0.28
100	<ld	6.49	<ld	<ld	0.07	<ld	1.66	0.12	<ld	94.23	<ld	0.55	0.17
101	<ld	16.85	<ld	<ld	0.04	<ld	3.65	0.12	0.14	57.19	9.49	<ld	0.09

Table 12.2 Continued

No.	Ag	Ca	Cr	Co	Cu	Fe	Mg	Mn	Mo	P	S	Zn	Al
102	<ld	14.13	<ld	0.25	0.06	<ld	3.23	0.29	0.72	99.54	11.46	<ld	0.15
103	<ld	13.31	<ld	0.21	0.05	<ld	2.83	0.34	0.75	98.33	<ld	<ld	0.26
104	<ld	8.01	<ld	<ld	0.06	<ld	1.59	0.16	<ld	89.06	<ld	<ld	<ld
105	<ld	13.59	<ld	0.06	0.05	<ld	3.81	0.24	0.16	55.44	<ld	<ld	0.10
106	0.54	10.52	0.51	<ld	0.06	<ld	2.53	0.14	0.68	101.9	<ld	0.44	0.11
107	<ld	15.65	<ld	<ld	0.05	<ld	3.22	0.36	0.23	63.84	<ld	<ld	0.10
108	<ld	28.23	0.11	<ld	0.07	<ld	6.96	1.22	<ld	67.79	21.86	<ld	<ld
109	<ld	12.75	<ld	0.03	0.06	<ld	4.78	<ld	<ld	54.44	15.65	0.36	0.19
110	<ld	8.70	<ld	<ld	0.05	<ld	1.91	0.11	<ld	94.48	<ld	<ld	0.05
111	<ld	9.85	<ld	<ld	0.03	<ld	3.03	0.21	0.34	64.98	<ld	<ld	<ld
112	<ld	19.68	<ld	<ld	0.06	<ld	6.83	0.57	<ld	73.84	15.39	<ld	0.07
113	<ld	22.41	<ld	<ld	0.08	0.43	4.95	0.74	0.61	105.2	12.22	0.49	0.21
114	<ld	64.54	<ld	<ld	0.12	<ld	15.00	2.93	<ld	148.7	<ld	0.72	0.25
115	<ld	107.8	<ld	0.20	0.11	4.63	18.00	3.08	<ld	154.3	23.83	0.66	0.37
116	<ld	9.79	<ld	<ld	0.04	0.43	2.69	<ld	<ld	57.96	<ld	<ld	<ld
117	<ld	28.33	<ld	<ld	0.08	0.63	6.44	1.97	<ld	116.1	<ld	<ld	0.51
118	<ld	16.88	<ld	<ld	0.06	0.73	3.26	0.32	<ld	101	<ld	<ld	0.40
119	<ld	11.61	<ld	0.21	0.07	0.40	2.73	0.18	0.65	100.4	10.93	0.56	0.16
120	<ld	23.94	<ld	<ld	0.10	0.65	5.16	0.21	<ld	116.2	<ld	<ld	<ld
121	<ld	27.41	<ld	<ld	0.07	0.72	4.73	0.78	<ld	69.80	6.56	<ld	0.42
122	<ld	37.52	<ld	0.08	0.31	0.61	14.64	0.35	0.16	110.5	23.59	<ld	<ld
123	2.16	7.96	0.50	<ld	0.06	<ld	2.48	0.09	<ld	96.83	8.10	<ld	0.15
124	<ld	21.28	<ld	0.19	0.08	0.55	4.25	0.31	0.49	104.5	<ld	<ld	0.30
125	<ld	21.92	<ld	<ld	0.08	<ld	4.61	0.20	<ld	118.4	<ld	0.69	<ld
126	<ld	2.98	<ld	<ld	0.05	<ld	10.14	0.18	0.62	100.4	11.61	0.58	0.15
127	0.57	14.55	0.52	<ld	0.07	<ld	3.46	0.24	0.67	96.61	9.89	0.39	0.11
128	<ld	9.57	<ld	<ld	0.05	3.48	2.96	1.23	<ld	55.16	<ld	<ld	0.09
129	<ld	48.15	<ld	0.09	0.09	5.24	12.65	0.87	<ld	96.70	25.88	0.66	0.25
130	<ld	28.95	<ld	<ld	0.06	<ld	4.17	0.19	0.64	100.7	<ld	<ld	0.17
131	<ld	23.47	<ld	<ld	0.08	<ld	7.01	0.37	<ld	129.1	<ld	<ld	0.30
132	<ld	52.40	<ld	<ld	<ld	<ld	12.14	0.34	0.25	62.24	22.73	<ld	<ld
133	0.61	33.96	0.30	0.07	<ld	1.15	10.26	0.19	<ld	61.56	20.13	0.43	<ld
134	<ld	31.58	<ld	<ld	0.34	1.02	8.88	0.35	<ld	56.20	20.48	<ld	0.38
135	<ld	26.71	<ld	<ld	0.19	0.64	8.02	0.33	<ld	50.20	19.31	0.60	0.65
136	<ld	14.52	<ld	<ld	<ld	0.59	2.95	0.25	<ld	33.45	10.51	<ld	0.20
137	<ld	38.04	<ld	0.08	<ld	0.92	13.04	1.03	0.10	72.40	27.99	0.73	<ld
138	<ld	43.32	<ld	<ld	<ld	1.37	5.98	3.28	0.24	49.06	17.97	1.74	<ld
139	<ld	37.65	<ld	<ld	0.30	1.39	9.61	1.55	<ld	49.99	21.99	0.76	0.78
140	<ld	30.04	<ld	0.06	0.33	0.69	12.28	0.79	0.26	59.34	22.82	<ld	<ld
141	<ld	6.87	<ld	0.03	<ld	<ld	3.25	0.25	0.55	47.04	<ld	0.11	<ld
142	<ld	29.86	<ld	<ld	<ld	0.24	9.32	0.45	0.49	53.20	<ld	0.04	<ld
143	<ld	11.66	<ld	<ld	0.06	3.19	2.59	0.24	<ld	97.36	4.95	<ld	0.11
144	<ld	11.32	<ld	<ld	0.04	3.06	3.69	0.27	<ld	64.96	<ld	<ld	<ld
145	<ld	8.35	<ld	<ld	0.05	0.16	2.31	0.22	0.17	69.35	<ld	<ld	0.09
146	1.03	8.77	<ld	<ld	0.45	1.02	1.43	0.64	0.36	93.40	6.37	0.92	<ld
147	0.52	21.34	0.47	<ld	0.05	0.64	3.70	0.20	<ld	101.8	<ld	<ld	0.17
148	<ld	9.83	<ld	<ld	0.06	<ld	1.98	<ld	<ld	98.58	<ld	<ld	<ld
149	<ld	14.24	<ld	<ld	<ld	0.38	5.01	0.43	0.10	40.78	<ld	<ld	0.23
150	<ld	27.73	<ld	<ld	<ld	0.69	9.16	0.44	<ld	59.13	22.75	0.44	0.45

Note
*<ld = Less than the limit of detection. For the values see text.

with concentrations of 0.06, 0.06, 0.04, 0.24, and 0.02 mg/kg, respectively. Finally, nickel (Ni), mercury (Hg), cadmium (Cd), and lead (Pb), which are particularly hazardous for biota and are indicators of industrial pollution, were not detected in the 150 honeys (Table 12.1). This is particularly surprising because about 50 percent of the samples were collected in hives located in polluted areas. Even if we can assume that some hives were misclassified by the beekeepers, the descriptions provided for most of them clearly show that numerous hives were undoubtedly located near sources of industrial pollution (e.g. highways, petroleum industries). In addition, the detectable presence of some elements such as Ag or Cr clearly reveals that some honey samples were collected in polluted areas. Information on the level of contamination of French honeys by heavy metals and related pollutants is scarce. Recently, Fléché and co-workers [7] revealed that, between 1986 and 1996, among the routine analyses performed by the CNEVA (Centre National d'Etudes Vétérinaires et Alimentaires – National Center for Veterinary and Alimentary Studies) on honeys of various origins, only 97 were focused on the detection of heavy metals, while 615 analyses were carried out for detecting pesticides and 341 were performed to find the level of contamination of honeys in antibiotics. In addition, among these 97 analyses, while the presence of Pb was investigated systematically (with 10.3 percent positive response (p.r.)) and that of Cd was searched in 83 samples (1.2 percent p.r.), the contamination in Hg was only investigated in four honey samples (0 percent p.r.). Fléché *et al.* [7] also emphasized that in the framework of their annual control of the quality of honeys, in 1994, the CNEVA analyzed 122 French honeys and 28 foreign honeys for their concentrations of Pb and Cd. While Pb was not detected in the former group, 43 percent of the latter were contaminated by detectable concentrations of this element with a mean concentration of 3.8 ppm. Conversely, Cd was not detected in the foreign honeys while 3 percent of the French honeys were contaminated by detectable amounts of Cd with a mean concentration of 0.07 ppm [7]. However, in these analytical results, the type of honey was not given even though it is well known that this parameter widely influences the levels of contamination found in samples gathered in the same geographical area. Thus, for example, in a recent study, Barisic and co-workers [24] showed that the concentrations of Pb in meadow honey, mixed meadow and honeydew honey, and honeydew honey from Gorski Kotar (Croatia) were 0.80 ± 0.64, 1.08 ± 0.59, and 3.38 ± 1.55 ppm, respectively.

In order to perform a rational analysis of Table 12.2 and provide a typology of the acacia honeys based on their detectable concentrations in metallic and nonmetallic elements, different linear multivariate analyses were performed on this 13×150 data matrix.

Multivariate analysis of the honey samples

Correspondence factor analysis

Background

Among the different linear multivariate methods that can be used to analyze Table 12.2, correspondence factor analysis (CFA) was selected because its χ^2 metrics permits work on data profiles and the natural biplot representation of the variables and objects which greatly facilitates the interpretation of the graphical displays [26]. In addition, CFA has been used successfully on similar data matrices for rationalizing (eco)toxicological information [27–30].

Analysis of the factorial map F_1F_2

CFA allows the dimensionality of the 13×150 data matrix (Table 12.2) to be significantly reduced since the six first axes (i.e. F_1 to F_6) account for about 93 percent of the total inertia of the system.

The factorial map F_1F_2 (Figure 12.2), which accounts for most of the variance of the system (i.e. 62.23 percent), clearly reveals an opposition between the presence or the absence of detectable concentrations of sulfur (S) in the samples. Thus, broadly speaking, the honey samples belonging to the compact cluster of points located on the right of Figure 12.2B do not have sulfur. Conversely, points located in the top left of Figure 12.2B deal with honey samples containing significant concentrations of sulfur. It is clear that CFA can be used to perform a more precise analysis of the points displayed on the factorial map. Thus, for example, sample number 41 does not contain a detectable concentration of sulfur but, in addition, it presents the highest concentration in zinc (i.e. 5.96 ppm). This explains its location as an outlier in the lower part of Figure 12.2B. Conversely, samples 85 and 86 contain fairly similar concentrations of sulfur but the former is also contaminated by Cr, Cu, and Al while the latter does not have detectable concentrations of these elements. In addition, sample 86 contains more Ca than sample number 85. These chemical differences explain their different locations on Figure 12.2B.

The strong opposition between the honeys with or "without" sulfur clearly reveals that this element has to be viewed as a contaminant. It is difficult to explain the origin of this contamination. It is assumed that environmental pollutions mainly explain the fairly high concentrations found in the honeys but direct human contamination cannot be excluded for some samples. Thus, for example, honey sample number 20 with 35.90 mg/kg of sulfur was collected near a highway, as were samples number 27 (S = 60.11 mg/kg), number 85 (S = 18.66 mg/kg), number 86 (S = 26.05 mg/kg), and others. In the same way, the honey sample number

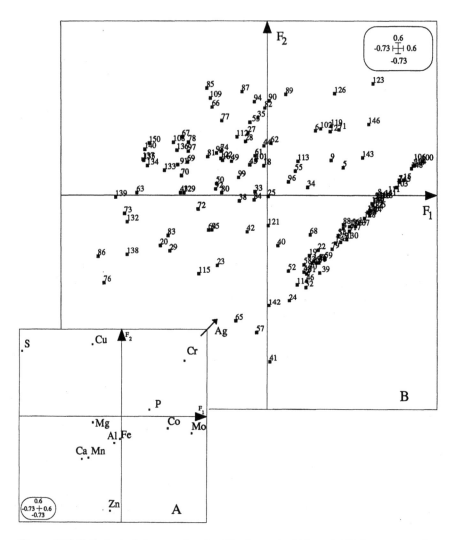

Figure 12.2 F_1F_2 factorial maps for the 13 elements (A) and 150 honey samples (B).

81 with 21.93 mg/kg of sulfur was collected in a hive located near a textile factory. Other honeys gathered near various industrial sites also contain substantial amounts of sulfur [e.g. 73, 132]. However, surprisingly, the highest concentration of sulfur was found in sample number 28 which was collected in a mountainous area apparently exempt from industrial pollution. Because this sample was provided jointly with sample number 27, collected in a polluted area, we cannot exclude human contamination introduced by the beekeeper, especially if we consider the very high or

fairly high concentrations found for most of the other elements in these two samples (Table 12.2).

Another trend which can be underlined in Figure 12.2B is the gradient determined by Cr and Ag (Figure 12.2A). Note that on Figure 12.2A, the true location of Ag was not indicated in order to have a scale yielding an optimal graphical display of the variables and objects. The joint reading of Figures 12.2A and 12.2B shows that sample number 123, which is located at the top right of Figure 12.2B, is the most contaminated in silver with 2.16 mg/kg and also contains a very high concentration of chromium (i.e. 0.50 mg/kg). This is not surprising because this sample was collected in a hive located within an urban area (more specifically in the center of a rural city of about 6000 inhabitants). A high concentration of Ag (i.e. 1.03 mg/kg) was also found in sample number 146 located in the upper right of Figure 12.2B. This sample was collected near a highway. Sample number 127 located in the vicinity of sample number 146 also presents a high concentration of Ag (i.e. 0.57 mg/kg) and is the most contaminated in Cr (i.e. 0.52 mg/kg). However, this sample was labeled by the beekeeper as being collected in a nonpolluted area. It is surprising because it is also contaminated by Cu, Mo, Zn, Al, and so on (Table 12.2).

P, Ca, and Mg are elements found in all the 150 honey samples but with various concentrations. Mn is also detected in most of the samples (Tables 12.1 and 12.2). Consequently, it is difficult to determine formal trends in relation to environmental pollutions for these elements. However, it is interesting to note that an inverse relationship can exist between the concentrations found for P and those recorded for Ca and Mn. Thus, because sample number 126 shows the lowest concentration of Ca (i.e. 2.98 mg/kg), a very low concentration of Mn (i.e. 0.18 mg/kg), and a fairly high concentration of P (i.e. 100.4 mg/kg), it is located in the upper right of Figure 12.2B. This is in accordance with the location of these three variables on Figure 12.2A. This sample, which was collected at a distance of 10 km from a city, also contains 0.05 mg/kg of Cu, 0.62 mg/kg of Mo, 11.61 mg/kg of S, and only 0.58 mg/kg of Zn. Undoubtedly, these concentrations also influence its location on Figure 12.2B.

Analysis of the factorial map F_1F_3

The factorial map F_1F_3 (Figure 12.3), which accounts for 56.86 percent of the total inertia of the system, emphasizes the fact that the presence of cobalt (Co) in the honey samples is correlated with that of molybdenum (Mo). Of the 46 samples containing detectable concentrations of Co (Table 12.1), only nine do not contain significant concentrations of Mo (Table 12.2). Consequently, these two elements form a cluster in the bottom right of Figure 12.3A. Note that the same situation occurs in Figure 12.2A but Figure 12.2B is more difficult to read than Figure 12.3B as regards honey samples containing Co and/or Mo. Conversely, it is easy

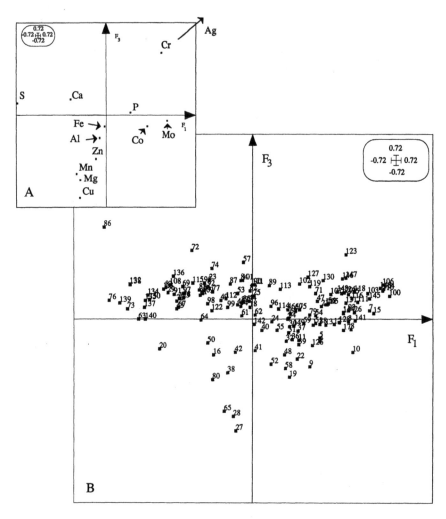

Figure 12.3 F₁F₃ factorial maps for the 13 elements (A) and 150 honey samples (B).

to see, for example, that sample number 10 which is located in the bottom right of Figure 12.3B presents a fairly high concentration of Co (i.e. 0.10 mg/kg) and the highest concentration of Mo (i.e. 0.81 mg/kg). Undoubtedly, Co and Mo are found mainly in the acacia honeys collected in polluted areas. However, exceptions can be found.

Figure 12.3B also highlights samples with specific contaminants. Thus, for example, samples 27 and 28 contain the highest concentrations of Cu. As indicated previously, sample number 123 is the most contaminated by Ag. Sample number 80 contains the highest concentration of Al (i.e. 1.44 mg/kg). Sample number 86 contains fairly high concentrations of Ca and S while Ag, Cr, Co, Cu, Mo, and Al have not been detected and the other elements are present in limited amounts. All these details can be readily deduced from Figures 12.3A and 12.3B.

Analysis of the factorial map F_2F_3

The factorial map F_2F_3 (Figure 12.4), which only accounts for 32.69 percent of the total inertia of the system, confirms the general trends stressed previously with the factorial maps F_1F_2 and F_1F_3. In addition, it

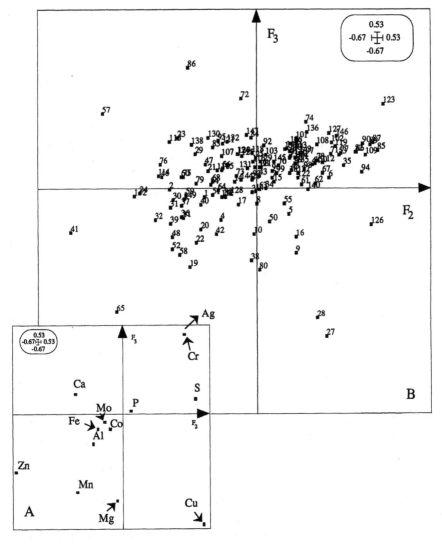

Figure 12.4 F_2F_3 factorial maps for the 13 elements (A) and 150 honey samples (B).

allows some chemical characteristics of the honey samples to be refined. Thus, for example, sample number 41, which was collected in a hive located near a paper pulp factory, presents the highest concentration of Zn (i.e. 5.96 mg/kg). This sample, located on the bottom left of Figure 12.4B, also contains a fairly high concentration of Al (i.e. 0.98 mg/kg). In the same way, sample number 65, which was collected near a highway, is also significantly contaminated by these two elements (Table 12.2). However, in neither of these two samples sulfur has been detected (Figure 12.2B). Sample number 19 contains substantial concentrations of Cr, Cu, and Co and a very high concentration of Mo. Similarly, sample number 52 contains measurable concentrations of Ag, Cr, Co, Mo, and Al. Sample number 32 contains no detectable concentrations of Ag and Al (Table 12.2) but is contaminated by Cr, Co, Mo, and Zn. Conversely, honey samples located in the upper left of Figure 12.4B generally do not contain these elements or are only contaminated by some of them. Thus, for example, sample number 57 only contains significant amounts of Ca and P, all the other elements are present in small quantities. The particular location of Al on Figure 12.4A, but also on Figures 12.2A and 12.3A, has to be related to the ubiquity of this pollutant. In the same way, Fe also presents a rather central location on Figures 12.2A, 12.3A, and 12.4A.

Principal component analysis and hierarchical cluster analysis

Because, as emphasized previously, Table 12.2 could be analyzed by other multivariate techniques, principal component analysis (PCA) [31] was also used to reduce the dimensionality of this data matrix. The PCA results (not shown) are broadly in accordance with those obtained from CFA. A hierarchical cluster analysis (HCA) was also carried out on Table 12.2. An aggregative procedure using a χ^2 distance and an average linkage algorithm were used [32, 33]. The results obtained with this type of multivariate method are difficult to compare directly with those produced by CFA or PCA. With PCA or CFA, the different variables and objects are explained on different factors, consequently, to draw conclusions, it is always necessary to consider different factorial maps accounting for different parts of the information. Conversely, with HCA, all the information of the data matrix is displayed through two dendrograms: one for the variables and another for the objects. Therefore, the comparison of the results obtained with a CFA and an HCA is not straightforward. Despite this point, on the dendrogram of the variables obtained from the HCA of Table 12.2, it has been possible to confirm the atypical position of Ag and the relative independence of the other elements except for Ca, P, and Mg and to a lesser extent S which form a cluster (figure not given). In the same way, the dendrogram of the objects clearly shows the existence of some important outliers [e.g. 41, 123] in contrast with samples organized in more or less strong clusters (figure not given).

Discussion

It is difficult to compare our results with those published in the literature because they generally deal with different types of honeys. In addition, other analytical methods and protocols have generally been used to quantify the concentrations of metallic and nonmetallic elements in the samples. Kump and colleagues [21], comparing the performances of radioisotope X-ray fluorescence spectrometry, total reflection X-ray fluorescence spectrometry, atomic absorption spectrometry, and inductively coupled plasma atomic emission spectrometry as methods for detecting contamination in metallic and nonmetallic elements in different types of honeys, have clearly addressed these problems. They have shown that in the acacia honey, the concentrations of most of the trace elements were lower than those generally found in the other honey varieties tested. Nevertheless, our results clearly reveal an absence of significant contamination of the French acacia honey by Ni, Cd, Hg, and Pb. In fact, because of the large number of samples collected in various contaminated sites located in different geographical regions (Figure 12.1), we can assume that the French acacia honey is not significantly contaminated by these elements. For comparison purposes, note that Rowarth [18] showed, by means of atomic absorption spectrometry, that the concentrations of Pb in 59 samples of New Zealand honey taken from several enterprises in three localities in the North Island, and from different stages, ranged from 0.009 to 1.131 ppm. With the same analytical technique, Cesco *et al.* [22] measured 1.84 ± 0.48 ppm of Pb in honeys collected in a polluted area located in the city of Portogruaro (Venice, Italy). The highest concentrations of Pb were found in propolis (13.7 ± 6.14 ppm) and royal jelly (13.1 ± 0.43 ppm). Cesco *et al.* [22] also found concentrations of Cd ranging from about 1 to 3 ppm, depending on the matrix analyzed. Barisic *et al.* [24] showed that the concentrations of Pb in meadow honey, mixed meadow and honeydew honey, and honeydew honey from Gorski Kotar (Croatia) ranged from 0.19 to 2.77, 0.22 to 2.62, and 0.84 to 6.78 ppm, respectively. In addition to their (eco)toxicological usefulness, these results are also very interesting from a methodological point of view. They clearly illustrate the difficulty in comparing honey samples due to intra (within) and inter (between) variability. Ni was found in lower amounts, with the highest concentrations measured for the meadow honey, mixed meadow and honeydew honey, and honeydew honey being 0.188, 0.211, and 0.472 ppm, respectively [24].

Conversely, some of the 150 acacia honeys analyzed are highly contaminated by Ag, Cr, Zn and/or other elements which are undoubtedly linked to human pollutions. However, the true source of the contamination is often difficult to determine. While samples collected in contaminated sites generally present the highest concentrations in these elements, exceptions can be found. Thus, honeys originating from apparently unpolluted sites can present a fairly high level of contamination for one or more of these elements. It is obvious that the role of the wind in transportation cannot be excluded for explaining the presence of contaminants far from their emission source. However, direct contamination induced by bee-keepers cannot be excluded. Thus, for example, as the pH of the acacia honey is equal to 3.9 [34], the contact of a sample with a galvanized surface will induce an elevated level of zinc in this sample. Even if the beekeepers were asked to provide samples collected directly in the honeycomb cells of the hives with the appropriate equipment to avoid contamination, we are aware that our protocol has not always been followed. Thus, changes have been already noted for three beekeepers (129/130/131, 134/135, 143/144). Sample number 129 was collected in a hive located near a dump while samples 130 and 131 originated from uncontaminated sites. While these samples have not been extracted directly from the honeycomb cells but after the honey harvesting, the different concentrations found in these samples (Table 12.2) are logical. Thus, we can assume that no bias has been introduced. This also seems to be the case for samples 134/135 and 143/144 which were apparently collected in nonpolluted areas. Because the analytical results of these seven samples were logical, they were kept to perform the multivariate analyses. However, we cannot certify that for all the other honey samples, the sampling protocol has been scrupulously respected by the beekeepers and hence, direct contamination during processing and/or storage cannot be excluded.

More generally, Tables 12.1 and 12.2 reveal a high variability in the concentrations found for most of the elements. Consequently, it is not surprising to see the scattering of the points (i.e. samples) on Figures 12.2 to 12.4. The variability in the concentrations of metallic and nonmetallic elements in honeys has been reported in numerous articles. However, generally these papers deal with honeys of different biological origins and/or collected according to various methods and/or not related to possible environmental contaminations. Thus, Bengsch [35] generally emphasized the large variations in the concentrations of K, P, Ca, S, Mg, Mn, Si, B, Fe, Zn, Cu, and Ba measured in honey samples by ICP-AES. However, while 14 different biological types of honeys were analyzed, acacia honey was excluded from his study. In addition, no relationships with direct or indirect human contaminations were considered. The same criticism can be made of the work of Lasceve and Gonnet [36] dealing with the comparison of light (*Robinia pseudoacacia, Lavandula*) and dark (*Abies pectinata, Calluna vulgaris*) honeys for their mineral composition measured by activation analysis with thermic neutrons. While the geographical origin of the samples was provided, it is obvious that on the basis of only 14 French samples analyzed (i.e. 4 *Rp*, 3 *L*, 4 *Ap*, 3 *Cv*), and without any indication of the levels of contamination found in the media in which these honeys were collected, no formal conclusions can be made. Tong *et al.* [14], from the analysis of 19 honey samples of various biological origins and collected with different protocols near zinc mines, industrial areas, or highways, also showed a high variability in the concentrations found for most of the metallic and nonmetallic elements. Because the concentrations of the elements were related to the sources of sampling, it is interesting to provide the ranges found by these authors for the 18 elements under study. These concentration ranges (ppm fresh weight) were the following: Ag (0.002–0.094), Ca (3–540), Cr (0.003–2.1), Co (0.002–0.50), Cu (0.13–3.3), Fe (0.41–40), Mg (2–370), Mn (0.18–12), Mo (0.003–0.10), P (5–500), S (0.9–390), Zn (0.18–5.6), Al (0.09–18), Li (not analyzed), Ni (0.011–0.83), Hg (<0.1 in all samples), Cd (<0.001–0.028), and Pb (0.03–0.28). While some of their samples were collected in the vicinity of highly polluted areas, it is surprising that the highest concentration of Ag was only 0.094 ppm if we consider that the highest concentration found for Cr was 2.1 ppm. Indeed, our study has clearly shown the relationship between these two elements. Conversely, even if Al was detected in about two-thirds of our samples, it is interesting to note that the highest concentration (i.e. 1.44 ppm, sample number 80 on Table 12.2) is about 12 times lower than the highest concentration found for this element by Tong *et al.* [14] (i.e. 18 ppm, sample collected near the New York State Thruway).

Concluding remarks

From 150 samples collected with comparable protocols, in various identified polluted and nonpolluted environments, all being located in France, it has been possible to show that acacia honey was not a good bioindicator of the environmental pollution by heavy metals and related elements. Indeed, it is true that generally the most contaminated honeys correspond to samples collected in hives located within polluted areas. However, contaminations can also be found in apparently uncontaminated areas. Even if some of these contaminations may be explained by acidic reactions of the honey with metallic surfaces during their processing, because some beekeepers were not able to respect our sampling protocol, the true source of contamination of the collected samples often remains difficult to determine.

More generally, our analytical results and multivariate analyses reveal that acacia honeys present a very high variability in their concentrations in metallic and nonmetallic elements. This makes it impossible to propose an average profile for characterizing the French acacia honeys from their elemental analysis. It would be worthwhile investigating whether similar conclusions can be drawn from the analysis of other types of French honeys.

Acknowledgment

This work has been funded partly by the EU Program supporting French beekeeping (Règlement du Conseil no. 1221/97, portant règles d'application pour des actions visant à l'amélioration de la production de miels).

References

1 Schnoor, J.L. (1992). *Fate of Pesticides and Chemicals in the Environment*. John Wiley, New York, p. 436.

2 Tardiff, R.G. and Goldstein, B.D. (1991). *Methods for Assessing Exposure of Human and Non-human Biota*. John Wiley, New York, p. 417.

3 Newman, P.J. (1992). Surface water quality indicators. In: *Environmental Impact Assessment* (Colombo, A.G., Ed.). Kluwer Academic Publishers, Dordrecht, The Netherlands, pp. 211–233.

4 Colombo, A.G. and Premazzi, G. (1990). *Workshop on Indicators and Indices for Environmental Impact Assessment and Risk Analysis*. Luxembourg, Office for Official Publications of the European Communities, EUR 13060 EN, p. 333.

5 Asta, J. (1995). Les lichens bioindicateurs de pollution atmosphérique: Exemples de cas étudiés en milieux perturbés dans les Alpes. In: *L'Ecotoxicologie du Milieu Aérien*. SEFA, pp. 191–196.

6 Pham-Delègue, M.H. (1998). *Abeilles*. La Martinière, Paris, p. 47.

7 Fléché, C., Clément, M.C., Zeggane, S. and Faucon, J.P. (1997). Contamination des produits de la ruche et risques pour la santé humaine: Situation en France. *Rev. Sci. Tech. Off. Int. Epiz.* **16**, 609–619.

8 Voget, M. (1989). Bees and beeproducts as biological indicators of environmental contamination: An economical alternative way of monitoring pollutants. *Toxicol. Environ. Chem.* **20–21**, 199–202.

9 Accorti, M., Guarcini, R. and Persano Oddo, L. (1991). L'ape: Indicatore biologico e insetto test. *Redia* **74**, 1–15.

10 Juberg, D.R., Kleiman, C.F. and Kwon, S.C. (1997). Position paper of the American council on science and health: Lead and human health. *Ecotoxicol. Environ. Saf.* **38**, 162–180.

11 Frederick, P.C., Spalding, M.G., Sepulveda, M.S., Williams, G.E., Nico, L. and Robins, R. (1999) Exposure of great egret (*Ardea albus*) nestlings to mercury through diet in the everglades ecosystem. *Environ. Toxicol. Chem.* **18**, 1940–1947.

12 Cobb, G.P., Sands, K., Waters, M., Wixson, B.G. and Dorward-King, E. (2000). Accumulation of heavy metals by vegetables grown in mine wastes. *Environ. Toxicol. Chem.* **19**, 600–607.

13 Outridge, P.M., Wagemann, R. and McNeely, R. (2000). Teeth as biomonitors of soft tissue mercury concentrations in beluga, *Delphinapterus leucas*. *Environ. Toxicol. Chem.* **19**, 1517–1522.

14 Tong, S.C.C., Morse, R.A., Bache, C.A. and Lisk, D.J. (1975). Elemental analysis of honey as an indicator of pollution. Forty-seven elements in honeys produced near highway, industrial, and mining areas. *Arch. Environ. Health* **30**, 329–332.

15 Bromenshenk, J.J., Carlson, S.R., Simpson, J.C. and Thomas, J.M. (1985). Pollution monitoring of Puget Sound with honey bees. *Science* **227**, 632–634.

16 Jones, K.C. (1987). Honey as an indicator of heavy metal contamination. *Water Air Soil Pollut.* **33**, 179–190.

17 Accorti, M., Guarcini, R., Modi, G. and Persano Oddo, L. (1990). Urban pollution and honey bees. *Apicoltura* **6**, 43–55.

18 Rowarth, J.S. (1990). Lead concentration in some New Zealand honeys. *J. Apic. Res.* **29**, 177–180.

19 Bromenshenk, J.J., Gudatis, J.L., Carlson, S.R., Thomas, J.M. and Simmons, M.A. (1991). Population dynamics of honey bee nucleus colonies exposed to industrial pollutants. *Apidologie* **22**, 359–369.

20 Raes, H., Cornelis, R. and Rzeznik, U. (1992). Distribution, accumulation and depuration of administered lead in adult honeybees. *Sci. Total Environ.* **113**, 269–279.

21 Kump, P., Nečemer, M. and Šnajder, J. (1996). Determination of trace elements in bee honey, pollen and tissue by total reflection and radioisotope X-ray fluorescence spectrometry. *Spectrochim. Acta B* **51**, 499–507.

22 Cesco, S., Barbattini, R. and Agabiti, M.F. (1994). Honey bees and bee products as possible indicators of cadmium and lead environmental pollution: An experience of biological monitoring in Portogruaro city (Venice, Italy). *Apicoltura* **9**, 103–118.

23 Leita, L., Muhlbachova, G., Cesco, S., Barbattini, R. and Mondini, C. (1996). Investigation on the use of honey bees and honey bee products to assess heavy metals contamination. *Environ. Monit. Ass.* **43**, 1–9.

24 Barišić, D., Vertačnik, A., Bromenshenk, J.J., Kezić, N., Lulić, S., Hus, M., Kraljević, P., Simpraga, M. and Seletković, Z. (1999). Radionuclides and selected elements in soil and honey from Gorski Kotar, Croatia. *Apidologie* **30**, 277–287.

25 ADE-4: http://pbil.univ-lyon1.fr/ADE-4.

26 Devillers, J. and Karcher, W. (1990). Correspondence factor analysis as a tool in environmental SAR and QSAR studies. In: *Practical Applications of Quantitative Structure–Activity Relationships (QSAR) in Environmental Chemistry and Toxicology* (Karcher, W. and Devillers, J., Eds). Kluwer Academic Publishers, Dordrecht, The Netherlands, pp. 181–195.

27 Devillers, J., Elmouaffek, A., Zakarya, D. and Chastrette, M. (1988). Comparison of ecotoxicological data by means of an approach combining cluster and correspondence factor analyses. *Chemosphere* **17**, 633–646.

28 Fortin, V., Doré, J.C., Poirier, F., Verhille, A.M., Carbonnier-Jarreau, M.C., Viel, C. and Cerceau-Larrival, M.T. (1995). Les profils de composition en oligoéléments des pollens, marqueurs potentiels en cladistique végétale. *Grana* **34**, 421–433.

29 Zakrgynska-Fontaine, V., Doré, J.C., Ojasoo, T., Poirier-Duchêne, F. and Viel, C. (1998). Study of the age and sex dependence of trace elements in hair by correspondence analysis. *Biol. Trace Element Res.* **61**, 151–168.

30 Siobud-Dorocant, E., Doré, J.C., Michelot, D., Poirier, F. and Viel, C. (1999). Multivariate analysis of metal concentration profiles in mushrooms. *SAR QSAR Environ. Res.* **10**, 315–370.

31 Thioulouse, J., Devillers, J., Chessel, D. and Auda, Y. (1991). Graphical techniques for multidimensional data analysis. In: *Applied Multivariate Analysis in SAR and Environmental Studies* (Devillers, J. and Karcher, W., Eds). Kluwer Academic Publishers, Dordrecht, The Netherlands, pp. 153–205.

32 Roux, M. (1991). Basic procedures in hierarchical cluster analysis. In: *Applied Multivariate Analysis in SAR and Environmental Studies* (Devillers, J. and Karcher, W., Eds). Kluwer Academic Publishers, Dordrecht, The Netherlands, pp. 115–135.

33 Roux, M. (1991). Interpretation of hierarchical clustering. In: *Applied Multivariate Analysis in SAR and Environmental Studies* (Devillers, J. and Karcher,

W., Eds). Kluwer Academic Publishers, Dordrecht, The Netherlands, pp. 137–152.

34 Persano Oddo, L., Sabatini, A.G., Accorti, M., Colombo, R., Marcazzan, G.L., Piana, M.L., Piaza, M.G. and Pulcini, P. (2000). *I Mieli Uniflorali Italiani Nuove Schede di Caratterizzazione.* Ministero Delle Politiche Agricole e Forestali, p. 105.

35 Bengsch, E. (1992). Connaissance du miel. Des oligo-éléments pour la santé. *Rev. Fr. Apic.* **521**, 383–386.

36 Lasceve, G. and Gonnet, M. (1974). Analyse par radioactivation du contenu minéral d'un miel. Possibilité de préciser son origine géographique. *Apidologie* **5**, 201–223.

13 The role of insect-resistant transgenic crops in agriculture

L. Jouanin and A.M.R. Gatehouse

Summary

Phytophagous insects are responsible for major losses in crops. For the past five decades pest control has been accomplished largely by the use of chemical pesticides, although some success has also been achieved towards producing plants with enhanced levels of endogenous resistance using conventional plant-breeding (i.e. host-plant resistance) and *in vitro* techniques. Recent technologies such as plant genetic engineering provide breeders with the opportunity for introducing resistance genes from foreign species into crop plants.

Different approaches have been considered to obtain such plants, through the expression of entomotoxic proteins. The main strategy to date has been based on the expression of endotoxins (Cry) originating from the soil bacterium *Bacillus thuringiensis* (*Bt*), with the commercialization of such crops in the USA since 1995. However, in order to enlarge the spectra of activity against insects and to co-express different toxins in transgenic crops, screenings for new entomotoxic proteins of plant, bacterial, and insect origin have become necessary and some genes encoding such toxins have already been introduced into crops and tested against selected insect pests. The state of the art of these different strategies is considered in this chapter.

Introduction

Pest control is accomplished largely by the use of chemical pesticides; however, losses in the major crops remain important [1]. In addition, major problems related to the use of these products have been reported, the most important being detrimental impacts on the environment, such as pollution of land and water tables, toxicity towards nontarget organisms, and accumulation in food chains. Thus, it is necessary to develop more environmentally benign methods of crop protection. The use of other types of pest control measures such as breeding for resistant varieties, modified agricultural practices, biological control, and biotechnology products must be developed. In this context, transgenic plants represent a very promising technology. The first transgenic plants were obtained in 1983 [2] and reports of the first applications to insect resistance were published in 1987 [3–6]. Many field trials have been performed in different countries during the following years, and in 1995 *B. thuringiensis* (*Bt*)-potatoes became the first *Bt*-expressing crop to be commercialized, soon to be followed by the commercialization and cultivation in 1996 of lepidopteran-insect-resistant cotton in the USA [7].

The expression of an insecticidal protein in plants presents many advantages over the exogenous application of chemicals. The "toxin," confined in the plant, is active at the early stages of insect attack and thus further reduces the level of damage. In addition, the "toxin" is only likely to have a direct effect on phytophagous insects feeding on the plant, although it may have indirect effects on insects which predate/parasitize these pest species. The expressed insecticidal gene product can be effective against insects feeding inside the plant (borers) as well as protecting parts of the plant which are difficult to treat with conventional pesticides (roots). The culture costs are reduced (but the seeds are more expensive) and the environment is more protected. Before introduction and expression in a transgenic plant, the gene(s) encoding the insecticidal protein must be identified. Since the insect gut is the prime target for the majority of insect resistance genes at present being utilized or developed, in order to confer the resistance trait, the "toxin" must be active after ingestion. This consideration has, up until now, excluded the use of neurotoxins. Insecticidal proteins can be of diverse origins and the most well known are derived from bacteria or plants. While the expression of endotoxins originating from the bacterium *B. thuringiensis* has been the most successful strategy for obtaining insect-resistant plants, many other strategies are also being developed; the different classes of insect resistance genes which have been expressed in transgenic crops are summarized in Table 13.1. The aim of this chapter is to summarize major studies carried out to date, and to discuss the potential problems posed by the use of this new technology. The reader is also referred to other recent reviews [7–9]. This chapter provides an introduction to two further chapters presented in this book (Chapters 14 and 15) which discuss, in detail, work carried out to evaluate the risks of entomotoxins expressed in transgenic plants on honey bees.

Entomotoxins introduced into plants by recombinant DNA technology

Bacillus thuringiensis δ-endotoxins

B. thuringiensis is a gram-positive bacterium that synthesizes insecticidal crystalline inclusions during sporulation. The crystalline structure of the inclusion is made up of protoxin subunits called δ-endotoxins. Most *B.*

Table 13.1 Classes of insect resistance genes expressed in transgenic crop plants

Source	Target pests
Microorganisms	
Bacillus thuringiensis (Bt)	Lepidoptera, Coleoptera
Isopentyl transferase (ipt)	Lepidoptera, Homoptera
Cholesterol oxidase	Lepidoptera, Coleoptera
Vegetative insectical proteins (Vips)	Lepidoptera
Plants	
Enzyme inhibitors (serine, cysteine, α-amylase)	Lepidoptera, Coleoptera, Homoptera
Lectins	Coleoptera, Homoptera, Lepidoptera
Chitinases	Homoptera
Anionic peroxidase	Lepidoptera, Coleoptera, Homoptera
Tryptophan decarboxylase (TDC)	Homoptera
Animals	
Protease inhibitors (insects)	Lepidoptera, Homoptera, Orthoptera
Chitinases (insects)	Lepidoptera
Avidin (chicken egg white)	Coleoptera, Lepidoptera

thuringiensis strains produce several crystal (Cry) proteins, each possessing a specific host range. The narrow host range of each individual toxin makes this group of insecticidal proteins very attractive with respect to both efficiency and environmental safety. The classification of the Cry proteins is based on hierarchical clustering using amino-acid sequence identity [10, http://epunix.biols.susx.ac.uk/Home/Neil_Crickmore/Bt/insdex.html]. A large number of the isolated and characterized genes encode toxins active against Lepidoptera (Cry1A, Cry1B, Cry1C, Cry2, Cry9) although others are toxic towards Coleoptera (Cry3), Diptera (Cry 4), and nematodes (Cry 5). Most of these proteins, even in the Cry1 subfamily, have a distinctive insecticidal spectrum. The size of most of these Cry proteins is about 130 kDa and they are produced in an inactive form. After ingestion, the alkaline environment of the insect midgut causes the crystals to dissolve and release their protoxins (several protoxins can be included in the same crystal). The protoxin is then cleaved by gut proteases to give a 65–70 kDa truncated form which is the active toxin. The toxin binds to specific receptors on the cell membranes and forms pores that destroy the epithelial cells by colloid osmotic lysis [11] resulting in the death of the insect. Specificity is, to a large extent, determined by a toxin–receptor interaction [12], although solubility of the crystal and protease activation also play a role [13].

B. thuringiensis was initially used as a bioinsecticide against different lepidopteran pests [14]; however, due to low field-persistence, the use of *Bt* sprays is relatively limited. The fact that *Bt* toxins have little effect on either nontarget organisms or mammals, together with their high and rapid toxicity towards target insects, as well as the availability of a large number of genes possessing different specificities, makes these toxins very interesting for introduction into plants.

The first published reports of the introduction and expression of *cry*1A genes into plants were published in 1987 [3, 4, 6]; in these early studies tobacco and tomato were used as model plants. *Bt* genes have now been transferred to a number of other crops such as cotton, maize, rice, and potato [reviewed in 15, 16]. Initially, both full-length (encoding the pro-toxin) and truncated (encoding the N-terminal part of the protein) *cry* genes were introduced into plants; only plants expressing truncated genes conferred protection against insect larvae. However, trials performed on these first-generation *Bt*-plants demonstrated low levels of protection under field conditions [16]. Subsequently, many attempts were made to increase the level of expression; however, the best improvement was observed by using partial or entirely synthetic genes (where the nucleotide sequences are modified without changing the amino-acid sequence [17]). A substantial increase in the amount of Cry protein expressed was observed after this gene modification and field trials of *Bt*-cotton demonstrated that the plants were completely protected against important lepidopteran pests [18]. Different synthetic Cry genes (Cry1Aa, b, c, Cry1C, cry9C) have been synthesized [reviewed in 15] and many reports of the successful introduction of these genes into various plants have been published together with the results of field trials [19]. Among the *Bt* δ-endotoxin genes cloned, several genes (Cry3A, B) encode toxins active against Coleoptera such as the colorado potato beetle (CPB, *Leptinotarsa decemlineata*). Synthetic Cry3A genes have also been designed and suc-cessfully introduced into potatoes. However, the activity spectra of coleopteran Cry-toxins is restricted to a limited number of insects from this order and there appear to be no published reports of Cry proteins with activity towards important insect pests such as the Southern- or Northern-corn rootworm or the boll weevil.

In order to increase the level of expression of the native *Bt* gene, the cry1Ab gene [20] and the cry2Aa2 gene [21] have been expressed in chloroplasts by homologous recombination. The large number of chloro-plasts in a cell leads to a very high level of toxin production (3–5 percent of soluble proteins) in tobacco. Nevertheless, chloroplast transformation is far from being routinely achieved and this technology needs to be adapted to crops.

Plant proteinase inhibitors

Plant proteinase inhibitors (PIs) are small proteins which are known to be involved in the natural defense of plants against herbivory [22]. Hydrolysis of dietary proteins in insects can involve different types of digestive proteinases – serine-, cysteine-, aspartic- and metallo-proteinases – and different proteinases predominate in the gut according to the insect order. Many different plant serine PIs have been characterized and cloned; they can be classified according to their sequence homology [23]. The most studied are the Bowman–Birk, the Kunitz, and the potato PI; fewer plant cysteine PIs have been characterized and cloned to date.

The mode of action of serine and cysteine PIs at the molecular level is known [24]. They are competitive inhibitors and form nonconvalent complexes with proteases. The antimetabolic action of these PIs against insects is not fully understood: direct inhibition of digestive enzymes or enzyme hypersecretion (to overcome the inhibition), inducing depletion in essential amino acids, is known to be involved [25].

Serine-like proteinases are predominant in lepidopteran larvae [26]. It has been shown that different serine PIs are able to inactivate lepidopteran proteases and to cause deleterious effects on development and growth when incorporated into artificial diets [reviewed in 23, 25]. The first constitutive expression of a PI in a plant was reported by Hilder *et al.* [5], who showed that a trypsin/trypsin inhibitor derived from cowpea (*Vigna unguiculata*), CpTI, conferred resistance against *Heliothis virescens* when expressed in tobacco. Many reports [reviewed in 8, 10, 11, 25] detail the production of transgenic plants expressing PIs of various origins and their antifeeding effects on different lepidopteran larvae. However, to be effective, the level of PI expression must be high [27]. In addition, insects can rapidly adapt to the ingestion of PI by overexpressing existing proteases or inducing the production of new types, less sensitive to the introduced PI [28–30]. In order to achieve durable resistance, crop protection strategies based on PIs will require further optimization, since lepidopteran larvae possess a diverse pool of serine proteases; information on the molecular interactions of the enzyme–inhibitor complex and the response of the insect to the presence of these inhibitors will be essential. This could be achieved by co-expressing PIs of different types and/or improving the affinity of introduced PIs for the target insect proteases [31, 32]. Until now, even if increased mortality and reduced growth of lepidopteran larvae have been observed after ingestion of serine PI-expressing plants, these effects have not been deemed sufficiently convincing to permit the commercialization of such crops.

Studies carried out on the protease content of the gut of different Coleoptera have shown the presence of cysteine proteases, which, in many cases, represent the major class of digestive proteases [33]. The cDNA of OC-I, a rice cysteine PI, has been constitutively expressed in different plant species. When expressed to a level of 1 percent of the soluble proteins in poplar, it causes an increase in insect mortality; however, this lethal effect is observed mainly at the end of the larval stages [34]. A significant growth reduction in Colorado potato beetle larvae was observed when OC-I was expressed in potatoes [35]. However, OC-I expression in oilseed rape failed to confer resistance towards several coleopteran species feeding on this plant [reviewed in 36]. As already observed with Lepidoptera, the lack of effects can be linked to a number of factors: the need for high expression levels (which was not obtained in oilseed rape), overexpression of cysteine proteases, compensation by serine proteases and degradation of the introduced PI by insensitive proteases [36]. The digestive complex of coleopteran insects involves proteases of different classes (serine, cysteine, aspartyl) and it may be difficult to obtain durable protection using PIs for this insect order, even if PIs of several types (serine and cysteine for example) are expressed simultaneously.

Plant lectins

Lectins are proteins containing at least one noncatalytic domain which binds reversibly to a specific mono- or oligosaccharide [37]. Lectins have been isolated from many plant tissues such as seeds, storage and vegetative tissues of dicots and monocots. On the basis of molecular and structural analyses, plant lectins can be classified into different families [38]. The role of lectins in the plant is not well characterized, but they are thought to be involved in different physiological processes such as storage proteins, sugar transport, cell-to-cell recognition, interaction with microorganisms, and defense against pests and pathogens. A role for lectins as defense proteins in plants against insect pests was first proposed by Janzen and Juster [39] who suggested that the lectin from the common bean (*Phaseolus vulgaris* PHA) was responsible for the resistance of these seeds to attack by coleopteran storage pests. Over the past few years, lectins from a wide variety of sources have been tested for their entomotoxic properties in intensive screening programs. These studies have shown that lectins belonging to different families and with different sugar specificities exert interesting effects on different insect genera. Effects included a delay in the rate of insect development, a decrease in fecundity, and mortality [reviewed in 40, 41]. The mechanism of action of lectins on insects is not well understood, but is thought to be complex. A prerequisite for lectin toxicity involves binding to specific "receptors," although binding in itself does not necessarily infer that a given lectin will be toxic. Many studies have demonstrated binding of lectins to the midgut epithelial cells of insects from different orders including Homoptera, Coleoptera, and Lepidoptera [42–45] and in some instances this binding has induced morphological changes such as disorganization of these cells, which in turn is thought to affect nutrient absorption. Further evidence that lectins affect digestion and absorption is provided by the recent findings that they can alter the activity of specific digestive enzymes within the insect gut or block glycoproteins involved in digestion or transport [40].

Not only do lectins exert their effects within the gut itself, but they are also known to confer systemic effects. They have been shown to be sequestered in the fat bodies of rice brown planthopper (*Nilaparvata lugens*; BPH) [44] and in the hemolymph of lepidopteran species such as tomato moth [45]. In addition to the toxic effects outlined above, lectins have also been implicated in altering insect behavior both in artificial diets [46] and when expressed in transgenic crops [47].

Lectins are currently receiving most interest as insecticidal agents for control of homopteran pests following the demonstration that they were toxic to planthoppers [48] and, to a lesser extent, aphids [49, 50]. Expression in transgenic plants of the mannose-specific lectin from snowdrop (*Galanthus nivalis* agglutinin, GNA) has been shown to be effective against homopteran pests [47, 51–55]. It is also effective against several lepidopteran pest species [56, 57]. However, to date, there are no published reports of field trials of plants expressing lectins.

Plant α-amylase inhibitors (α-AIs)

The common bean, *Phaseolus vulgaris*, contains a family of related seed proteins (PHA-E and -L, arcelin and α-AI). PHA-E and -L are classical lectins with strong agglutination activity while α-AI can complex insect α-amylases and is thought to play a role in plant defense; it has been shown to inhibit the α-amylases present in the midgut of coleopteran pests of stored products [58]. The common bean α-AI has been expressed in pea and in Azuki bean, where its expression confers resistance to the bruchid beetles, *Callosobruchus maculatus* and *C. chinensis* [59, 60]. As well as being active against pests of stored grain, Schroeder *et al.* [60] further demonstrated that the expression of this gene in pea confered resistance to *Bruchus pisorum*. In a recent study Morton *et al.* [61] demonstrated complete protection under field conditions of transgenic peas expressing the α-AI-1 against this pea weevil.

Other toxins of bacterial origin

In order to identify new insecticidal proteins, large screening programs of bacterial extracts have been initiated in different laboratories [7]. These programs have allowed the identification of new gene candidates for generating insect-resistant crops. Supernatants from exponential cultures of *B. thuringiensis* were shown to contain toxins active against Lepidoptera such as *Agrotis ipsilon* (black cutworm, BCW). Two of these toxins, vegetative insecticidal proteins (VIPs), with toxicity towards lepidopteran larvae, have been isolated [62]. Insecticidal proteins (VIP1 and VIP2) have also been isolated from supernatants of *Bacillus cereus* isolates [62]. *Streptomyces* cultures are known to secrete cholesterol oxidase (COX), an enzyme active against the boll weevil (*Anthonomus grandis*), a major cotton pest worldwide. This protein is active within the same range as *Bt* toxins [63] and has been expressed in tobacco protoplasts [64].

To date, while no reports of transgenic plants expressing these recently identified bacterial toxins have been published, Estruch *et al.* [7] have nevertheless described the use of these genes to generate a second generation of insecticidal plants.

Toxins of insect origin

In the search for new toxin genes, several studies have raised the possibility of altering/interfering with specific physiological processes within insects using proteinase inhibitors or chitinase of insect origin. For example, one serine PI isolated from the hemolymph of *M. sexta* adversely affects insect development when expressed in plants [65–67]. Chitin is present in insects, not only as exoskeletal material but also in the peritrophic membrane [68], and during molting there is known to be an increase in chitinase activity. In recent studies, constitutive expression of the *M. sexta* (tobacco hornworm) gene encoding this chitinase in tobacco was shown to cause a significant reduction in growth of tobacco budworm (*H. virescens*) larvae, whereas no differences were observed in tobacco hornworm (*M. sexta*) [69]. A synergistic effect was observed when this insect chitinase was used in combination with sublethal doses of *Bt* toxin, with detrimental effects being observed in the case of *M. sexta* [69].

Commercialization and risk assessment of insect-resistant transgenic crops

Commercialization

The first *Bt*-cotton field trial was reported in 1992 [18] and since 1996 only one *Bt*-cotton (Bollgard™, Monsanto) has been released. This plant expresses the Cry1Ac protein which protects it against several lepidopteran insect pests (*Heliothis virescens*, *Helicoverpa zea*, and *Pectinophora gossypiella*). In 1999, 27 percent of the total acreage of cotton was planted with *Bt*-cotton in the USA.

Similarly, *Bt*-maize has been developed with resistance to the European corn borer (ECB; *Ostrinia nubilabis*), with the first report of a field trial published by Koziel *et al.* [70]. The commercialized *Bt* varieties originate from five different transformation events which vary according to which gene is expressed (*cry*1Ab, *cry*1Ac, and *cry* 9C), and the promoter associated with the coding sequence (which affects the quantity and location of the Cry protein). In 1999, 30 percent of the cultivated area in the USA consisted of transgenic varieties. In 1995, *Bt*-potato (NewLeaf™, Monsanto) became the first *Bt*-crop to be commercialized. However, they are not, as yet, cultivated on large areas (4 percent acreage in 1999 in the USA). A summary of the global area of transgenic crops by country, crop, and trait is given in Figure 13.1.

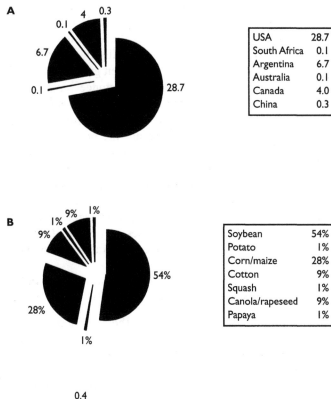

A

USA	28.7
South Africa	0.1
Argentina	6.7
Australia	0.1
Canada	4.0
China	0.3

B

Soybean	54%
Potato	1%
Corn/maize	28%
Cotton	9%
Squash	1%
Canola/rapeseed	9%
Papaya	1%

C

Herbicide tolerance	29.4
Herbicide & insect resistance	2.9
Insect resistance	9.1
Virus resistance	0.4

Figure 13.1 Global area of transgenic crops in 1999 by (A) country (millions of hectares); (B) crop; (C) trait (millions of hectares). Reference source: *Global Review of Commercialized Transgenic Crops* (1999). *ISAAA Briefs*, No. 12.

Insect resistance

The repeated and unmanaged use of chemical pesticides has led to the rapid evolution of resistant insect populations. However, development of resistance within insect populations is not just confined to chemicals since field uses of *B. thuringiensis*-based biopesticide products have led, in the case of one insect, *Plutella xylostella,* to the occurrence of resistant insect populations in Hawaii [71] and in other areas [reviewed in 72]. The important increase in the cultivation of transgenic insect-resistant crops could lead to the same problem. Most of the introduced genes work as monogenic traits and could therefore be readily overcome. For the most part, only crops expressing Cry genes have been grown in the field in large quantities and as yet no cases of insect resistance have been reported. However, there is no doubt that the potential for resistance is present [73]. In addition, under laboratory conditions many strains of Cry-resistant insects have been selected [72]. As a result, the potential for insect resistance to develop is a major consideration whenever large plantations of insect-resistant crops are planned [74].

Resistance management strategies are oriented towards a reduction of selection [reviewed in 19, 75, 76]. These strategies are of different types: tissue- or time-specific expression of toxins, transfer of multiple toxins with different modes of action, low doses in combination with natural enemies, high doses plus refuge, and other cultural practices.

Use of tissue- or time-specific promoters

In most cases, the toxin is expressed under the control of constitutive promoters such as the CaMV 35S promoter and its derivatives, or monocot ubiquitin or actin promoters. Tissue- and time-specific promoters can be used to limit toxin production to the tissues fed upon by the pest, or to periods when the pest attacks the plant. For example, to protect against seed-attacking insects, the promoter from the seed protein phytohemagglutinin from beans has been used to drive expression of the α-amylase inhibitor [59]. The rice sucrose synthase promoter which confers phloem-specific expression has been used to generate plants resistant to sap-sucking insects such as aphids and planthoppers [54, 77]. The use of inducible promoters allowing toxin expression only after wounding such as insect feeding has also been considered. Duan *et al.* [78] obtained lepidopteran-resistant transgenic rice lines expressing a potato PI under the control of its own promoter. Induction of expression by chemicals (salicylic acid) has also been observed using the tobacco promoter of the pathogenesis-related protein [79].

Gene pyramiding

The use of multiple resistance genes or gene-pyramiding (stacking) requires the incorporation into the plant genome of genes encoding two or more entomotoxins each possessing different modes of action. Increasing attention is now being devoted to the study of the co-expression of different genes. It is for this reason that it is important for the future to identify new toxins since, for many insects, the choice of genes available for transfer is limited.

High-dose and refuge strategy

The high-dose strategy is considered to be the most efficient and promising way of managing resistance in Bt crops, if used in conjunction with refuges [80]. Refuges are areas planted with nontransgenic plants where the pest population can survive and act as a reservoir of wild-type susceptible alleles. The success of this strategy depends upon the initial frequency of allele resistance [81, 82].

Risk assessment

When using transgenic plants or derived products, it is important to determine the entomotoxin toxicity towards other organisms. Three categories need to be considered: humans, animals, and nontarget insects. In this chapter, we will only consider the risks which are specific to insect-resistant transgenic plants, and not those relevant to all transgenic plants and which are more related to the biology of the plant itself (impact on biodiversity by crossing with wild relatives, pollen dispersion, etc.). Another point which will not be discussed here concerns the potential risks associated with the marker genes (coding for antibiotic or herbicide resistance) generally used to select the transformed cells at the first stages of the transformation procedure. Some studies have demonstrated the innocuity of the proteins encoded by such marker genes [83–85]. In addition, different strategies are now available which avoid or eliminate these marker genes in plants available commercially.

Risk for humans and animals

Potential risks must be considered in relation to the final use of the transgenic crop. For example, it will differ for cotton (industrial use), maize (use of derived products and animal feeds), and vegetables (human consumption), and as to whether it is eaten raw or after cooking. However, even if eaten raw, in the case of Bt, the ingested proteins are very rapidly degraded by the digestive enzymes and, in most cases, lose their activity and properties. *B. thuringiensis* sprays have been used for a long time and different studies have demonstrated its innocuity for humans and mammals. In the case of proteins of plant origin, most of them are already present in vegetables and fruits and are thus consumed on a regular basis. However, in some cases (proteinase inhibitors, α-amylase inhibitors, certain lectins), they are considered as anti-nutritional and vegetables containing them in large amounts should be cooked before consumption, as in fact is usually carried out for many vegetables such as potatoes, beans, etc.

Risk for non-target insects

A major advantage of insect-resistant plants (whether produced by conventional plant breeding or via recombinant DNA technology) is the confinement of the entomotoxin within the plant, thus restricting exposure of the toxin to insects feeding on the plant. However, secondary pests, predators, or parasites of pests could ingest or come into contact with the toxin. Natural enemies of pest species are an important component of integrated pest management (IPM) and, therefore, it is imperative to investigate possible adverse effects upon natural biological agents [86]. Apart from Cry-expressing crops, most of the studies on nontarget insects have been performed under laboratory conditions and must be considered as the "worst-case scenario" [87].

Even if the main target of a toxin is an insect which causes considerable damage to the crop, very often other insects can feed on the plant. If they are sensitive to the expressed toxin, they will also be affected which, of course, is advantageous in terms of crop protection. However, some insects could be affected in a nonintended way. An example of this is the monarch butterfly (*Danaus plexippus*), a mythic butterfly of North America. Losey *et al.* [88] observed a higher mortality rate in butterfly larvae fed milkweed coated with *Bt*-maize pollen as compared to larvae fed leaves coated with nontransformed maize pollen or with leaves free of pollen. However, it is important to note that this study was performed under artificial laboratory conditions which do not reflect most of the characteristics of the monarch way of life [89]. In a very recent report, the EPA (September 22, 2000), on the basis of further trials, concluded "that monarch butterflies were at very little risk from *Bt* corn products, contrary to widely published reports." EPA further found that "In fact, some authors are predicting that the widespread cultivation of *Bt* crops may have huge benefits for monarch butterfly survival."

Potential risk for beneficial insects

If transgenic insect-resistant crops are to play a useful role in decreasing pesticide usage, it is apparent that they must be compatible with other components of IPM. Indeed, the recommended practices for deploying transgenic crops are all based on IPM. Ideally, genes expressed in transgenic plants for control of pests should at the same time produce no directly deleterious effects on predators or parasitoids, which may play an important role in biological control. In this context, it is important to distinguish between indirect effects, resulting from a decreased food supply or reduced food quality, i.e. as a consequence of controlling the pest (host) species, and direct effects where the transgene product is toxic to the beneficial insect.

The high level of specificity shown by Bt toxins suggests that the encoding genes are unlikely to cause deleterious effects on predators when expressed in transgenic plants. Many studies have now been carried out both in laboratory trials and in the field and, in the main, this assumption has been shown to be the case [90]. For example, plants expressing Bt toxins were used as hosts for aphids (toward which the toxin has no protective effect) and shown to have no deleterious effects on ladybirds feeding on those aphids [91]. Other studies found no deleterious effects on beneficial insects in transgenic cotton [18], potatoes [92], or corn [93]. On the other hand, other studies have reported Bt to be toxic to lacewing, a beneficial predator [94]. In the case of transgenes whose products do not cause complete, or almost complete, mortality of the target pest, the situation is different, and in these situations, natural enemies may form an important component of crop protection. Much interest is therefore being placed on the effects of transgenes, including lectins and PIs on both predators and parasitoids. Recent studies showed that when adult 2-spot ladybirds (*Adalia bipunctata*) were fed on aphids (*Myzus persicae*) colonizing transgenic potato plants expressing GNA, ladybird fecundity, egg viability, and adult longevity were adversely affected, although no acute toxicity was observed [95]. More recently, neonate larvae of 2-spot ladybird have been reared to adulthood on either GNA-fed or control-fed *M. persicae*, using an artificial diet. Under these conditions, GNA failed to show any deleterious effects on ladybird survival or development [96]. In these studies it was noted that the ladybird larvae consumed more GNA-dosed aphids which were significantly smaller. The cowpea trypsin inhibitor (CpTI), similarly, did not affect ladybird survival or development [97].

The effects of GNA and CpTI on the ability of the gregarious ectoparasitoid wasp *Eulophus pennicornis* to parasitize lepidopteran larvae have also been investigated recently. The pest *Lacanobia oleracea* was selected for study since transgenic potato plants expressing GNA were shown to be significantly resistant to attack [57]. In these studies, using both artificial diet and GNA-expressing potato plants, no deleterious effects were observed on any of the measured biological parameters of the parasitoid (survival, development, egg load, fecundity, F1 generation) [45]. However, in the case of CpTI expressed in transgenic potato, indirect adverse effects on the parasitoid were observed since the pest larvae did not grow to a sufficient size for parasitism to take place; in the few instances when the parasitoid was able to parasitize the pest, its subsequent development was not affected [98].

Potential effects of transgene products on pollinating insects such as honey bees and bumble bees, which play a major role in seed production and fruit set of many crops, are of great importance. They feed on pollen and nectar and therefore it is necessary to determine the toxicity of the entomotoxin expressed in insect-resistant transgenic plants both in the short and long term. To date such studies with honey bees have been performed predominantly using artificial diets where the entomotoxins are incorporated at different doses; these types of experiments must be considered as a "worst-case scenario." When deleterious effects are observed, even under such artificial conditions, it is important to try and avoid expression of the given toxin in pollen and nectar. Detailed discussions of experiments of this kind will be addressed in Chapter 14. However, the final evaluation must be performed on transgenic plants grown under natural conditions; this topic will be considered fully by Pham-Delègue *et al.* in Chapter 15.

Conclusion and perspectives

To increase the yield and reduce the use of chemicals in modern agriculture, it is important to develop new approaches to crop protection, including the use of recombinant DNA technology. This technology has opened up new avenues for obtaining crops resistant to their major insect pests. Expression of bacterial *Bacillus thuringiensis* Cry endotoxins is the most advanced of these strategies. *Bt*-expressing maize, cotton, and potatoes are already commercialized in some countries. However, to date, they have been grown mainly in industrial countries and it is of importance that this technology be extended to developing countries [99, 100].

To avoid problems with the emergence of resistance within insect populations it is important to cultivate these crops under resistance-management regimes. In the long term, and with the aim of extending the range of insect pests to be controlled, it is important to increase the number of genes which can be expressed in plants. Many studies, currently at the laboratory stage, are being performed with this objective in mind. Another factor which will affect the future of these crops is the public acceptance of products derived from transgenic plants [101–103]. Better consumer information is necessary to allow a well-informed decision to be made based on the comparison of the potential benefits of using transgenic plants as against the continued reliance on chemical insecticides.

References

1 Oerke, E.C. (1994). Estimated crop losses due to pathogens, animal pests and weeds. In: *Crop Production and Crop Protection: Estimated Losses in Major Food and Cash Crops* (Oerke E.C., Dehne, H.W., Schönbeck, F. and Weber, A., Eds). Elsevier, Amsterdam, pp. 72–88.

2 Horsch, R.B., Fry, J.E., Hoffman, N.L., Eichholtz, D., Rogers, S.G. and Fraley, R.T. (1985). A simple and general method for transferring genes into plants. *Science* **227**, 1229–1231.

3 Barton, K.A., Whiteley, H.R. and Yang, N.-S. (1987). *Bacillus thuringiensis* δ-endotoxin expressed in transgenic *Nicotiana tabaccum* provides resistance to Lepidopteran insects. *Plant Physiol.* **85**, 1103–1109.

4 Fischhoff, D.A., Bowdish, K.S., Perlak, F.J., Marrone, P., McCormick, S., Niedermeyer, J., Dean, D., Kusano-Kretzmer, K., Mayer, E., Rochester, D.,

Rogers, S. and Fraley, R. (1987). Insect tolerant transgenic tomato plants. *Bio/Technology* **5**, 807–813.

5 Hilder, V.A., Gatehouse, A.M.R., Sheerman, S.E., Barker, R.F. and Boulter, D. (1987). A novel mechanism of insect resistance engineered into tobacco. *Nature* **333**, 160–163.

6 Vaeck, M., Reynaert, A., Höfte, H., Jansens, S., De Beuckeleer, M., Dean, C., Zabeau, M., van Montagu, M. and Leemans, J. (1987). Transgenic plants protected from insect attack. *Nature* **328**, 33–37.

7 Estruch, J.J., Carozzi, N.B., Desai, N., Duck, N.B., Warren, G.W. and Koziel, M.G. (1997). Transgenic plants: An emerging approach to pest control. *Nat. Biotechnol.* **15**, 137–141.

8 Jouanin, L., Bonadé-Bottino, M., Girard, C., Morrot, G. and Giband, M. (1998). Transgenic plants for insect resistance. *Plant Sci.* **131**, 1–11.

9 Schuler, T.H., Poppy, G.M., Kerry, B.R. and Denholm, I. (1998). Insect-resistant transgenic plants. *Trends Biotechnol.* **16**, 147–196.

10 Crickmore, N., Zeigler, D.R., Feitelson, J., Schnepf, E., Van Rie, J., Lereclus, D., Baum, J. and Dean, D.H. (1998). Revision of the nomenclature for the *Bacillus thuringiensis* pesticidal crystal proteins. *Microbiol. Mol. Biol. Rev.* **6**, 807–813.

11 Knowles, B.H. and Dow, J.A.T. (1993). The crystal delta-endotoxins of *Bacillus thuringiensis* – models for their mechanism of action in the insect gut. *BioEssays* **15**, 469–476.

12 Van Rie, J., McGaughey, W.H., Johnson, D.E., Barnett, B.D. and van Mellaert, H. (1990). Mechanism of resistance to the microbial insecticide *Bacillus thuringiensis*. *Science* **247**, 72–74.

13 de Maagd, R.A., Bosch, D. and Stiekema, W. (1999). *Bacillus thuringiensis* toxin-mediated insect resistance in plants. *Trends Plant Sci.* **4**, 9–13.

14 Lambert, B. and Peferoen, M. (1992). Insecticidal promise of *Bacillus thuringiensis*. Facts and mysteries about a successful biopesticide. *BioScience* **42**, 112–122.

15 Mazier, M., Pannetier, C., Tourneur, J., Jouanin, L. and Giband, M. (1997). The expression of *Bacillus thuringiensis* toxin genes in plant cells. *Biotechnol. Ann. Rev.* **3**, 313–347.

16 Peferoen, M. (1997). Insect control with transgenic plants expressing *Bacillus thuringiensis* crystal proteins. In: *Advance in Insect Control: The Role of Transgenic Plants* (Carozzi, N. and Koziel, M., Eds). Taylor & Francis, London, pp. 21–48.

17 Perlak, F.J., Fuchs, D.A., Dean, R.L., McPherson, S.L. and Fishhoff, D.A. (1991). Modification of the coding sequence enhances plant expression of insect control genes. *Proc. Natl. Acad. Sci. USA* **88**, 3324–3328.

18 Wilson, W.D., Lfint, H.M., Deaton, R.W., Fischhoff, D.A., Perlak, F.J., Armstrong, T.A., Fuchs, R.L., Parks, N.J. and Stapp, B.R. (1992). Resistance of cotton lines containing a *Bacillus thuringiensis* toxin to pink bollworm (Lepidoptera: Gelechiidae) and other insects. *J. Econom. Entomol.* **85**, 1516–1521.

19 Peferoen, M. (1997). Progress and prospects for field use of *B. thuringiensis* genes in crops. *Trends Biotechnol.* **15**, 173–177.

20 McBride, K.E., Svab, Z., Schaaf, D.J., Hogan, P.S. and Maliga, P. (1995). Amplification of a chimeric *Bacillus* gene in chloroplasts leads to an extraordinary level of an insecticidal protein in tobacco. *Bio/Technology* **13**, 362–365.

21 Kota, M., Daniell, H., Varma, S., Garczynski, S.F., Gould, F. and Moar, W.J. (1999). Overexpression of the *Bacillus thuringiensis* (Bt) Cry2Aa2 protein in chloroplasts confers resistance to plants against susceptible and Bt-resistant insects. *Proc. Natl. Acad. Sci. USA* **96**, 1840–1845.

22 Ryan, C.A. (1990). Proteinase inhibitors in plants: Genes for improving defenses against insects and pathogens. *Annu. Rev. Phytopathol.* **28**, 839–943.

23 Boulter, D. (1993). Insect pest control by copying nature using genetically engineered crops. *Phytochemistry* **34**, 1453–1466.

24 Bode, W. and Huber, R. (1992). Natural protein proteinase inhibitors and their interactions with proteinases. *Eur. J. Biochem.* **204**, 433–451.

25 Reeck, G.R., Kramer, K.J., Baker, J.E., Kanost, R., Fabrick, J.A. and Brehnke, C.A. (1997). Proteinase inhibitors and resistance of transgenic plants to insects. In: *Advance in Insect Control: The Role of Transgenic Plants* (Carozzi, N. and Koziel, M., Eds). Taylor & Francis, London, pp. 157–183.

26 Christeller, J.T., Laing, W.A., Markwick, N.P. and Burgess, E.P.J. (1992). Midgut protease activities in 12 phytophagous lepidopteran larvae – Dietary and protease inhibitor interactions. *Insect Biochem. Mol. Biol.* **22**, 735–746.

27 De Leo, F., Bonadé-Bottino, M., Ceci, L.R., Gallerani, R. and Jouanin, L. (1998). Opposite effects on *Spodoptera littoralis* larvae of low and high expression level of a trypsin proteinase inhibitor in transgenic plants. *Plant Physiol.* **118**, 997–1004.

28 Bown, D., Wilkinson, H.S. and Gatehouse, J.A. (1997). Differentially regulated inhibitor-sensitive and insensitive protease genes from the phytophagous insect pest, *Helicoverpa armigera*, are members of complex multigene families. *Insect Biochem. Mol. Biol.* **27**, 625–638.

29 Jongsma, M.A., Bakker, P.L., Peters, J., Bosch, D. and Stiekema, W.J. (1995). Adaptation of *Spodoptera exigua* larvae to plant proteinase inhibitors by induction of gut proteinase activity insensitive to inhibitors. *Proc. Natl. Acad. Sci. USA* **92**, 8041–8045.

30 Jongsma, M.A. and Bolter, C. (1997). The adaptation of insects to plant protease inhibitors. *J. Insect Physiol.* **4**, 885–895.

31 Jongsma, M.A., Stiekema, W.J. and Bosch, D. (1996). Combating inhibitor-insensitive proteases of insect pests. *Trends Biotechnol.* **14**, 331–333.

32 Michaud, D. (1997). Avoiding protease-mediated resistance to herbivorous pests. *Trends Biotechnol.* **15**, 4–6.

33 Murdock, L.L., Brookhart, G., Dunn, P.E., Foard, D.E., Kelley, S., Kitch, L., Shade, R.E., Shukle, R.H. and Wolfson, J.L. (1987). Cysteine digestive proteinases in coleoptera. *Comp. Biochem. Physiol.* **87B**, 783–787.

34 Leplé, J.C., Bonadé-Bottino, M., Augustin, S., Pilate, G., Dumanois-LêTân, V., Delplanque, A. and Jouanin, L. (1995). Toxicity to *Chrysomela tremulae* (Coleoptera: Chrysomelidae) of transgenic poplars expressing a cysteine proteinase inhibitor. *Mol. Breed.* **1**, 319–328.

35 Lecardonnel, A., Chauvin, L., Jouanin, L., Beaujean, A., Prévost, G. and Sangwan-Norreel, B. (1999). Effects of the rice cystatin I expression in transgenic potato on colorado potato beetle larvae. *Plant Sci.* **140**, 71–79.

36 Jouanin, L., Bonadé-Bottino, M., Girard, C., Lerin, J. and Pham-Delègue, M.-H. (2000). Expression of protease inhibitors in rapeseed. In: *Recombinant Protease Inhibitors in Plant* (Michaud, D., Ed.). Academic Press, pp. 182–194.

37 Peumans, W.J. and Van Damme, E.J.M. (1995). Lectins as plant defense proteins. *Plant Physiol.* **109**, 347–352.
38 Van Damme, E.J.M., Peumans, W.J., Barre, A. and Rougé, P. (1998). Plant lectins: A composite of several distinct families of structurally and evolutionary related proteins with diverse biological roles. *Crit. Rev. Plant Sci.* **17**, 575–692.
39 Jansen, D.H. and Juster, H.B. (1976). Insecticidal action of the phytohemagglutinin in black beans on a bruchid beetle. *Science* **192**, 795–796.
40 Czapla, H. (1997). Plant lectins as insect control proteins in transgenic plants. In: *Advance in Insect Control: The Role of Transgenic Plants* (Carozzi, N. and Koziel, M., Eds). Taylor & Francis, London, pp. 123–138.
41 Gatehouse, A.M.R., Powell, K.S., Van Damme, E.J.M., Peumans, W.J. and Gatehouse, J.A. (1995). Insecticidal properties of plant lectins: Their potential in plant protection. In: *Lectins: Biomedical Perspectives* (Pusztai, A.J. and Bardocz, S., Eds). Taylor & Francis, London, pp. 35–57.
42 Habidi, J., Backus, E.A. and Czapla, T.H. (1998). Subcellular effects and localization of binding sites of phytohemagglutinin in the potato leafhopper, *Empoasca* fabae (Insecta: Homoptera: Cicadellidae). *Cell Tissue Res.* **294**, 561–574.
43 Habidi, J., Backus, E.A. and Huesing, J.E. (2000). Effect of phytohemagglutinin (PHA) on the midgut epithelial cells and localization of its binding sites in western tarnished plant bug, *Lygus hesperus* Knight. *J. Insect Physiol.* **46**, 611–619.
44 Powell, K.S., Spence, J., Bharathi, M., Gatehouse, J.A. and Gatehouse, A.M.R. (1998). Immunohistochemical and developmental studies to elucidate the mechanism of action of the snowdrop lectin on the rice brown planthopper, *Nilaparvata lugens* (Stal). *J. Insect Physiol.* **44**, 529–539.
45 Bell, H.A., Fitches, E.C., Down, R.E., Marris, G.C., Edwards, J.P., Gatehouse, J.A. and Gatehouse, A.M.R. (1999). The effect of snowdrop lectin (GNA) delivered *via* artificial diet and transgenic plants on *Eulophus pennicornis* (Hymenoptera: Eulophidae), a parasitoid of the tomato moth *Lacanobia oleracea* (Lepidoptera: Noctuidae). *J. Insect Physiol.* **45**, 983–991.
46 Powell, K.S., Gatehouse, A.M.R., Hilder, V.A., Van Damme, E.J.M., Peumans, W.J., Boonjawat, J., Horsham, K. and Gatehouse, J.A. (1995). Different antimetabolic effects of related lectins towards nymphal stages of *Nilaparvata lugens. Entomol. Exp. Appl.* **75**, 61–65.
47 Foissac, X., Thi Loc, N., Christou, P., Gatehouse, A.M.R. and Gatehouse, J.A. (2000). Resistance to green leafhopper (*Nephotettix virescens)* and brown planthopper (*Nilaparvata lugens*) in transgenic rice expressing snowdrop lectin (*Galanthus nivalis* agglutinin; GNA). *J. Insect Physiol.* **46**, 573–583.
48 Powell, K.S., Gatehouse, A.M.R., Hilder, V.A. and Gatehouse, J.A. (1993). Antimetabolic effects of plant lectins and plant and fungal enzymes on the nymphal stages of two important rice pests, *Nilaparvata lugens* and *Nephotettix nigropictus. Entomol. Exp. Appl.* **66**, 119–126.
49 Rahbé, Y. and Febway, G. (1993). Protein toxicity to aphids: An *in vitro* test on *Acyrthosiphon pisum. Entomol. Exp. Appl.* **67**, 149–160.
50 Sauvion, N., Rahbé, Y., Peumans, W.J., Van Damme, E., Gatehouse, J.A. and Gatehouse, A.M.R. (1996). Effects of GNA and other mannose binding lectins on development and fecundity of the peach-potato aphid. *Entomol. Exp. Appl.* **79**, 285–293.

51 Down, R.E., Gatehouse, A.M.R., Hamilton, W.D.O. and Gatehouse, J.A. (1996). Snowdrop lectin inhibits development and decreases fecundity of the glasshouse potato aphid (*Aulacorthum solani*) when administrated *in vitro* and *via* transgenic plants in laboratory and glasshouse trials. *J. Insect Physiol.* **42**, 1035–1045.

52 Gatehouse, A.M.R., Down, R.E., Powell, K.S., Sauvion, N., Rahbé, Y., Newell, C.A., Merryweather, A., Hamilton, W.D.O. and Gatehouse, J.A. (1996). Transgenic potato plants with enhanced resistance to the peach-potato aphid *Myzus persicae. Entomol. Exp. Appl.* **79**, 295–307.

53 Hilder, V.A., Powell, K.S., Gatehouse, A.M.R., Gatehouse, J.A., Gatehouse, L.N., Shi, Y., Hamilton, W.D.O., Merryweather, A., Newell, C.A., Timans, J.C., Peumans, W.J., van Damme, E. and Boulter, D. (1995). Expression of snowdrop lectin in transgenic tobacco plants results in added protection against aphids. *Transgenic Res.* **4**, 18–25.

54 Rao, K.V., Rathore, K.S., Hodges, T.K., Fu, X., Stoger, E., Sudhakar, D., Williams, S., Christou, P., Bharathi, M., Bown, D.P., Powell, K.S., Spence, J., Gatehouse, A.M.R. and Gatehouse, J.A. (1998). Expression of snowdrop lectin (GNA) in transgenic rice plants confers resistance to rice brown plant-hopper. *Plant J.* **15**, 469–477.

55 Stoger, E., Willians, S., Christou, P., Down, R. and Gatehouse, J.A. (1999). Expression of the insecticidal lectin from snowdrop (*Galanthis nivalis* agglutinin; GNA) in transgenic wheat plants: Effects on predation by the grain aphid *Sitobion avenae. Mol. Breed.* **5**, 63–73.

56 Fitches, E., Gatehouse, A.M.R. and Gatehouse, J.A. (1997). Effects of snow-drop lectin (GNA) delivered *via* artificial diet and transgenic plants on the development of tomato moth (*Lacanobia oleracea*) larvae in laboratory and glasshouse trials. *J. Insect Physiol.* **44**, 1213–1224.

57 Gatehouse, A.M.R., Davison, G., Newell, C.A., Merryweather, A., Hamilton, W.D.O., Burgess, E.J., Gilbert, R.J.C. and Gatehouse, J.A. (1997). Transgenic potato plants with enhanced resistance to the tomato moth, *Lacanobia oler-acea. Mol. Breed.* **3**, 49–53.

58 Shade, R.E., Schroeder, H.E., Pueyo, J.J., Tabe, L.M., Murdock, L.L., Higgins, T.J.V. and Chrispeels, M.J. (1994). Transgenic pea seeds expressing the α-amylase inhibitor of the common bean are resistant to bruchid beetles. *Bio/Technology* **12**, 793–796.

59 Ishimoto, M., Sato, T., Chrispeels, M.J. and Kitamura, K. (1996). Bruchid resistance of transgenic azuki bean expressing seed α-amylase inhibitor of common bean. *Entomol. Exp. Appl.* **79**, 309–315.

60 Schroeder, H.E., Gollasch, S., Moore, A., Tabe, L.M., Craing, S., Hardie, D., Chrispeels, M.J., Spencer, D. and Higgins, T.J.V. (1995). Bean α-amylase inhibitor confers resistance to the pea weevil (*Bruchus pisorum*) in transgenic peas (*Pisun sativum* L). *Plant Physiol.* **111**, 393–401.

61 Morton, R.L., Schroeder, H.E., Bateman, K.S., Chrispeels, M.J., Armstrong, E. and Higgins, T.J.V. (2000). Bean α-amylase inhibitor 1 in transgenic peas (*Pisum sativum*) provides complete protection from pea weevil (*Bruchus pisorum*) under field conditions. *Proc. Natl. Acad. Sci. USA* **97**, 3820–3825.

62 Estruch, J.J., Warren, G., Mullins, M.A., Nye, G.J., Craing, J.A. and Koziel, M.G. (1996). Vip3A, a novel *Bacillus thuringiensis* vegetative insecticidal

Insect-resistant transgenic crops 287

protein with a wide spectrum of activities against lepidopteran insects. *Proc. Natl. Acad. Sci. USA* **93**, 5389–5394.

63 Purcell, J.P., Greenplate, J.T., Jennings, M.G., Ryerse, J.S., Pershing, J.C., Sims, S.R., Prinsen, M.J., Corbin, D.R., Tran, M., Sammons, R.D. and Stonard, R.J. (1993). Cholesterol oxidase: A potent insecticidal protein active against boll weevil larvae. *Biochem. Biophys. Res. Commun.* **196**, 1406–1413.

64 Corbin, D.R., Greenplate, J.T., Wong, E.Y. and Purcell, J.P. (1994). Cloning of an insecticidal cholesterol oxidase gene and its expression in bacteria and in plant protoplasts. *Appl. Environ. Microbiol.* **6**, 4239–4244.

65 Thomas, J.C., Wasmann, C.C., Echt, C., Dunn, R.L. and Bohnert, H.J. (1994). Introduction and expression of an insect proteinase inhibitor in alfalfa. *Plant Cell Rep.* **14**, 31–36.

66 Thomas, J.C., Adams, D.G., Kepenne, V.D., Wasmann, C.C., Brown, J.K., Kanost, M.R. and Bohnert, H.J. (1995). *Manduca sexta* encoded protease inhibitors expressed in *Nicotiana tabacum* provide protection against insects. *Plant Physiol. Biochem.* **33**, 611–614.

67 Thomas, J.C., Adams, D.G., Kepenne, V.D., Wasmann, C.C., Brown, J.K., Kanost, M.R. and Bohnert, H.J. (1995). Protease inhibitors of *Manduca sexta* expressed in cotton. *Plant Cell Rep.* **14**, 758–762.

68 Kramer, K.J., Muthukrishnan, S., Johnson, L. and White, F. (1997). Chitinases for insect control. In: *Advances in Insect Control: The Role of Transgenic Plants* (Carozzi, N. and Koziel, M., Eds). Taylor & Francis, London, pp. 185–193.

69 Ding, X., Gopalakrishnan, B., Johnson, L.B., White, F.F., Wang, X. and Muthukrishnan, S. (1998). Insect resistance in transgenic tobacco expressing an insect chitinase gene. *Transgenic Res.* **7**, 77–84.

70 Koziel, M.G., Beland, G.L., Bowman, C., Carozzi, N., Crenshaw, R., Crossland, L., Dwason, J., Desai, N., Hill, M., Kadwell, S., Launis, K., Lewis, K., Maddox, D., McPherson, K., Meghji, M.R., Merlin, E., Rhodes, R., Warrren, G.W., Wright, M. and Evola, S.V. (1993). Field performance of elite transgenic maize plants expressing an insecticidal protein derived from *Bacillus thuringiensis*. *Bio/Technology* **11**, 194–200.

71 Tabashnik, B.E., Cushing, N.L., Finson, N. and Johnson, M.W. (1990). Field development of resistance to *Bacillus thuringiensis* in diamondback moth (Lepidoptera: Pyrallidae). *J. Econ. Entomol.* **83**, 1671–1676.

72 Frutos, R., Rang, C. and Royer, M. (1999). Managing insect resistance to plants producing *Bacillus thuringiensis* toxins. *Crit. Rev. Biotechnol.* **19**, 227–276.

73 Gould, F., Anderson, A., Jones, A., Sumerford, D., Heckel, D.G., Lopez, J., Micinski, S., Leonard, R. and Laster, M. (1997). Initial frequency of alleles for resistance to *Bacillus thuringiensis* in field populations of *Heliothis virescens*. *Proc. Natl. Acad. Sci. USA* **94**, 3519–3523.

74 Riebe, J.F. (1999). The development and implementation of strategies to prevent resistance to *B.t.* expressing crops. *Can. J. Plant Pathol.* **21**, 101–105.

75 McGaughey, W.H. and Whalon, M.E. (1992). Managing insect resistance to *Bacillus thuringiensis* toxins. *Science* **258**, 1451–1455.

76 Tabashnik, B.E. (1989). Managing resistance with multiple pesticide tactics: Theory, evidence and recommendations. *J. Econ. Entomol.* **82**, 1263–1269.

77 Shi, Y., Wang, M.B., Powell, K.S., Van Damme, E., Hilder, V.A., Gatehouse,

A.M.R., Boulter, D. and Gatehouse, J.A. (1994). Use of the rice sucrose syn-thase-1 promoter to direct phloem-specific expression of beta-glucuronidase and snowdrop lectin in transgenic tobacco plants. *J. Exp. Bot.* **45**, 623–631.

78 Duan, X., Li, X., Xue, Q., Abo-El-Saad, M., Xu, D. and Wu, R. (1996). Trans-genic rice plants harboring an introduced potato proteinase inhibitor II gene are insect resistant. *Nat. Biotechnol.* **14**, 494–498.

79 Williams, S., Friedrich, L., Dincher, S., Carozzi, N., Kessmann, H., Ward, E. and Ryals, J. (1993). Chemical regulation of *Bacillus thuringiensis* delta-endo-toxin expression in transgenic plants. *Bio/Technology* **7**, 194–200.

80 Roush, R.T. (1996). Can we slow adaptation by pests to insect transgenic crops? In: *Biotechnology and Integrated Pest Management* (Persley G.L., Ed.). CAB International, Cambridge, pp. 242–263.

81 Liu, Y.-B. and Tabashnik, B.E. (1997). Experimental evidence that refuges delay insect adaptation to *Bacillus thuringiensis*. *Proc. R. Soc. London* **264**, 605–610.

82 Liu, Y.-B., Tabashnik, B.E., Dennehy, T.J., Patin, A.L. and Bartlett, A.C. (1999). Development time and resistance to *Bt* crops. *Nature* **400**, 519.

83 Flavell, B., Dart, E., Fuchs, R.L. and Fraley, R.T. (1992). Selectable marker genes: Safe for plants. *Bio/Technology* **10**, 141–144.

84 Fuchs, R.L., Ream, J.E., Hammond, B.G., Naylor, M.W., Leimgruber, R.M. and Berberich, S.A. (1998). Safety assessment of the neomycin phosphotrans-ferase II (NPTII) protein. *Bio/Technology* **11**, 1543–1547.

85 Metz, L.J., Stiekema, W.J. and Nap, J.P. (1998). A transgene-centered approach to the biosafety of transgenic phosphinothricin-tolerant plants. *Mol. Breed.* **4**, 335–341.

86 Schuler, T.H., Poppy, G.M., Kerry, B.R. and Denholm, I. (1999). Potential side effects of insect-resistant transgenic plants on arthropod natural enemies. *Trends Biotechnol.* **16**, 210–216.

87 Poppy, G. (2000). GM crops: Environmental risks and non-targets effects. *Trends Plant Sci.* **5**, 4–6.

88 Losey, J.E., Rayor, L.S. and Carter, M.E. (1999). Transgenic pollen harms monarch larvae. *Nature* **399**, 214.

89 Hodgson, J. (1999). Monarch *Bt*-corn paper questioned. *Nat. Biotechnol.* **17**, 627.

90 Schuler, T.H., Potting, R.P., Denholm, I. and Poppy, G.M. (1999). Parasitoid behaviour and *Bt* plants. *Nature* **400**, 825.

91 Dogan, E.B., Berry, R.E., Reed, G.L. and Rossignol, P.A. (1996). Biological parameters of convergent lady beetle (Coleoptera: Coccinellidae) feeding on aphids (Homoptera: Aphididae) on transgenic potatoes. *J. Econ. Entomol.* **89**, 1105–1108.

92 Riddick, E.W. and Barbosa, P. (1998). Impact of Cry3-intoxicated *Leptino-tarsa decemlineata* (Coleoptera: Chrysomelidae) and pollen on consumption, development and fecundity of *Coleomegilla maculata* (Coleoptera: Coccinelli-dae). *Ann. Entomol. Soc. Am.* **91**, 303–307.

93 Pilcher, C.D., Obrycki, J.J., Rice, M.E. and Lewis, L.C. (1997). Preimaginal development, survival, and field abundance of insect predators on transgenic *Bacillus thuringiensis* corn. *Environ. Entomol.* **26**, 446–454.

94 Hilbeck, A., Baumgartner, M., Fried, P.M. and Bigler, F. (1998). Effect of transgenic *Bacillus thuringiensis* corn-fed prey on mortality and development

time of immature *Chrysoperla carnae* (Neuroptera: Chrysopidae). *Environ. Entomol.* **27**, 480–487.

95 Birch, A.N.E., Geoghegan, I.E., Majerus, M.E.N., McNicol, J.W., Hackett, C.A., Gatehouse, A.M.R. and Gatehouse, J.A. (1999). Tri-trophic interactions involving pest aphids, predatory 2-spot ladybirds and transgenic potatoes expressing snowdrop lectin for aphid resistance. *Mol. Breed.* **5**, 75–83.

96 Down, R.E., Ford, L., Woodhouse, S.D., Raemaekers, R.J.M., Leitch, B., Gatehouse, J.A. and Gatehouse, A.M.R. (2000). Snowdrop lectin (GNA) has no acute toxic effects on the beneficial insect predator, the 2-spot ladybird (*Adalia bipunctata* L.). *J. Insect Physiol.* **46**, 379–391.

97 Walter, A.J., Ford, L., Majerus, M.E.N., Geoghegan, I.E., Birch, A.N.E., Gatehouse, J.A. and Gatehouse, A.M.R. (1998). Characterization of the proteolytic activity of the larval gut of two-spot ladybird (*Adalia punctata* L) and its sensitivity to proteinase inhibitors. *Insect Biochem. Mol. Biol.* **28**, 173–180.

98 Bell, H.A., Fitches, E.C., Down, R.E., Marris, G.C., Edwards, J.P., Gatehouse, J.A. and Gatehouse, A.M.R. (2001). Dietary Cowpea trypsin inhibitor (CpTI) affects growth and development of the tomato moth *Lacanobia oleracea* (Lepidoptera: Noctuidae) and the success of the gregarious ectoparasitoid *Eulophus pennicornis* (Hymenoptera: Eulophidae). *Pest Manage. Sci.* **57**, 57–66.

99 Serageldin, I. (1999). Biotechnology and food security in the 21st century. *Science* **285**, 387–389.

100 Toenniessen, G.H. (1995). Plant biotechnology and developing countries. *Trends Biotechnol.* **13**, 404–409.

101 Boulter D. (1995). Plant biotechnology: Facts and public perception. *Phytochemistry* **40**, 1–9.

102 Gaskell, G., Bauer, M.W., Durant, J. and Allum, N.C. (1999) Worlds apart? The reception of genetically modified foods in Europe and the US. *Science* **285**, 384–387.

103 Ruibal-Mendieta, N.L. and Lints, F.A. (1998). Novel and transgenic food crops: Overview of scientific versus public perception. *Transgenic Res.* **7**, 379–386.

14 Using proteins to assess the potential impacts of genetically modified plants on honey bees

L.A. Malone and M.H. Pham-Delègue

Summary

Genetically modified plants manifest new traits via the expression of foreign proteins encoded by inserted transgenes. For example, cotton modified to contain a *Bacillus thuringiensis* (Bt) gene and expressing Bt toxin in its leaves and buds will be protected from bollworm attack. Since the protein products of many transgenes can be purified, these "active ingredients" of genetically modified plants can be used in experiments to assess the likely impacts of such plants on bees. Such tests have a number of advantages: they can be conducted prior to the lengthy process of plant modification, the effects of the proteins can be quantified and some tests may be conducted with bees outside strict quarantine conditions. The shortcomings of this approach are that indirect impacts of genetically modified plants on bees, such as pleiotropic effects resulting from changes in plant phenotype, cannot be assessed and that the test conditions may be somewhat artificial, for example keeping the bees in cages in an incubator. This chapter summarizes current results from bioassays with bees and purified transgene products. Effects of a range of proteins, Bt toxins, protease inhibitors, chitinases, glucanases, and biotin-binding proteins, on adult bee gut physiology, food consumption, olfactory learning behavior, and longevity are presented.

Introduction

Genetically modified (GM) plants are becoming an increasingly common component of agro-ecosystems throughout the world. For example, between 1997 and 1998 the global acreage planted in commercial GM crops, excluding China, increased 250 percent to almost 70 million acres [1]. Over 90 percent of this acreage was planted with either herbicide-resistant (71 percent) or Bt (*Bacillus thuringiensis*) insect-resistant (28 percent) crops. Most GM crops have been planted in industrialized nations, with the United States accounting for nearly three-quarters of the total. Herbicide-resistant soybeans were the most commonly planted GM crop in 1998 (52 percent), followed by Bt-corn (24 percent), herbicide-resistant oilseed rape (or canola) (9 percent), Bt- and herbicide-resistant cotton (9 percent), and herbicide-resistant corn (6 percent) [1]. Techniques have now been developed for genetically modifying a huge array of crop species. Field trials have been approved and are under way for field crops (e.g. wheat, rice, barley, tobacco), flowers, trees (e.g. poplar, spruce, sweetgum), oil crops (e.g. sunflower, peanut), grasses, sugar crops (beet and cane), fruits (e.g. apple, cranberry, grape, melon, strawberry), and vegetables (e.g. tomato, potato, broccoli, carrot, eggplant, lettuce, pea) [2].

In recent years there has also been increasing interest in and concern about the potential ecological impacts of GM plants, both in scientific circles [e.g. 3, 4] and in the popular media [e.g. 5]. Recent research has sought to generate information with which to quantify the ecological risks that may be associated with widespread use of GM plants. Key issues include the dispersal of transgenes to related weedy plants [6, 7], to soil microbes [8–10], and perhaps to bee gut microflora [11], effects on plant decomposition rates [12, 13], persistence of transgene products in the environment [14, 15], the development of insect resistance [16], and effects on nontarget organisms. This final category encompasses tests of the effects of GM plants on mammals [17–20], on tri-trophic interactions [21–25], on attractive insects such as butterflies [26–28], on soil biota [29, 30], and on pollinators such as the honey bee. Many GM crop species depend upon or have yields improved by honey bee pollination and many represent important nectar sources for honey production [31]. Thus there is significant interest in this beneficial insect and the impacts that GM plant technology may have upon it.

GM plants may have direct or indirect effects on bees. Direct effects may be defined as those that arise when a bee ingests the protein that a transgene encodes, for example when it is expressed in pollen. Indirect effects may be defined as those which arise if the process of introducing the transgene into the plant results in inadvertent changes to plant phenotype affecting its attractiveness or nutritive value to bees. Direct effects can be examined in experiments using purified proteins. Since the process of isolating and characterizing genes for incorporation into GM plants usually involves the expression of the gene in a bacterial or other expression system [32], it is often possible to obtain reasonable quantities of the proteins they encode. These can then be used in tests with both target and nontarget organisms. Alternatively, many such proteins can be obtained in purified form from their original sources, e.g. proteinase inhibitors from potatoes [33, 34] or Bt toxins from bacterial cultures [35]. Since the production of GM plants is labor-intensive, time-consuming, and may require strict containment conditions, it is often advantageous to conduct bioassays using the purified proteins in advance of beginning the process of plant transformation. This is a standard approach when identifying potential pest-resistance genes. For example, a range of protease inhibitors may be bioassayed before one is selected for incorporation into a GM plant and then tested *in planta* [e.g. 36, 37]. The same approach has been employed with honey bees and the results of such bioassays are the subject of this chapter.

Recent research results

Bt Cry proteins

The insecticidal properties of delta endotoxin proteins produced in crystals formed by the soil-dwelling bacterium, *Bacillus thuringiensis* or Bt, are well known [38]. These crystal (Cry) proteins are toxic only to insects and many have activity specific to insects of a particular Order, e.g. Cry1 proteins for Lepidoptera, Cry3 proteins for Coleoptera. Because of this, and the ease with which Bt can be cultured, preparations consisting of Bt cells, spores, and crystals have been used as biopesticides since 1961 [39]. These have a good safety record with honey bees, especially when compared to many chemical insecticides. For example, Bt products such as Dipel® or Foray® are described as having "very low toxicity" to bees, whereas Orthene® (an organophosphate), Carbaryl® (a carbamate), and Ripcord® (a synthetic pyrethroid) are all listed as "toxic to bees" [40].

More recently, the genes encoding various Cry proteins have been isolated, characterized, modified for plant expression and introduced successfully into a wide range of crop plants. Expression of a Cry protein in the leaves or roots effectively protects the plant against attack by insects susceptible to that protein. Recent commercial examples include Bt-corn resistant to European corn borer (*Ostrinia nubilalis*), Bt-cotton resistant to the cotton bollworm (*Helicoverpa armigera*), and Bt-potatoes resistant to the Colorado potato beetle (*Leptinotarsa decemlineata*) [2]. These plants present single Cry proteins to the insect in a pure and "activated" form, whereas the biopesticide preparations, containing whole bacteria and spores, usually present the insects with mixtures of toxins that need to be activated by conditions in the insect's gut. Because of this, additional testing needs to be undertaken to ensure the safety of GM Bt-plants to beneficial insects such as bees. Fortunately, Bt toxins can be purified and activated to resemble the state in which they are expressed in transgenic plants [41–43], and these can be used in trials with bees.

Purified Cry1Ac (formerly CryIA(c)) toxin (lepidopteran-active) fed at a concentration of 20 µg/ml to 1–3-day-old larvae and adults of *Apis mellifera* had no significant effect on the survival of these insects [44]. This toxin concentration was more than "100 times the concentration of Cry1Ac protein found in the field as present in pollen and nectar of transgenic cotton" [44], but the authors did not give details of these gene expression measurements. Similarly, purified Cry3Ba (formerly CryIIIB) toxin (coleopteran-active) fed in sugar syrup at concentrations of 0.066 or 0.332

percent to colonies of honey bees over a 2-month period had no effect on larval survival or pupal dry weight [45]. Purified Cry1Ba toxin (lepidopteran-active), mixed into a pollen-based food at 10, 2.5, or 0.25 mg/g and fed to adult honey bees for 7 days post-emergence, had no significant effect on the rate at which each food was consumed (Figure 14.1a) or on the longevity of the bees (Figure 14.1b) [46]. A similar lack of

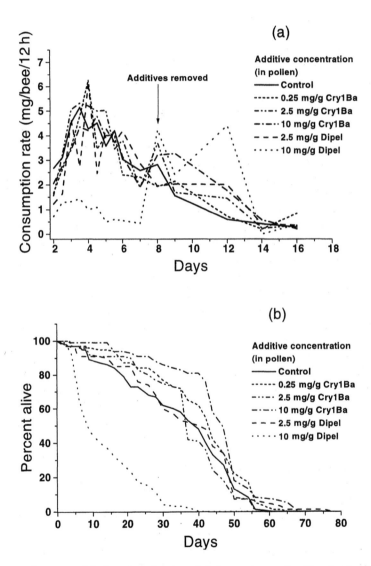

Figure 14.1 (a) Consumption by adult honey bees of pollen-food with 0.25, 2.5, or 10 mg/g Bt toxin (Cry1Ba) or with 2.5 or 10 mg/g Dipel® Bt biopesticide added. After 7 days all bees were given pollen-based food without additive. (b) Survival of these bees.

effect was noted with two Bt biopesticide preparations (Foray® 48B and Dipel® 2X) fed to bees in the same experiment at 2.5 mg/g. However, an extremely high concentration of Dipel® (10 mg/g) resulted in significantly reduced food consumption and survival, although whether this was due to the Bt toxin or some of the "inert" ingredients in the preparation was not ascertained [46]. Purified Cry1Ba protein (625 µg/g in pollen-based food) has also been fed for 7 days to newly-emerged adult bees that were tagged and then returned to outdoor hives and monitored. Bees fed with Cry1Ba did not differ significantly from control bees fed plain pollen-food in the timing of their first flights, the period during which flights took place or in estimated longevity [47].

In response to public concerns raised by recent monarch butterfly/Bt-corn pollen studies [26, 27], tests are also under way to compare the respective effects of purified lepidopteran-active Bt toxins, of a Bt biopesticide preparation, and of a conventional chemical pesticide commonly used on corn and potentially present when pollen is produced (M.H. Pham-Delègue, unpublished data). Such work is expected to assess the biosafety correlates of the different methods used for crop protection, rather than simply comparing the effects of GM plants versus control plants, which is not agronomically realistic.

Bee larval tests with Bt Cry proteins have been carried out in-house by biotech companies (S. Sims, personal communication), but there are no published reports of such work. Laboratory-based methods for rearing bee larvae have been established [48] and these could be adapted for use with Bt and other pest-resistance proteins, especially for the later larval instars which are known to ingest pollen as well as the glandular secretions of adult bees.

Results so far suggest that the specificities of different Bt toxins are retained in their activated form and, with the possible exception of those derived from hymenopteran-active Bt strains [49], Bt transgene products are very likely to be safe for honey bees.

Protease inhibitors

Protease inhibitors (PIs) represent a second class of proteins that may be expressed at insecticidal levels in GM plants [e.g. 37, 50–52], although none has been commercialized for this application as yet [53]. PIs can be isolated and purified from many different plants, animals, and microbes. When ingested by insects, some PIs can inhibit their digestive proteolytic enzymes, causing starvation and death [e.g. 54–60].

PIs vary in their ability to inhibit specific proteases. For example, cysteine proteases respond to one set of PIs and serine proteases to another. Some PIs bind strongly to only one type of protease; others have dual specificity. The impact of a PI on a particular insect will depend on the insect's gut protease profile and the specific activity (or activities) of the PI in question. Because their mechanism of action involves molecule-to-molecule binding, the impacts of PIs on insects are often dose-dependent [e.g. 55].

Honey bees use proteolytic enzymes, including trypsin, chymotrypsin, elastase, and leucine aminopeptidase, to digest dietary protein [61–64]. It is not surprising, therefore, that some PIs at some concentrations have been demonstrated to have effects on these insects. Bioassay methods for determining the effects of purified PIs incorporated into artificial diets on various pest insects are well established [e.g. 55, 58]. These methods have been adapted for use with adult honey bees, by mixing PIs at various concentrations into either sugar syrup or a protein food and presenting these to bees kept in small cages in an incubator.

Purified Bowman–Birk soybean trypsin inhibitor (BBI) fed to foraging (older) honey bees at concentrations of 1, 0.1, 0.01, or 0.001 mg/g of sugar syrup had no effect on bee survival over 4 days [65]. However, trypsin activity levels in foraging bees fed three different concentrations of BBI in syrup for 3.5 days were significantly different from those in control bees. The lowest BBI concentration (0.001 mg/g) resulted in a slight but significant increase in trypsin activity, while two other concentrations (0.1 and 1 mg/g) resulted in significant reductions in activity. *In vitro* tests, in which enzyme extracts from control bee guts were incubated with BBI at a range of concentrations, showed an 80 percent reduction in non-specific protease activity and a 100 percent reduction in trypsin activity.

Some other studies on the direct effects of PIs on bees have used newly-emerged adult bees [63, 64, 66]. It is only during the first week or so of adulthood that honey bees consume and need to digest significant amounts of protein-rich pollen [67], so the impacts of PIs would be expected to be greater at that time. When fed to young adult bees, four different serine endopeptidase inhibitors had dose-dependent effects on bee survival and many of the PI treatments significantly altered protease activity levels in the midguts of these bees [63, 64, 66].

Aprotinin (also known as bovine pancreatic trypsin inhibitor or BPTI) and SBTI (also known as SKTI or soybean Kunitz trypsin inhibitor) both significantly reduce the survival of bees fed these PIs *ad libitum* in sugar syrup at 10, 5, or 1 mg/ml, but not at 0.1 or 0.01 mg/ml [63, 66] (Figure 14.2a). *In vivo* activity levels of three midgut endopeptidases (trypsin, chymotrypsin, and elastase) and the exopeptidase leucine aminopeptidase (LAP) were determined for these bees at two time points: Day 8 after emergence and when 75 percent of bees had died. LAP activity levels increased significantly in bees fed with either inhibitor at all concentrations. At Day 8, bees fed BPTI at all concentrations had significantly reduced levels of trypsin, chymotrypsin, and elastase (Figure 14.3a). At the time of 75 percent mortality, bees fed BPTI at each concentration had reduced trypsin levels, but only those fed the inhibitor at the highest concentration had reduced chymotrypsin or elastase activity. At both time

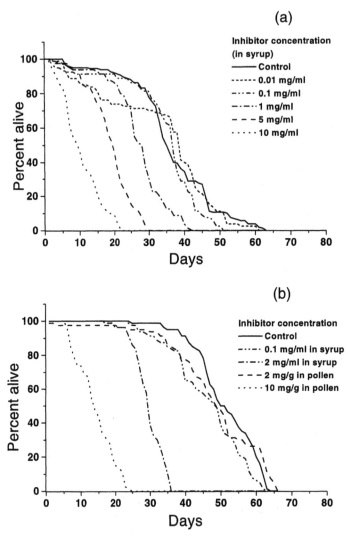

Figure 14.2 Survival of adult honey bees fed with (a) aprotinin (0.01, 0.1, 1, 5, or 10 mg/ml in sugar syrup, continuously) or with (b) potato protease inhibitor 1 (POT-1) (0.1 or 2 mg/ml in syrup, continuously, or 2, 10 mg/g in pollen-based food for 8 days).

points, only bees fed SBTI at the highest concentration had lowered trypsin, chymotrypsin, and elastase activities. These results suggest that the observed reductions in bee survival at the higher PI concentrations are in fact the result of a disruption in their ability to digest protein. We may also speculate that the increased levels of LAP represent some kind of compensatory mechanism to make up for the loss of proteolytic function in the gut.

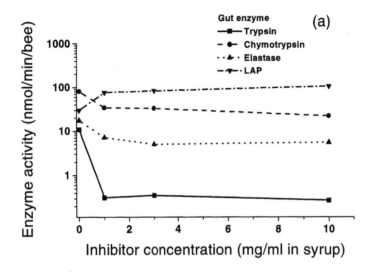

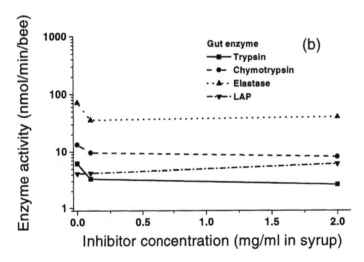

Figure 14.3 In vivo activities of four different digestive enzymes in adult bees after 8 days of feeding *ad libitum* on syrup with (a) 1, 3, or 10 mg/ml aprotinin or (b) 0.1 or 2 mg/ml potato protease inhibitor 1 (POT-1) added. Control bees were fed plain syrup without additive.

Aprotinin (2.5 mg/g in pollen-based food) has also been fed for 7 days to newly-emerged adult bees which were tagged and then returned to outdoor hives and monitored. These bees began to fly and also died about 3 days sooner than control bees fed with plain pollen-food, in accordance with the survival effects noted in some of the laboratory bioassays reported above. The period during which flights took place was not altered by the PI treatment [47].

Similar laboratory bioassay results were obtained with bees fed potato proteinase inhibitor I (POT-1 or PI-I) and potato proteinase inhibitor II (POT-2 or PI-II) [64] (Figure 14.2b). Newly-emerged bees were fed each PI in either sugar syrup (2 or 0.1 mg/ml) administered *ad libitum* or in a pollen-based food (10 or 2 mg/g) which was replaced with control food after 8 days. *In vivo* activities of trypsin, chymotrypsin, elastase, and LAP were determined at Day 3 and at Day 8 (Figure 14.3b). Enzyme activities were significantly lower at Day 8 than at Day 3, except for elastase, which did not change. Potato PI-II significantly reduced the activity of all endopeptidases at both time points, regardless of the dose level or the medium in which the inhibitor was administered. Potato PI-I acted in a similar manner, except that 0.1 mg/ml potato PI-I in syrup had no effect on bees. There was no consistent trend in changes in LAP activity. Survival of bees fed either inhibitor at 10 mg/g in pollen or at 2 mg/ml in syrup was significantly reduced, with the effect of the pollen treatment being greater than the syrup treatment. Survival of bees fed potato PI-I or potato PI-II at 2 mg/g in pollen or 0.1 mg/ml in syrup was similar to that of control bees.

Acute toxicity tests similar to those used to test chemical pesticides, in which 10-day-old adult honey bees were either fed or injected with cowpea trypsin inhibitor (CpTI), showed that an oral dose of 11 µg per bee and an injected dose of 0.5 µg per bee had no effect on bee survival after 24 or 48 hours [68].

Tests of short- and long-term toxicity of BBI, oryzacystatin I (OC-I), and chicken egg white cystatin to honey bees have also been carried out [69]. In the short-term test, 15-day-old worker bees were supplied with 11 µg of PI each over a period of 24 hours, and then given control syrup. None of the treatments resulted in significant bee mortality at 24, 48, or 96 hours. In the long-term test, 2-day-old bees were given a continuous supply of syrup with 26 µg/ml PI added and their longevity recorded. There was considerable variability in bee longevity in this test, but no significant effects could be attributed to the ingestion of these PIs at this low concentration and bees taken from the long-term test at 15–16 days had levels of midgut proteolytic activity that did not differ from the controls.

Sandoz [70] conducted further long-term tests with SBTI, OC-I, BBI, and a mixture of OC-I and BBI fed continuously to 2-day-old bees at concentrations of 1, 0.1, or 0.01 mg/ml. Significant mortality occurred only for bees fed SBTI, BBI, or the OC-I/BBI mixture at the highest dose level. These findings were confirmed by Jouanin *et al.* [71] who reported that OC-I (1, 0.1, or 0.01 mg/ml) had no effect on short- or long-term honey bee mortality. BBI at 1 mg/ml, however, reduced bee survival, altered olfactory learning performance and resulted in overproduction of the gut proteases, trypsin and chymotrypsin.

Additional work has been conducted on the effects of the two serine proteinase inhibitors, BBI and SBTI [72]. These experiments have shown that, compared to a control diet containing a neutral protein (bovine serum albumin), diets containing these PIs at 1 mg/ml, and at 0.1 mg/ml to a lesser extent, significantly increased the probability of bee death at a given time. Bee gut proteolytic activities were increased when BBI and SBTI were ingested at 1 or 0.1 mg/ml, with trypsin activity being increased at both concentrations, and other activities only at the higher concentration. Interestingly, new forms of proteinases that were still sensitive to BBI and SBTI were produced. This suggests that bees ingesting high doses of BBI or SBTI will overproduce proteinases and thus will require large quantities of amino acids derived from body proteins. Such mobilization of body proteins might explain the reduced longevity and lower behavioral responses of bees fed high doses of BBI or SBTI.

The impact of exposure to sublethal doses of PIs on adult honey bees is not yet known, but some studies of one component of foraging behavior, olfactory learning, have been carried out with honey bees that have consumed PIs. Addition of CpTI at 1, 5, or 10 μg/ml to the reward syrup offered in a conditioned proboscis extension assay significantly reduced the ability of bees to learn this response [68]. By contrast, addition of BBI or cystatin at the same concentrations did not affect short- or long-term learning ability in 15-day-old bees [69]. Furthermore, the learning performances of bees that had been fed *ad libitum* with syrup containing 26 μg/ml of either OC-I or BBI for about 13 days prior to the proboscis extension assay were unaltered by this treatment [69]. When bees were fed with SBTI, OC-I, BBI, or a OC-I/BBI mixture at 1, 0.1, or 0.01 mg/ml for 15 days prior to testing, their learning ability was significantly impaired only with the 1 mg/ml BBI treatment [71, 72].

Recently a study was designed to evaluate bees' intrapopulation variability in response to BBI exposure, and to relate these responses to the genetic background of the bees (M.H. Pham-Delègue, unpublished data). Worker bees from the same hive were caged and fed 1 μg/ml BBI for 15 days prior to testing in the conditioned proboscis assay. Individual learning performances were recorded. Then each individual was subjected to a genetic analysis using microsatellite markers, so that the paternal origin of each bee could be defined. Samples of bees from these lines were analyzed for protease content in both treated and control bees. Interline differences appeared in the learning abilities and in the protease content of the control groups. BBI treatment induced a significant change in the learning performances, especially during the extinction phase, which was consistent with previous data [72]. This effect was found for all bee lines, which suggests that although the genetic background is different, the need to compensate the ingestion of BBI will result in a general metabolic reaction leading to a change in the resistance to extinction of a learnt response. In parallel, an increase in the amount of digestive proteases was shown, with new forms, which were all sensitive to BBI, being produced (M.H. Pham-Delègue, unpublished data). Further studies are still needed to establish whether there are interline differences in the response to BBI ingestion at the level of the digestive physiology.

Thus, research with PIs and bees so far suggests that adult bee gut protease activities may be reduced, with a resultant impact on bee longevity, when bees ingest these proteins. However, the effects will depend on the specificity of the particular inhibitor and the concentration to which the bee is exposed. PI effects on bee larvae have not yet been ascertained.

Few data are available to evaluate long-term exposure of bee colonies to transgene products. Colonies fed sucrose solutions supplemented with BBI (6 µg/ml) over a 2-month period were compared with a control colony fed with standard sucrose solution. There were no differences between the two, either in the amount of stored food and brood or in the foraging activity and learning abilities of the treated bees. Additionally, the proteolytic activities measured in adult bees and larvae sampled at the end of the exposure period were similar for control and BBI-treated hives (M.H. Pham-Delègue, unpublished data).

There are as yet few published measurements of PI expression levels in pollen. It is therefore difficult to extrapolate from the results of experiments with purified transgene products to making predictions about the effects of GM plants on bees. However, if we assume that the bees in the bioassays described above received a diet which was 25 percent protein, then the doses of PIs administered ranged from 0.004 to 4 percent of total protein received. GM PI-plants that are effectively protected from pest attack typically have leaf expression levels ranging from 0.05 to 2.5 percent of total protein. For example, rice expressing 0.5 to 2 percent of a potato PI was resistant to pink stem borer [73], *Spodoptera litura* were killed by feeding on leaves of tobacco expressing 0.4 to 1 percent soybean trypsin inhibitor [37], rice expressing 0.05 to 2.5 percent soybean trypsin inhibitor had improved resistance to brown planthopper [74], and *Wiseana* spp. growth was reduced on white clover expressing 0.07 percent aprotinin [75]. Thus we may expect that, if pollen PI expression levels are identical to those in leaves, then some GM PI-plants may have the potential to have an impact on adult bee longevity. However, such effects will undoubtedly be vastly less than the impacts that many currently-used chemical pesticides, especially those with high contact toxicity, have on bees.

Chitinases

Genes encoding chitin-degrading enzymes have been isolated from a number of sources, including plants, insects, and entomopathogenic microorganisms [e.g. 76–81]. As chitin is an important structural component in fungi and insects, chitinase genes have been engineered into plants in order to protect them from fungal infection and pest attack [e.g. 82–84]. As with other insects, chitin is an important component of the cuticle of honey bees. Thus bees might be affected by ingesting chitinases expressed in transgenic plants.

Acute toxicity tests with 10-day-old adult honey bees fed sugar solution containing a chitinase purified from tomato ($11\,\mu g$ per bee) showed that this transgene product had no significant impact on bee survival after 24 or 48 hours [68]. Bees injected with $1.69\,\mu g$ of chitinase were similarly unaffected.

Using a standard conditioned proboscis extension assay in individual restrained bees, it was demonstrated that concentrations of 1, 5, or $10\,\mu g/ml$ chitinase added to the sugar reward delivered during the training period did not affect olfactory learning performance [68]. Complementary studies were conducted at the colony level in a flight room, using an artificial flower device [85]. Sucrose solutions, either pure or combined with $1.3\,mg/ml$ chitinase diluted either 100 or 1000 times, were presented in a choice situation. There was no evidence of discrimination in the weights of solution collected. However, the number of visits was lower by a factor of four on the protein-added sources, compared to the control solution.

Results so far suggest that bees will not be directly affected by the chitinases tested, although ranges of doses have not yet been tested and effects on bee larvae have not yet been ascertained.

β-1,3-Glucanases

Glucanase genes have been isolated from a number of different plants, where they form an important part of the plant's response to attack from fungal pathogens [e.g. 86–88]. They have also been isolated from microorganisms [e.g. 89, 90]. Transgenic plants expressing β-1,3-glucanase have demonstrated enhanced resistance to fungal pathogens [91]. This protein is highly unlikely to be harmful to bees, since its substrate, β-1,3-glucan, has not been found in insects.

Purified β-1,3-glucanase had no effect on the 24- or 48-hour survival of adult bees fed with $11\,\mu g$ per bee or injected with $0.3\,\mu g$ of this transgene product [68].

As with chitinase, the effects of 1, 5, or $10\,\mu g/ml$ β-1,3-glucanase on bee olfactory learning abilities has also been tested using the conditioned proboscis extension assay [68]. With this protein, a lower resistance to the extinction of the conditioned response was found, i.e. after training, bees stopped exhibiting the proboscis extension response to unrewarded presentations of the olfactory stimulus more rapidly than in the control group. At the colony level, the number of visits to feeders of an artificial flower device set in a flight room, filled with sucrose solution mixed with $110\,\mu g/ml$ β-1,3-glucanase diluted between 100 and 10000 times, was lower as the concentration increased. However, there were no differences in the amounts of solution collected that could be attributed to the type of feeder solution presented [85].

Lectins

Lectins are one class of entomotoxins under investigation for use against Homopteran pests and the snowdrop lectin, *Galanthus nivalis* agglutinin (GNA), has been introduced into several plants, including tobacco, wheat, and potatoes, to confer resistance against Homoptera [92]. The mode of action of GNA is still relatively unknown but it has been shown to bind to insect gut cells, including those of aphids [93, 94], consequently inducing a disruption in nutrient assimilation.

The acute toxicity of GNA at three concentrations (500, 800, $1000\,\mu g/ml$) was tested. No additional mortality was observed after 24 and 48 hours for any of the treated groups. Chronic toxicity of 10, 100 and $1000\,\mu g/ml$ GNA-added solutions, over a 2-month period, was not significantly different from controls. Consistently, 2 weeks' exposure to the protein at these concentrations did not affect bees' learning responses in the conditioned proboscis extension procedure (M.H. Pham-Delègue, unpublished data).

Biotin-binding proteins

Proteins that bind to vitamins, such as biotin, represent another category of potential pest-resistance transgene products [95–97]. Genes encoding two such proteins, avidin and streptavidin, have been isolated from chickens [98] and a bacterium [99], respectively. The avidin gene has recently been incorporated into GM plants, which are then insecticidal to a variety of insects [100, 101] Since avidin functions by binding to biotin and creating a vitamin deficiency (which is reversible) in the insect eating it, predictions about its likely effects can be made with some knowledge of the relative molar concentrations of both compounds in the insect's diet. Quantitative information about the insect's biotin requirements is also helpful, but for many species, including the honey bee, this is lacking.

Multifloral New Zealand bee-collected pollen contains $1.85 \pm 0.08\,\mu M$ biotin and New Zealand bee-bread contains $1.83 \pm 0.36\,\mu M$ biotin (J.T. Christeller, personal communication). This suggests that pollen processing by the bees does not introduce additional biotin. Pollen avidin expression levels have not yet been measured (J.T. Christeller, personal communication). However, it is possible that avidin levels may exceed biotin levels in GM pollen. We have fed newly-emerged adult honeybees *ad libitum* with multifloral pollen with 6.7 or $20\,\mu M$ avidin added and observed no significant change in the amounts of food consumed or in bee longevity, suggesting that such bees either have no need for biotin or can biosynthesize it (L.A. Malone, unpublished data). Further tests with bee larvae are under way.

Even if further bioassays and experiments with whole GM plants were to show that avidin may have an impact on bee nutrition, feeding a biotin supplement to reverse its effects would not be difficult, as biotin is inexpensive and beekeeping in most countries involves routine feeding and medication of colonies.

Glufosinate resistance proteins

Herbicide resistance is at present one of the most commonly used traits in commercial cultivars of GM crop plants [1]. Since this resistance operates via the production of an enzyme to break down the herbicide and bees lack such substrates, they are extremely unlikely to be harmed by these plants. Bioassays with the purified products of such genes have not been carried out, but some experiments with GM plants have (see Chapter 15).

Concluding remarks

Laboratory bioassay and in-hive test results so far suggest that the effects of ingestion of transgene products by honey bees will be relatively predictable. These will depend on the nature of the protein, its specificity and mode of action, and its concentration in the bee's diet. For example, tests to date with lepidopteran-specific Bt Cry proteins and with β-1,3-glucanase have shown that these proteins have no effect on honey bees. This is not surprising as bees probably lack the necessary Bt receptors and do not contain β-1,3-glucan. In contrast, PIs capable of inhibiting some of the proteases known to be active in adult bee guts may have a negative impact on bee longevity. As with other insects [55], these effects appear to be dose dependent.

With better information on transgene expression levels in pollen and nectar, laboratory assays could be designed to deliver purified proteins at realistic concentrations to bees and thus provide useful information for predicting direct effects of GM plants on adult and larval bees.

One potential disadvantage with this methodology is that the effects on the social interactions that occur among bees in the hive are excluded in caged-bee experiments. Even behaviors that are not obviously connected to social aspects of bee life, such as the duration of pollen feeding, may differ somewhat between bees in cages and in hives [67]. However, these are perennial problems for honey bee researchers. Laboratory experiments may sacrifice the "social reality" of the beehive, but allow better control over the factors that may influence the bees in the experiment. The replication necessary to obtain statistically significant results can be attained more easily with caged bees in the laboratory than with field experiments where often the unit of replication is the whole hive rather than a cage containing 25 bees. Furthermore, while purified proteins can often be expensive to obtain, they are inevitably less expensive and more readily obtained than GM plants, which require time, skill, labor, and glasshouse space or land to grow.

Bioassays with purified proteins provide useful information for broadly determining the areas that require further study. Information obtained may also assist with decisions about how appropriate a particular genetic modification will be for a particular plant species. In the future this methodology could usefully be extended to look at other transgene products under development for expression in GM plants.

Acknowledgments

L. Malone would like to thank the following people for assistance with the experimental work which generated the data shown in the figures and described as previously unpublished work: E.P.J. Burgess, J.T. Christeller, H.S. Gatehouse, A. Gunson, M. Lester, B.A. Philip and E.L. Tregidga, all of The Horticulture and Food Research Institute of New Zealand Ltd. She also wishes to acknowledge E.L. Tregidga's assistance in the preparation of the manuscript. M.H. Pham-Delègue and her French co-workers wish to acknowledge support from the EU Biotechnology program, BIO4-CT96-0365.

References

1 James, C. (1998). *Global Review of Commercialized Transgenic Crops: 1998, International Service for the Acquisition of Agri-biotech Applications (ISAAA).* Briefs No. 8, Ithaca, New York, pp. vi and 43.

2 Hagedorn, C. (1997). Commercial status of transgenic crops and microorganisms. www.ext.vt.edu/news/0021/cals/cses/chagedor/crplist.html.

3 Poppy, G. (1998). Transgenic plants and bees: The beginning of the end or a new opportunity? *Bee World* **79**, 161–164.

4 Beringer, J.E. (2000). Releasing genetically modified organisms: Will any harm outweigh any advantage? *J. Appl. Ecol.* **37**, 207–214.

5 Huttall, N. (1998). Genetically altered plants could wreak environmental havoc. *The Times*, July 13.

6 Dale, P.J., Scheffler, J.A., Schmidt, E.R. and Hankeln, T. (1996). Gene dispersal from transgenic crops. In: *Transgenic Organisms and Biosafety: Horizontal Gene Transfer, Stability of DNA, and Expression* (Schmidt, E.R., Ed.). Springer Verlag, Berlin, pp. 85–93.

7 Hill, J.E. (1999). Concerns about gene flow and the implications for the development of monitoring protocols. In: *Gene Flow and Agriculture: Relevance for Transgenic Crops.* BCPC Symp. Proc. 72, Keele, UK, pp. 217–224.

8 Tirodimos, I., Pretorius-Guth, I.M., Priefer, U., Tsaftaris, A. and Tsiftsoglou A.S. (1993). Methods of assessment of transfer of a gentamycin-resistance gene (aacC1) from genetically engineered microorganisms into agriculturally important soil bacteria. *Appl. Microbiol. Biotechnol.* **38**, 526–530.

9 Nielsen, K.M., Gebhard, F., Smalla, K., Bones, A.M. and van Elsas, J.D. (1997). Evaluation of possible horizontal gene transfer from transgenic plants to the soil bacterium *Acinetobacter calcoaceticus* BD413. *Theor. Appl. Genet.* **95**, 815–821.

10 Nielsen, K.M., van Elsas, J.D. and Smalla, K. (2000). Transformation of *Acinetobacter* sp. strain BD41 3 (pFG4DELTAnptII) with transgenic plant DNA in soil microcosms and effects of kanamycin on selection of transformants. *Appl. Environ. Microbiol.* **66**, 1237–1242.

11 Barnett, A. (2000). New research shows genetically modified genes are jumping species barrier. *NewsCenter*, available at www.commondreams.org/headlines/052800–01.htm.

12 Donegan, K.K., Seidler, R.J., Fieland, V.J., Schaller, D.L., Palm, C.J., Ganio, L.M., Cardwell, D.M. and Steinberger, Y. (1997). Decomposition of genetically engineered tobacco under field conditions: Persistence of the proteinase inhibitor I product and effects on soil microbial respiration and protozoa, nematode and microarthropod populations. *J. Appl. Ecol.* **34**, 767–777.

13 Glandorf, D.C.M., Bakker, P.A.H.M. and van Loon, L.C. (1997). Influence of the production of antibacterial and antifungal proteins by transgenic plants on the saprophytic soil microflora. *Acta Bot. Neerl.* **46**, 85–104.

14 Sims, S.R. and Ream, J.E. (1997). Soil inactivation of the *Bacillus thuringiensis* subsp. *kurstaki* CryIIA insecticidal protein within transgenic cotton tissue: Laboratory microcosm and field studies. *J. Agric. Food Chem.* **45**, 1502–1505.

15 Saxena, D., Flores, S. and Stotzky, G. (1999). Insecticidal toxin in root exudates from Bt corn. *Nature* **402**, 480.

16 Gould, F. (1998). Sustainability of transgenic insecticidal cultivars: Integrating pest genetics and ecology. *Annu. Rev. Entomol.* **43**, 701–726.

17 Kough, J.L. (1994). US EPA considerations for mammalian health effects presented by transgenic plant pesticides. In: *Food Safety Evaluation. Proc. OECD-Sponsored Workshop.* OECD, pp. 156–161.

18 Hammond, B.G., Vicini, J.L., Hartnell, G.F., Naylor, M.W., Knight C.D., Robinson, E.H., Fuchs, R.L. and Padgette, S.R. (1996). The feeding value of soybeans fed to rats, chickens, catfish and dairy cattle is not altered by genetic incorporation of glyphosate tolerance. *J. Nutrit.* **126**, 717–727.

19 Noteborn, H.P.J.M., Kuiper, H.A. and Jones, D.D. (1994). Safety assessment strategies for genetically modified plant products: A case study of *Bacillus thuringiensis*-toxin tomato. In: *Proc. 3rd Int. Symp on The Biosafety Results of Field Tests of Genetically Modified Plants and Organisms* (Young, A.L. and Economidis, I., Eds). USDA, Monterey, USA, pp. 199–207.

20 Hashimoto, W., Momma, K., Yoon, H.J., Ozawa, S., Ohkawa, Y., Ishige, T., Kito, M., Utsumi, S. and Murata, K. (1999). Safety assessment of transgenic potatoes with soybean glycinin by feeding studies in rats. *Biosci. Biotechnol. Biochem.* **63**, 1942–1946.

21 Dogan, E.B., Berry, R.E., Reed, G.L. and Rossignol, P.A. (1996). Biological parameters of convergent lady beetle (Coleoptera: Coccinellidae) feeding on aphids (Homoptera: Aphididae) on transgenic potato. *J. Econ. Entomol.* **89**, 1105–1108.

22 Lozzia, G.C., Furlanis, C., Manachini, B. and Rigamonti, I.E. (1998). Effects of Bt corn on *Rhopalosiphum padi* L. (Rhynchota Aphididae) and on its predator *Chrysoperla carnea* Stephen (Neuroptera Chrysopidae). *Boll. Zool. Agrar. Bachic.* **30**, 153–164.

23 Birch, A.N.E., Geoghegan, I.E., Majerus, M.E.N., McNicol, J.W., Hackett, C.A., Gatehouse, A.M.R. and Gatehouse, J.A. (1999). Tri-trophic interactions

involving pest aphids, predatory 2-spot ladybirds and transgenic potatoes expressing snowdrop lectin for aphid resistance. *Mol. Breed.* **5**, 75–83.

24 Schuler, T.H., Poppy, G.M., Potting, R.P.J., Denholm, I. and Kerry, B.R. (1999). Interactions between insect tolerant genetically modified plants and natural enemies. In: *Gene Flow and Agriculture: Relevance for Transgenic Crops. British Crop Protection Council Symposium Proceedings No. 72* (Lutman, P.J.W., Ed.). British Crop Protection Council, Staffordshire, UK, pp. 197–202.

25 Jørgensen, H.B. and Lövei, G.L. (1999). Tri-trophic effect on predator feeding: Consumption by the carabid *Harpalus affinis* of *Heliothis armigera* caterpillars fed on proteinase inhibitor-containing diet. *Entomol. Exp. Appl.* **93**, 113–116.

26 Losey, J.E., Rayor, L.S. and Carter, M.E. (1999). Transgenic pollen harms monarch larvae. *Nature* **399**, 214.

27 Jesse, L.C.H. and Obrycki, J.J. (2000). Field deposition of Bt transgenic corn pollen: Lethal effects on the monarch butterfly. *Oecologia*, available at athene.em.springer.de.

28 Wraight, C.L., Zangeri, A.R., Carroll, M.J. and Berenbaum, M.R. (2000). Absence of toxicity of *Bacillus thuringiensis* pollen to black swallowtails under field conditions. *Proc. Natl. Acad. Sci. USA* **97**, 7700–7703.

29 Lozzia, G.C. (1999). Biodiversity and structure of ground beetle assemblages (Coleoptera Carbidae) in Bt corn and its effects on non target insects. *Boll. Zool.a Agrar. Bachic.* **31**, 37–50.

30 Griffiths, B.S., Geoghegan, I.E. and Robertson, W.M. (2000). Testing genetically engineered potato, producing the lectins GNA and Con A, on non-target soil organisms and processes. *J. Appl. Ecol.* **37**, 159–170.

31 Crane, E. and Walker P. (1984). *Pollination Directory for World Crops.* International Bee Research Association, Gerrards Cross, UK, p. 183.

32 Meyers, R.A. (1995). *Molecular Biology and Biotechnology: A Comprehensive Desk Reference.* VCH, New York, pp. xxxviii and 1034.

33 Melville, J.C. and Ryan, C.A. (1972). Chymotryptic inhibitor I from potatoes. Large scale preparation and characterisation of its subunit components. *J. Biol. Chem.* **247**, 3445–3453.

34 Bryant, J., Green, T.R., Gurusaddaiah, T. and Ryan, C.D. (1976). Proteinase inhibitor II from potatoes: Isolation and characterisation of its promoter components. *Biochemistry* **15**, 3418–3424.

35 Höfte, H., de Greve, H., Seurinck, J., Jansens, S., Mahillon, J., Ampe, C., Vandekerckhove, J., Vanderbruggen, H., Van Montagu, M., Zabeau, M. and Vaeck, M. (1986). Structure and functional analysis of a cloned delta-endotoxin of *Bacillus thuringiensis* Berliner 1715. *Eur. J. Biochem.* **161**, 273–280.

36 McManus, M.T. and Burgess, E.P.J. (1995). Effects of the soybean (Kunitz) trypsin inhibitor on growth and digestive proteases of larvae of *Spodoptera litura. J. Insect Physiol.* **41**, 731–738.

37 McManus, M.T. and Burgess, E.P.J. (1999). Expression of the soybean (Kunitz) trypsin inhibitor in transgenic tobacco: Effects on feeding larvae of *Spodoptera litura. Transgenic Res.* **8**, 383–395.

38 Schnepf, H.E. (1995). *Bacillus thuringiensis* toxins: Regulation, activities and structural diversity. *Curr. Opin. Biotechnol.* **6**, 305–312.

39 Bryant, R., Bite, M.G. and Hopkins, W.L. (1999). *Global Insecticide Direc-*

tory, Second Edition, A Global Reference of Major Experimental and Commercial Insecticide Compounds, The Companies That Discovered Them and The Markets for Their Application. Agranova, UK, p. 208.

40 Walton, T. (2000). *New Zealand Agrichemical Manual*. WHAM Chemsafe Limited, Wellington, New Zealand, p. 636.

41 Hofmann, C. and Luthy, P. (1986). Binding and activity of *Bacillus thuringiensis* delta-endotoxin to invertebrate cells. *Arch. Microbiol.* **146**, 7–11.

42 Hofmann, C., Luthy, P., Hutter, R. and Pliska, V. (1988). Binding of the delta-endotoxin from *Bacillus thuringiensis* to brush-border membrane vesicles of the cabbage butterfly (*Pieris brassicae*). *Eur. J. Biochem.* **173**, 85–91.

43 Simpson, R.M., Burgess, E.P.J. and Markwick, N.P. (1997). *Bacillus thuringiensis* delta-endotoxin binding sites in two Lepidoptera, *Wiseana* spp. and *Epiphyas postvittana*. *J. Invertebr. Pathol.* **70**, 136–142.

44 Sims, S.R. (1995). *Bacillus thuringiensis* var. *kurstaki* (CryIA(c)) protein expressed in transgenic cotton: Effects on beneficial and other non-target insects. *Southwest. Entomol.* **20**, 493–500.

45 Arpaia, S. (1996). Ecological impact of Bt-transgenic plants: 1. Assessing possible effects of CryIIIB toxin on honey bee (*Apis mellifera* L.) colonies. *J Genet. Breed.* **50**, 315–319.

46 Malone, L.A., Burgess, E.P.J. and Stefanovic, D. (1999). Effects of a *Bacillus thuringiensis* toxin, two *Bacillus thuringiensis* biopesticide formulations, and a soybean trypsin inhibitor on honey bee (*Apis mellifera* L.) survival and food consumption. *Apidologie* **30**, 465–473.

47 Malone, L.A., Burgess, E.P.J., Gatehouse, H.S., Voisey, C.R., Tregidga, E.L. and Philip, B.A. (2001). Effects of ingestion of a *Bacillus thuringiensis* toxin and a trypsin inhibitor on honey bee flight activity and longevity. *Apidologie* **32**, 57–68.

48 Peng, Y.S.C., Mussen, E., Fong, A., Montague, M.A. and Tyler, T. (1992). Effects of chlortetracycline of honey bee worker larvae reared *in vitro*. *J. Invertebr. Pathol.* **60**, 127–133.

49 Benz, G. and Joeressen, H.J. (1994). A new pathotype of *Bacillus thuringiensis* with pathogenic action against sawflies (Hymenoptera, Symphyta). *Bull. OILB-SROP* **17**, 35–38.

50 Boulter, D., Edwards, G.A., Gatehouse, A.M.R., Gatehouse, J.A. and Hilder, V.A. (1990). Additive protective effects of different plant-derived insect resistance genes in transgenic tobacco plants. *Crop Prot.* **9**, 351–354.

51 Hilder, V.A., Gatehouse, A.M.R., Sherman, S.E., Barker, R.F. and Boulter, D. (1987). A novel mechanism for insect resistance engineered into tobacco. *Nature* **330**, 160–163.

52 Johnson, R., Narvaez, J., An, G. and Ryan, C. (1989). Expression of proteinase inhibitors I and II in transgenic tobacco plants: Effects on natural defense against *Manduca sexta* larvae. *Proc. Natl. Acad. Sci. USA* **86**, 9871–9875.

53 Hilder, V.A. and Boulter, D. (1999). Genetic engineering of crop plants for insect resistance – a critical review. *Crop Prot.* **18**, 177–191.

54 Burgess, E.P.J., Stevens, P.S., Keen, G.K., Laing, W.A. and Christeller, J.T. (1991). Effects of protease inhibitors and dietary protein level on the black field cricket *Teleogryllus commodus*. *Entomol. Exp. Appl.* **61**, 123–130.

55 Burgess, E.P.J., Main, C.A., Stevens, P.S., Christeller, J.T., Gatehouse,

A.M.R. and Laing, W.A. (1994). Effects of protease inhibitor concentration and combinations on the survival, growth and gut enzyme activities of the black field cricket, *Teleogryllus commodus. J. Insect Physiol.* **40**, 803–811.

56 Gatehouse, A.M.R., Gatehouse, J.A., Dobie, P., Kilminster, A.M. and Boulter, D. (1979). Biochemical basis of insect resistance in *Vigna unguiculata. J. Sci. Food Agric.* **30**, 948–958.

57 Johnston, K.A., Lee, M., Gatehouse, J.A. and Anstee, J.H. (1991). The partial purification and characterisation of serine protease activity in the midgut of larval *Helicoverpa armigera. Insect Biochem.* **21**, 389–397.

58 Johnston, K.A., Gatehouse, J.A. and Anstee, J.H. (1993). Effects of soybean protease inhibitors on the growth and development of larval *Helicoverpa armigera. J. Insect Physiol.* **39**, 657–664.

59 Johnston, K.A., Lee, M., Brough, C., Hilder, V.A., Gatehouse, A.M.R. and Gatehouse, J.A. (1995). Protease activities in the larval midgut of *Heliothis virescens*: Evidence for trypsin and chymotrypsin-like enzymes. *Insect Biochem. Mol. Biol.* **25**, 375–383.

60 Steffens, R., Fox, F.R. and Kassel, B. (1978). Effect of trypsin inhibitors on growth and metamorphosis of corn borer larvae *Ostrinia nubilalis. J. Agric. Food Chem.* **26**, 170–174.

61 Dahlmann, B., Jany, K.D. and Pfleiderer, G. (1978). The midgut endopeptidases of the honey bee (*Apis mellifica*): Comparison of the enzymes in different ontogenetic stages. *Insect Biochem.* **8**, 203–211.

62 Moritz, B. and Craildheim, K. (1987). Physiology of protein digestion in the midgut of the honeybee (*Apis mellifera* L.). *J. Insect Physiol.* **33**, 923–931.

63 Burgess, E.P.J., Malone, L.A. and Christeller, J.T. (1996). Effects of two proteinase inhibitors on the digestive enzymes and survival of honey bees (*Apis mellifera*). *J. Insect Physiol.* **42**, 823–828.

64 Malone, L.A., Burgess, E.P.J., Christeller, J.T. and Gatehouse H.S. (1998). *In vivo* responses of honey bee midgut proteases to two protease inhibitors from potato. *J. Insect Physiol.* **44**, 141–147.

65 Belzunces, L.P., Lenfant, C., Di Pasquale, S. and Colin, M.E. (1994). *In vivo* and *in vitro* effects of wheat germ agglutinin and Bowman–Birk soybean trypsin inhibitor, two potential transgene products, on midgut esterase and protease activities from *Apis mellifera. Comp. Biochem. Physiol.* **109B**, 63–69.

66 Malone, L.A., Giacon, H.A., Burgess, E.P.J., Maxwell, J.Z., Christeller, J.T. and Laing, W.A. (1995). Toxicity of trypsin endopeptidase inhibitors to honey bees (Hymenoptera: Apidae). *J. Econ. Entomol.* **88**, 46–50.

67 Crailsheim, K. and Stolberg, E. (1989). Influence of diet, age and colony condition upon intestinal proteolytic activity and size of the hypopharyngeal glands in the honeybee (*Apis mellifera* L.). *J. Insect Physiol.* **35**, 595–602.

68 Picard-Nizou, A.L., Grison, R., Olsen, L., Pioche, C., Arnold, G. and Pham-Delègue, M.H. (1997). Impact of proteins used in plant genetic engineering: Toxicity and behavioral study in the honeybee. *J. Econ. Entomol.* **90**, 1710–1716.

69 Girard, C., Picard-Nizou, A.L., Grallien, E., Zaccomer, B., Jouanin, L. and Pham-Delègue, M.H. (1998). Effects of proteinase inhibitor ingestion on survival, learning abilities and digestive proteinases of the honeybee. *Transgenic Res.* **7**, 239–246.

70 Sandoz, G. (1996). Étude des effets d'inhibiteurs de protéases sur un insecte

pollinisateur, l'abeille domestique *Apis mellifera* L. Diplôme d'Agronomie Approfondie, Institut National Agronomique Paris-Grignon, p. 32.

71 Jouanin, L., Girard, C., Bonadé-Bottino, M., Le Metayer, M., Picard Nizou, A., Lerin, J. and Pham-Delègue, M. (1998). Impact of oilseed rape expressing proteinase inhibitors on coleopteran pests and honeybees. *Cah. Agric.* **7**, 531–536.

72 Pham-Delègue, M.H., Girard, C., Le Métayer, M., Picard-Nizou, A.L., Hennequet, C., Pons, O. and Jouanin, L. (2000). Long-term effects of soybean protease inhibitors on digestive enzymes, survival and learning abilities of honeybees. *Entomol. Exp. Appl.* **95**, 21–29.

73 Duan, X., Li, X., Xue, Q., Abo-el-Saad, M., Xu, D. and Wu, R. (1996). Transgenic rice plants harboring an introduced potato proteinase inhibitor II gene are insect resistant. *Nat. Biotechnol.* **14**, 494–498.

74 Lee, S.I., Lee, S.H., Koo, J.C., Chun, H.J., Lim, C.O., Mun, J.H., Song, Y.H. and Cho, M.J. (1999). Soybean Kunitz trypsin inhibitor (SKTI) confers resistance to the brown planthopper (*Nilaparvata lugens* Stal.) in transgenic rice. *Mol. Breed.* **5**, 1–9.

75 Voisey, C.R., Nicholls, M.F., Skou, B.J., Broadwell, A.H., Chilcott, C.N., Frater, C.M., Wigley, P.J., Burgess, E.P.J., Philip, B.A. and White, D.W.R. (1999). Expression of pest resistance genes in white clover. In: *Proc. 11th Australian Plant Breeding Conference* (Langridge, P., Barr, A., Auricht, G., Collins, G., Granger, A. and Paull, J., Eds). CRC for Molecular Plant Breeding, Waite Campus, University of Adelaide, Glen Osmond, South Australia, pp. 201–202.

76 Bogo, M.R., Rota, C.A., Pinto, H., Ocampos, M., Correa, C.T., Vainstein, M.H. and Schrank, A. (1998). A chitinase encoding gene (chit1 gene) from the entomopathogen *Metarhizium anisopliae*: Isolation and characterization of genomic and full-length cDNA. *Curr. Microbiol.* **37**, 221–225.

77 Gatehouse, A.M.R., Davison, G.M., Newell, C.A., Merryweather, A., Hamilton, W.D.O., Burgess, E.P.J., Gilbert, R.J.C. and Gatehouse, J.A. (1997). Transgenic potato plants with enhanced resistance to the tomato moth, *Lacanobia oleracea*: Growth room trials. *Mol. Breed.* **1**, 49–63.

78 Girard, C. and Jouanin, L. (1999). Molecular cloning of a gut-specific chitinase cDNA from the beetle *Phaedon cochleariae*. *Insect Biochem. Mol. Biol.* **29**, 549–556.

79 Kang, W.K., Tristem, M., Maeda, S., Crook, N.E. and O'Reilly, D.R. (1998). Identification and characterization of the *Cydia pomonella* granulovirus cathepsin and chitinase genes. *J. Gen. Virol.* **79**, 2283–2292.

80 Kim, M.G., Shin, S.W., Bae, K.S., Kim, S.C. and Park, H.Y. (1998). Molecular cloning of chitinase cDNAs from the silkworm, *Bombyx mori*, and the fall webworm, *Hyphantria cunea*. *Insect Biochem. Mol. Biol.* **28**, 163–171.

81 Kramer, K.J. and Muthukrishnan, S. (1997). Insect chitinases: Molecular biology and potential use as biopesticides. *Insect Biochem. Mol. Biol.* **11**, 887–900.

82 Ding, X.F., Gopalakrishnan, B., Johnson, L.B., White, F.F., Wang, X.R., Morgan, T.D., Kramer, K.J. and Muthukrishnan, S. (1998). Insect resistance of transgenic tobacco expressing an insect chitinase gene. *Transgenic Res.* **7**, 77–84.

83 Gatehouse, A.M.R., Down, R.E., Powell, K.S., Sauvion, N., Rahbe, Y.,

Newell, C.A., Merryweather, A., Hamilton, W.D.O. and Gatehouse, J.A. (1996). Transgenic potato plants with enhanced resistance to the peach-potato aphid *Myzus persicae*. *Entomol. Exp. Appl.* **79**, 295–307.

84 Wang, X., Ding, X., Gopalakrishnan, B., Morgan, T.D., Johnson, L., White, F.F., Muthukrishnan, S. and Kramer, K.J. (1996). Characterisation of a 46 kDa insect chitinase from transgenic tobacco. *Insect Biochem. Mol. Biol.* **10**, 1055–1064.

85 Picard, A.L., Pham-Delègue, M.H., Douault, P. and Masson, C. (1991). Transgenic rapeseed (*Brassica napus* L. var. *oleifera* Metzger): Effect on the foraging behaviour of honeybees. *Proc. Symp. Pollination, Acta Hort.* **288**, 435–439.

86 Chang, M.M., Hadwiger, L.A. and Horovitz, D. (1992). Molecular characterization of a pea beta-1,3-glucanase induced by *Fusarium solani* and chitosan challenge. *Plant Mol. Biol.* **20**, 609–618.

87 Gottschalk, T.E., Mikkelsen, J.D., Nielsen, J.E., Neilsen, K.K. and Brunstedt, J. (1998). Immunolocalization and characterization of a beta-1,3-glucanase from sugar beet, deduction of its primary structure and nucleotide sequence by cDNA and genomic cloning. *Plant Sci. Lim.* **132**, 153–167.

88 Neuhaus, J.M., Flores, S., Keefe, D., Ahi-Goy, P. and Meins, F., Jr. (1992). The function of vacuolar beta-1,3-glucanase investigated by antisense transformation. Susceptibility of transgenic *Nicotiana sylvestris* plants to *Cercospora nicotianae* infection. *Plant Mol. Biol.* **19**, 803–813.

89 Haapalainen, M.L., Kobets, N., Piruzian, E. and Metzler, M.C. (1998). Integrative vector for stable transformation and expression of a beta-1,3-glucanase gene in *Clavibacter xyli* subsp. *cynodontis*. *FEMS-Microbiol. Lett.* **162**, 1–7.

90 Okada, H., Tada, K., Sekiya, T., Yokoyama, K., Takahashi, A., Tohda, H., Kumagai, H. and Morikawa, Y. (1998). Molecular characterization and heterologous expression of the gene encoding a low-molecular-mass endoglucanase from *Trichoderma reesei* QM9414. *Appl. Environ. Microbiol.* **64**, 555–563.

91 Jongedijk, E., Tigelaar, H., Van Roekel, S.C., Bres-Vloemans, S.A., Dekker, I., Van Den Elzen, P.J.M., Cornelissen, J.C., Melchers, L.S. and Jones, P.W. (1995). Synergistic activity of chitinases and beta-1,3-glucanases enhances fungal resistance in transgenic tomato plants. In: *Eucarpia, Genetic Manipulation in Plant Breeding Section Meeting, Cork, Irish Republic, 11–14 September 1994, Euphytica* (Cassells, A.C., Ed.). pp. 173–180.

92 Gatehouse, A.M.R., Down, R.E., Powell, K.S., Sauvion, N., Rabhe, Y., Newell, C.A., Merryweather, A., Hamilton, W.D.O. and Gatehouse, J.A. (1996). Transgenic potato plants with enhanced resistance to the peach-potato aphid *Myzus persicae*. *Entomol. Exp. Appl.* **79**, 295–307.

93 Sauvion, N., Rabhe, Y., Peumans, W.J., Vandamme, E.J.M., Gatehouse, J.A. and Gatehouse, A.M.R. (1996). Effects of GNA and other mannose-binding lectins on the development and fecundity of the peach-potato aphid *Myzus persicae*. *Entomol. Exp. Appl.* **79**, 285–293.

94 Powell, K.S., Spence, J., Bharathi, M., Gatehouse, J.A. and Gatehouse, A.M.R. (1998). Immunohistochemical and developmental studies to elucidate the mechanism of action of the snowdrop lectin on the rice brown planthopper, *Nilaparvata lugens* (Stal). *J. Insect Physiol.* **44**, 529–539.

95 Bruins, B.G., Scharloo, W. and Thorig, G.E.W. (1991). The harmful effect of light on *Drosophila* is diet-dependent. *Insect Biochem.* **21**, 535–539.

96 Morgan, T.D., Oppert, B., Czapla, T.H. and Kramer, K.J. (1993). Avidin and streptavidin as insecticidal and growth inhibiting dietary proteins. *Entomol. Exp. Appl.* **69**, 97–108.

97 Markwick, N.P., Christeller, J.T., Docherty, L.C. and Lilley, C.M. (2001). Insecticidal activity of avidin and streptavidin against four species of pest Lepidoptera. *Entomol. Exp. Appl.* **98**, 59–66.

98 Keinanen, R.A., Wallen, M.J., Kristo, P.A., Laukkanen, M.O., Toimela, T.A., Helenius, M.A. and Kulomaa, M.S. (1994). Molecular cloning and nucleotide sequence of chicken avidin-related genes, 1–5. *Eur. J. Biochem.* **220**, 615–621.

99 Argarana, C.E., Kuntz, I.D., Birken, S., Axel, R. and Cantor, C.R. (1986). Molecular cloning and nucleotide sequence of the streptavidin gene, *Nucl. Acids Res.* **14**, 1871–1882.

100 Christeller, J., Sutherland, P., Murray, C., Markwick, N., Phung, M., Burgess, E. and Malone, L. (1999). Chimeric polypeptides allowing expression of plant-noxious proteins. New Zealand Patent: PCT/NZ99/00110.

101 Kramer, K.J., Morgan, T.D., Throne, J.E., Dowell, F.E., Bailey, M. and Howard, J.A. (2000). Transgenic avidin maize is resistant to storage insect pests. *Nat. Biotechnol.* **18**, 670–674.

15 Direct and indirect effects of genetically modified plants on the honey bee

M.H. Pham-Delègue, Lise Jouanin, and J.C. Sandoz

Summary

In this chapter we consider genetically modified (GM) oilseed rape–honey bee interactions, and some factors that could affect plant attractiveness to bees. We report observations on the foraging behavior of honey bees in situations of choice between GM oilseed rape expressing different genes and untransformed ones. Studies were conducted under controlled, semi-field, and field conditions, and no differential behavior was found between GM and control genotypes. To evaluate the risk of direct exposure, we investigated the amounts of gene products expressed in nectar and pollen. In the plant material under test, no transgene proteins were detected, which indicates that the risk of exposure to the proteins is reduced. Differences were found between GM and control genotypes in nectar and floral odor composition. However, it was shown that foragers did not discriminate among the genotypes, and that they could learn the olfactory signals from GM plants as well as from control plants. From these studies, it appears that even though the bees can be exposed to the gene products or subjected to secondary changes in the plant chemistry, these changes do not lead to noticeable modifications in the behavior of the honey bee for the genotypes tested.

Introduction

Mutual benefits between plants and pollinators such as honey bees rely on the ability of bees to discover flowers providing nectar and pollen, to memorize plant characteristics (floral color and shape, and chemical cues), and to communicate information within the hive leading to the recruitment of new foragers. These interactions can be affected by the genetic transformation of melliferous plants. In order to assess possible risks of genetically modified (GM) plants on bees, two types of effects must be considered: bees could be affected by direct exposure to the gene product either when foragers feed on contaminated nectar or pollen or when hive bees feed on stored food, corresponding to short-term and long-term exposure, respectively. In addition, the genetic transformation process itself may induce phenotypic modifications including changes in the nutritional quality of the plant and/or its attractiveness to bees. Risk assessment schemes for conventional insecticides involve a three-tiered approach [1]: first tier would correspond to small-scale laboratory bioassays, the second tier to extended laboratory or semi-field tests under more realistic conditions, and the third tier to large-scale field studies. Such a tiered approach could also be used for the risk assessment of GM plants on beneficial insects [2]. Tests using gene products would preferentially be conducted at the first-tier level, in a worst-case scenario where bees are exposed to high doses of proteins, whereas the transformed plants would be more suitable for testing under more natural conditions. Direct effects can be assessed by using both purified protein products of the transgenes and whole GM plants, but indirect effects should be evaluated mainly using the plants themselves.

In this chapter, we focus on the effects of whole plants on the behavior of honey bees. We also investigate the risk of direct exposure to the transgene products in the nectar and/or pollen of GM plants, and the possible changes in the secondary metabolism of the plants (nectar quality, floral volatile composition).

Honey bee–GM plant interactions

Few experiments have been conducted to assess the behavior of bee populations on GM plants on a large scale, most probably because of the rather drastic regulatory conditions imposed of the production of pre-commercialized GM plants in the field. However, some observations of bees exposed to transformed plants have been reported.

Studies on isolated plants set in indoor or outdoor cages

The first extensive study of the impact of GM plants on the foraging behavior of honey bees was performed under confined conditions in an indoor flight room (about 2.5×2 m) and in an outdoor flight cage (same size) in a more natural environment [3]. The plants under study were two oilseed rape genotypes modified to increase fungal disease resistance (developed by Sanofi Elf-BioRecherche Company) and the corresponding untransformed genotypes, with plants being grown in individual pots. The number of visits of foragers was similar on GM and control genotypes, as well as under indoor and outdoor conditions. More detailed behavioral analyses were conducted from video recordings, and confirmed that no change was induced by plant transformation for any of the variables considered (such as time spent on the plant or on isolated flowers, and number of nectar collection trials). However, differences appeared between the pairs of genotypes considered, one pair of GM/control genotypes being more attractive than the other. Differences were also found for a given pair of genotypes according to the environmental conditions, the number of visits to the plants being higher in indoor conditions. Interestingly, parallel nectar analyses conducted on the studied genotypes showed that for one pair of GM/control genotypes, the GM plants secreted more nectar and had a higher sugar content than the untransformed ones (see Table 15.1). Therefore, the conclusion of the study was that the foraging behavior of the bees was not markedly different on the fungi disease-resistant genotypes and on the control genotype, even though nectar volumes and sugar composition revealed differences between the plants, these differences being in favor of the transformed plants in terms of nectar quality.

More recently, a similar study was conducted on other GM genotypes with a chitinase gene for fungi resistance, coded as G genotypes (developed by Rustica-Prograin Génétique Company) [4]. Five pots of GM and control plants produced in greenhouses were set in an indoor flight room. Foragers from a hive placed in the flight room could visit the flowers for 15 minutes. Then the plants were removed, the flowers counted, and new plants were introduced for another observation period, up to a total of 10 replicates. The mean number of visits per 50 flowers was 72.35 ± 27.16 for the controls and 65.43 ± 21.71 for the GM genotypes, without any significant difference. Individual foraging sequences were videotaped and analyzed [5]. Behavioral items were investigated such as the location of the bee on the plant (flowers or green parts, rank from the top of the flower visited on the plant), or the type of behavior (exploration of the flower, foraging for nectar, scratching of stamina, pollen pellet gathering, cleaning, etc.). The mean duration of some items such as scratching the stamina or nectar foraging could vary among genotypes, but no drastic change in the foraging strategy on both types of plants could be clearly shown. Again, the parallel analyses of nectars did not show any significant difference in volume or content of sugar in GM and control genotypes (see Table 15.1).

Similar experiments were conducted with insect-resistant GM plants, expressing a cysteine protease inhibitor oryzacystatin I (OCI, developed by INRA) [6]. Foragers were given a choice between five GM and five control plants at the same flowering stage, in a flight room under controlled conditions. No differences between genotypes were found, either in the number of bees visiting each genotype or in individual foraging sequences analyzed from videotapes. From all these studies under confined or outdoor small-scale conditions, carried out with various GM plants expressing different gene products, no difference in the behavior of honey bees was found. However, in these experiments, plants were cultivated under artificial conditions, and the observations of plant–honey bee interactions were carried out in rather unnatural situations. Therefore, complementary experiments under more natural conditions are needed

Table 15.1 Volume and sugar content of nectars secreted by GM and control oilseed rape flowers

Type of resistance (protein expressed)	Genotype	Name (or code)	Nectar volume (μl/flower)	Sugar concentration (g/100 ml)	Ref.
Fungi disease (chitinase)	GM	1T	0.16 ± 0.08	57.0 ± 18.6	[3]
	Control	1	0.16 ± 0.12	60.7 ± 17.5	
	GM	76T	0.61 ± 0.21	55.1 ± 14.4	
	Control	76	0.32 ± 0.19	37.3 ± 15.4	
	GM	G	0.63 ± 0.15	57.01 ± 7.0	[4]
	Control	T	0.67 ± 0.18	64.91 ± 9.34	
Herbicide (pat protein)	GM	Falcon pat	1.05 ± 0.22	31.5 ± 2.3	[10]
	Control	Falcon	1.04 ± 0.10	31.0 ± 2.1	
	GM	Artus LL	1.00 ± 0.66	15.8 ± 7.8	[7]
	Control	Artus	0.87 ± 0.66	12.9 ± 6.9	
Insect (protease inhibitor)	GM	OCI (cysteine PI)	1.34 ± 0.38	40.5 ± 7.83	[3]
		CII (serine PI)	0.66 ± 0.05	72.17 ± 27.74	
		OCI × CII	0.91 ± 0.18	71.13 ± 2.13	
	Control	Drakkar	0.80 ± 0.18	76.84 ± 2.09	

before drawing any conclusions with confidence about the effect of plant genetic transformation on the honey bees' behavior.

Studies on crops under tunnels

An experiment was carried out under semi-field conditions to study the impact of a transgenic herbicide-resistant oilseed rape genotype tolerant to the herbicide Glufosinate on honeybee colonies [7]. The experiment consisted of two types of tunnels (6 × 17 m): mono-crop tunnels with either control or transgenic oilseed rape, and choice tunnels containing two parcels of transgenic plants and two parcels of control plants. The genotype of oilseed rape tested was transformed for resistance to Glufosinate (Artus LL, AgrEvo). The control genotype was the untransformed oilseed rape variety, Artus. The GM oilseed rape was treated with Glufosinate and the control with the usual herbicides. Honey bee colonies were introduced into the tunnels 3 days before the beginning of the experiment.

The results showed that the GM genotype tended to reach full bloom later than the control, although the number of flowers available to foragers was not different. In the choice tunnels, mortality was low. In the mono-crop tunnels mortality was positively correlated with the size of the colonies, but did not depend on the genotype. When having a choice between the two genotypes, bees did not show any foraging preference (Figure 15.1). The development of the colonies observed in the mono-crop tunnels was variable in terms of population size and brood surface, depending on the initial state of the colonies. However, this was not correlated with the plant genotypes to which the bees were exposed.

The foraging activity on the GM and control genotypes was tentatively

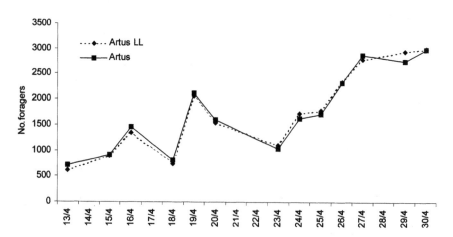

Figure 15.1 Density of foragers on herbicide-resistant oilseed rape (Artus LL) and untransformed oilseed rape (Artus) in the field during the flowering period.

correlated with the amount and sugar composition of the nectar, and with the residues of herbicide or the amount of *pat* protein potentially detected in the nectar and pollen. These analyses are still in progress but preliminary data indicate that no deleterious effects to bees would result from these plant characteristics. This semi-field experiment did not show any difference in the behavior or health of colonies foraging either on Artus LL herbicide-resistant oilseed rape or on its control Artus. The protocol developed in this work proved to be robust as long as variability between tunnels and bee colonies' needs is reduced as much as possible. The study of detailed effects of GM crops requires this kind of extensive study, including the monitoring of parameters such as flowering stage, weather conditions, assessing a large range of data relevant to the biology and behavior of bees.

To complete this study, herbicide residues and the presence of recombinant proteins have to be analyzed.

Field studies

Few studies have been carried out on a large scale to investigate the environmental impact of GM plants. Herbicide-resistant oilseed rape plants have been evaluated mainly to assess the gene flows within species or to weed species closely related to oilseed rape [8, 9]. Regarding the pollinating entomofauna only two studies have been achieved recently.

Observations have been undergone with two genotypes of transformed/nontransformed herbicide-resistant winter oilseed rape: Artus LL/Artus [10]. The transgene codes for the PAT protein which confers tolerance to Glufosinate. Four parcels of 22×22 m, with two parcels of each type, were sown in the South-west of France (Spring 2000). From the beginning of the flowering, the diversity of the pollinators was evaluated by counting the foragers visiting the crop and by classifying them into four groups (honey bee *Apis mellifera*, bumble bees *Bombus* sp., solitary bees, diptera). The results expressed as the number of insects per 1000 flowers indicated no difference in the number of foragers on both genotypes, with a mean of 8 insects per 1000 flowers per observation, the number of insects fluctuating according to environmental conditions (temperature mainly) (Figure 15.2). However, when considering honey bees alone, a significant difference was found, the density of foragers being slightly higher on the GM plants. This could not be related directly to the availability of the nectar collected in 2000 from the tested genotypes, since no differences were found either in the volume secreted or in the amounts of constitutive sugars. However, prior nectar analyses conducted on the same genotypes in 1999 indicated a tendency to higher secretion and sugar quantity in the GM genotype. This tendency seems to be a general trait of GM plants as similar results were found in other paired GM–control oilseed rape genotypes (see Table 15.1). As for the occurrence of the different insect taxa, it

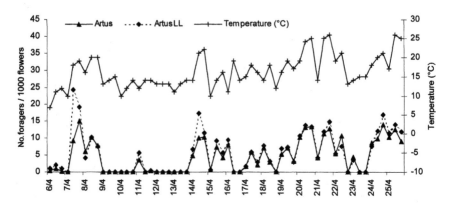

Figure 15.2 Density of foragers on herbicide-resistant oilseed rape (Artus LL) and untransformed oilseed rape (Artus) under tunnels during the flowering period, and corresponding temperature.

appeared that the great majority of pollinators were honey bees (more than 80 percent), the other groups being nearly equally represented. No significant difference in the representation of insect taxa was seen between plant genotypes.

Parallel to the Artus/Artus LL experimentation, another study was conducted in Brittany [10], on another transformed genotype, Falcon *pat,* with the same transgene conferring tolerance to Glufosinate. The experimental design was made up of two parcels (6 × 30 m) of Falcon *pat* and its control genotype Falcon, separated by 24 m, with a hive set between the parcels. In addition to the same observations as were performed on Artus, more detailed recordings of foraging postures and of crossings between the two parcels were carried out. Nectar as well as pollen samples were collected on both genotypes. No difference was found in the diversity and density of the pollinating insect population, or in the foraging behavior strategy between genotypes. No secondary changes in pollen and nectar production were noted, which could account for the fact that bees did not differentiate between the two genotypes.

Potential direct effects of GM plants on honey bees

Direct effects may derive from the ingestion by bees of the protein encoded by a transgene. Honey bees feed exclusively on pollen, nectar, and resins. To be ingested by honey bees and to induce direct deleterious effects, the transgene product must be present in these secretions of transgenic plants. There are surprisingly few published measurements of transgene expression levels in the pollen or nectar of GM plants and none for the resins, gums, or exudates that bees collect for propolis manufacture. The level of expression of a transgene (reported in percent soluble proteins, percent dry or fresh weight) is generally evaluated in the green plant tissues on which the target pest insects feed. Therefore, this information does not provide pertinent insights regarding the potential exposure of pollinating insects. Of the plant products that bees collect, pollen represents the most likely vehicle for a transgene product. Pollen is a plant tissue composed of 8 to 40 percent protein [11], whereas nectar and resin are plant secretions without significant protein content [12, 13]. Data available on the gene product content of plant pollen are scarce. GM corn (N4640) containing a Bt gene controlled by a pollen-specific promoter was found to have pollen containing 260–418 ng of Bt toxin per mg of total soluble protein [14]. However, GM corn plants containing the same Bt gene on a different promoter (cauliflower mosaic virus, or CaMV 35S) produced reduced quantities of the toxin in pollen. Bt-cotton plants (commercial genotype, Bollgard™, with *cry*1Ac gene driven by CaMV 35S promoter) had 0.6 μg of Bt toxin in their pollen (per gram fresh weight), whereas the petals of the same plants contained 3.4 μg of toxin per gram [15]. GM oilseed rape plants containing a gene encoding the protease inhibitor OCI, under the control of the CaMV 35S promoter, had measurable quantities of this transgene product in their leaves (0.2–0.4 percent of total soluble protein) but not in their pollen [16]. This finding was confirmed by Jouanin *et al.* [17], who also noted that Bowman–Birk soybean trypsin inhibitor (BBI) could not be detected in the nectar or pollen of GM oilseed rape plants which had measurable expression levels in leaves (gene also on the CaMV 35S promoter).

The choice of the promoter used in the GM plant construct seems to be essential in the control of the protein expression in the pollen. In many transgenic plants, the transgene is expressed under the control of the CaMV 35S promoter or derivatives (double enhancer sequences). Recent studies have shown that this promoter is inactive in pollen of *Arabidopsis* [18], oilseed rape [19], cotton, maize [reviewed in 20], and potatoes (A.M.R. Gatehouse, personal communication). However, it is not possible to generalize to all plants since CaMV 35S activity has been detected in tobacco pollen, although at a low level [18]. In addition, other promoters such as wounded inducible or tissue specific promoters can be used [20]. For example, the potential insecticide activity of pollen of a specific transgenic maize line expressing the δ-endotoxin of *Bacillus thuringiensis* (Bt N4640) against the monarch larvae [21] is due to the fact that the Bt gene is driven by a pollen/leaf specific promoter and is therefore present at a high dose in pollen. In the future, the range of promoters used to direct expression in given tissues or conditions will be enlarged. When pollinators are to be considered (in the case of plants attracting pollinating insects), studies must be performed on these promoters to determine the level of accumulation of toxins in the pollen. In addition, it has been shown that pollen proteins can be stable in honey [22], and therefore can be active in the hive a long time after being collected. To avoid the presence of transgene product in pollen, Bt genes were expressed in chloroplasts by homologous recombination [23, 24]. Chloroplasts are transmitted in the progeny via the female gametes, thus the pollen of the transgenic plants does not contain the toxin. This technology is a new way to be explored since chloroplast transformation is far from being routinely achieved for crops.

In conclusion, there are two possibilities to avoid risk for honey bees: the nonexpression of the toxin in the tissues bees feed on, or the innocuousness of the toxin for bees. The risk assessment of the expressed protein in a transgenic plant must be considered case by case.

Potential indirect effects of GM plants on honey bees

The introduction of the transgene into the plant may result in secondary changes in plant phenotype affecting its attractiveness or nutritive value to bees. Insertional mutagenesis is one such change. In this case, the random positioning of the transgene in the plant's genome interferes with a gene or suite of genes needed for a "normal" phenotype. For example, an insertional mutagenesis event that resulted in plants without flowers would have a definite negative impact on bees. Less obvious changes, such as alterations in nectar quality or volume, would be more difficult, but not impossible, to detect. Effects due to insertional mutagenesis will vary among different lines of plants derived from separate transformation events and can be eliminated easily by line selection. Pleiotropic effects represent a second type of inadvertent phenotypic change. In this case, it is not the position of the transgene, but its product, which interferes unexpectedly with a biochemical pathway in the plant to create a phenotypic change. Such changes would occur in all lines of the GM plant and could not be remedied by line selection. Indirect effects have been tested on the two main plant products mediating honey bees' attraction to plants, i.e. nectar and floral odors.

Nectar analyses

In order to investigate possible indirect pleiotropic effects on plant characteristics mediating honey bee–plant relationships, in most studies the nectar quantity and quality were compared between GM and control plant genotypes. Oilseed rape, expressing various types of resistance, has been the main GM plant under investigation.

As a general procedure, the nectar was sampled from both GM and control plants, parallel to behavioral observations of bees foraging on both genotypes. Nectar was collected at a uniform flowering stage, on the same dates, using glass pipettes. The number of flowers sampled to fill the pipettes ($5\,\mu l$) were counted to evaluate the volume secreted per flower. The sugar composition of nectar was analyzed using high-performance liquid chromatography according to a standard method [25], modified for oilseed rape nectars [26]. The main constitutive sugars for all conventional oilseed rape nectars analyzed to date are glucose and fructose [27]. The data obtained from the many studies on GM oilseed rape and the corresponding controls can be summarized as follows (Table 15.1).

Differences appear in the amounts of nectar secreted, and correlatively in the amount of constitutive sugars (the sugar concentration is higher when volumes are smaller). These differences depend on the date of collection (climatic conditions, physiological stage), the environmental and breeding conditions (indoor/outdoor, pots/field), and the genotype, as already shown for conventional oilseed rape varieties [27]. When considering studies on GM plants, all samplings have been done simultaneously on the GM and the control genotypes, environmental conditions were similar for both genotypes, and the transformed and untransformed genotypes are closely related genetically, when not completely isogenic except for the gene of interest. Therefore, it may be assumed that if differences arise between GM and control plants, they are the consequence of pleiotropic effects. Interestingly, among the studies listed in Table 15.1, significant differences were reported, e.g. for ArtusLL/Artus [7] and 76T/76 [3], with more abundant secretion and more concentrated nectar in the transgenic genotype. Although available data are still insufficient to conclude whether this could be a general trait of the transformation, it suggests that pleiotropic effects noticeable on the nectar secretion are not negative regarding the attractiveness of these plants for bees.

Floral odor analyses

To assess whether the effect of a genetic transformation of oilseed rape could imply changes in secondary plant metabolites, and consequently in the behavior of the bee, combined behavioral and chemical studies were conducted (Sandoz, unpublished data). The ability of honeybees to learn the odor of transformed and control oilseed rape was compared. The GM genotype under testing was expressing a cysteine protease inhibitor (OCI), and the control was Drakkar. Behavioral recordings were based on the conditioned proboscis extension (CPE) bioassay, where restrained bees learn to associate an odor (here from oilseed rape flowers) with a sugar reward. To stimulate the bees with the odor from intact plants, a stimulation system was developed, with racemes of oilseed rape enclosed in an airtight glass chamber. Air was flown through the chamber to stimulate the bees. In such conditions, bees learned rapidly and with the same efficiency odors from transformed and control oilseed rape (Figure 15.3).

Complementarily, after being conditioned to the odor of one genotype, bees were found to respond to the odor of the other genotype as well. Furthermore, in a differential conditioning procedure, where bees are

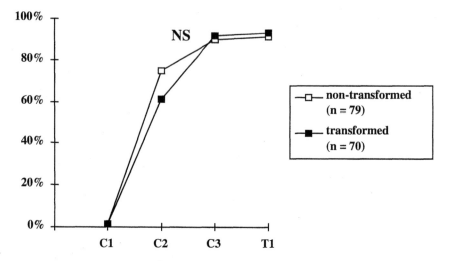

Figure 15.3 Percentage of conditioned responses obtained in the conditioned proboscis extension paradigm by stimulating the bees with the floral
volatiles of transformed (OCI variety expressing a cystein proteinase
inhibitor gene) or control (Drakkar variety) oilseed rape.

stimulated alternately with odors from each genotype, one being rewarded
and the other being unrewarded, they responded equally to both odors.

At the chemical level, the characterization of the compounds used by
bees to recognize the whole floral blend of transformed or control oilseed
rape was carried out. Air entrainment of floral odors was trapped on tenax
polymers and the constitutive components of the odor mixture were separated by optic gas chromatography (GC). Bees previously conditioned to
the floral odor of an oilseed rape genotype were tested in a combined
GC–CPE procedure [28], the effluents of the gas chromatograph being
directed to the bees. This method provided simultaneous recordings from
chemical (gas chromatograph) and biological (honey bee) detectors.
Therefore, individual compounds eliciting behavioral responses could be
identified. For both plant genotypes, two compounds (linalool and phenyl
acetaldehyde) elicited most of the activity of the conditioned bees (Figure
15.4). This study showed that bees did not differentiate between the odor
of transformed and control oilseed rape and suggests that they rely on the
same key compounds to recognize these complex odors. Finally, these data
indicate that even though qualitative or quantitative differences in the
chemical composition of floral odors may occur between transformed and
control plants, these differences are not detected by bees, or do not induce
discriminative behavior. It may be assumed that under more natural conditions, bees facing transformed or conventional oilseed rape would not
differentiate between them on the basis of their respective floral odors.

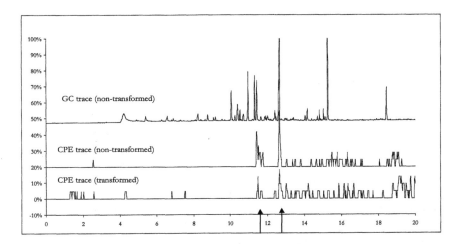

Figure 15.4 Simultaneous recordings of chemical (gas chromatography, GC) and biological (conditioned proboscis extension, CPE) responses. The upper line shows the volatile components of control oilseed rape flowers, and the lower lines the CPE responses of bees previously conditioned either to the control or to the transformed (OCI) floral volatiles, and tested for the individual components of the blend separated at the effluent of the chromatograph. Arrows indicate the main compounds (linalool and phenyl acetaldehyde) eliciting most of the behavioral activity.

Conclusion

As in toxicity studies of chemical pesticides, the evaluation of the impact of gene products potentially expressed in GM plants can be based on a three-tiered approach where laboratory acute toxicity tests and observations under more natural conditions are combined. Although parallels can be drawn in the methodologies used in the study of the sublethal effects of chemical pesticides and the risk assessment of GM plants, the main difference relies on the fact that the evaluation of GM plant implies specifically the study of secondary changes in plant metabolites mediating their attractiveness for honey bees.

This chapter reports the studies dealing with honey bee–plant interactions under semi-field or field conditions. These studies have tentatively established relationships between the observed behaviors and the transformed plant characteristics in terms of gene product expression or secondary changes in attraction cues. Direct observations of honey bees foraging on GM plants are still scarce, and have been reported mainly for oilseed rape expressing insecticide, herbicide, or disease resistance. Bt maize pollen was also tested on bees, but whole plants were not. In recent work (not yet published in peer-reviewed journals but reported in newspapers such as *The Observer*, May 2000) by Dr Hans-Heinrich Kaatz (Iena University, Germany), bees were allowed to forage on herbicide-resistant oilseed rape (Agrevo-Aventis). Pollen pellets taken back to the hive were then trapped, and used to feed young honey bees under laboratory conditions. It was shown that the herbicide-resistant genes of the oilseed rape had transferred across to the bacteria and yeast inside the intestines of the young bees. If confirmed, these data open a new area of risk to be assessed, to control whether genes used to modify crops can in fact "jump" the species barrier without external engineering as needed to transfer the foreign genes in a plant genome. Until now, available published data gave no evidence of a negative effect on the foraging behavior of bees or on the population development of pollinators when visiting GM plants. The possibility exists for GM pollen to express foreign proteins at levels sufficient to alter the diet of honey bees foraging on these plants. However, there are as yet insufficient experimental data to make generalizations about this or the effects that it might have on the bee. As for the changes in secondary plant metabolites, such as nectar or floral odors, even though quantitative or qualitative differences have been found, they did not seem to affect the attractiveness of the plants for honey bees. However, few plant species and genes have been evaluated yet, and large-scale studies are still lacking. Methodologies are now available both in laboratory, semi-field, and field conditions. They should be extended to new GM plants potentially visited by pollinators, following a case-by-case approach.

Acknowledgments

The authors are grateful to the students who contributed to the experiments on oilseed rape–honeybee interactions, namely A.-L. Picard-Nizou, D. Marsault, and N. Châline. J. Pierre (INRA Rennes) contributed to the field work, L.J. Wadhams (IACR Rothamsted) to the floral odor analyses, and L. Malone (Horticultural and Food Research Institute, New Zealand) and A. Couty (IACR Rothamsted) to the manuscript. Part of this work was funded by the EU in the Biotechnology Program of the 4th Framework, and by the CETIOM within the framework of an inter-institute study.

References

1 Barrett, K.L., Grandy, N., Harrison, E.G., Hassan, S.A. and Oomen, P.A. (1994). *Guidance Document on Regulatory Testing Procedures for Pesticides and Non-target Arthropods*. SETAC Europe, British Library Publisher, p. 51.

2 Poppy, G. (2000). GM crops: Environmental risks and non-target effects. *Trends Plant Sci.* **5**, 4–6.

3 Picard-Nizou, A.L., Pham-Delègue, M.H., Kerguelen, V., Douault, P., Maril-

leau, R., Olsen, L., Grison, R., Toppan, A. and Masson, C. (1995). Foraging behaviour of honey bees (*Apis mellifera* L.) on transgenic oilseed rape (*Brassica napus* L. var. *oleifera*). *Transgenic Res.* **4**, 270–276.

4 Picard-Nizou, A.L. and Pham-Delègue, M.H. (2000). *Environmental Impact of Transgenic Oilseed Rape on Beneficial Insects: Effects on Honey Bees and Bumble Bees*. Final report of the EU Biotechnology Program BIO4-CT96-0365 1996–1999, p. 90.

5 Bailez, O. and Pham-Delègue, M.H. (1996). Analyse de la structure du comportement de butinage de l'abeille *Apis mellifera* L. sur colza. *Actes Coll. Insectes Sociaux* **10**, 153–156.

6 Grallien, E., Marilleau, R., Pham-Delègue, M.H., Picard-Nizou, A.L., Jouanin, L. and Marion-Poll, F. (1995). Impact of pest insect resistant oilseed rape on honey bees. In: *Proceedings of the 9th International Rapeseed Congress on "Rapeseed Today and Tomorrow"*, Cambridge, UK, 4–7 July 1995, pp. 784–786.

7 Châline, N., Decourtye, A., Marsault, D., Lechner, M., Champolivier, J., Van Waetermeulen, X., Viollet, D. and Pham-Delègue, M.H. (1999). Impact of a novel herbicide resistant oilseed rape on honey bee colonies in semi-field conditions. In: *Proceedings of the 5th International Conference on Pests in Agriculture*, Montpellier, France, 7–9 December 1999, pp. 905–912.

8 Chèvre, A.M., Eber, F., Baranger, A. and Renard, M. (1997). Gene flow from transgenic crops. *Nature* **389**, 924.

9 Lavigne, C., Klein, E., Vallée, P., Pierre, J., Godelle, B. and Renard, M. (1998). A pollen dispersal experiment with transgenic oilseed rape. Estimation of the average pollen dispersal on an individual plant within a field. *Theor. Appl. Genet.* **96**, 886–896.

10 Pierre, J., Marsault, D., Genecque, E., Renard, M. and Pham-Delègue, M.H. (2000). Effects of herbicide tolerant transgenic oilseed rape genotypes on honey bees and other pollinating insects under field conditions (*Entomol. Exp. Applic.*, submitted).

11 Herbert, E.W., Jr. (1992). Honey bee nutrition. In: *The Hive and the Honey Bee* (Graham, J.M., Eds). Hamilton, IL, pp. 197–224.

12 Baker, H.G. and Baker, I. (1977). Intraspecific constancy of floral nectar amino acid complements. *Bot. Gaz.* **138**, 183–191.

13 Schmidt, J.O. and Buchmann, S.L. (1992). Other products of the hive. In: *The Hive and the Honey Bee* (Graham, J.M., Ed.). Hamilton, IL, pp. 928–977.

14 Koziel, M.G., Beland, G.L., Bowman, C., Carozzi, N.B., Crenshaw, R., Crossland, L., Dawson, J., Desai, N., Hill, M., Kadwell, S., Launis, K., Lewis, K., Maddox, D., McPherson, K., Meghji, M.R., Rhodes, R., Warren, G., Wright, M. and Evola, S.V. (1993). Field performance of elite transgenic maize plants expressing an insecticidal protein derived from *Bacillus thuringiensis*. *BioTechnology* **11**, 194–200.

15 Greenplate, J. (1997). Response to reports of early damage in 1996 commercial Bt transgenic cotton (Bollgard™) plantings. *Soc. Invertebr. Pathol. Newsletter* **29**, 15–18.

16 Bonadé-Bottino, M., Girard, C., Le Métayer, M., Picard-Nizou, A.L., Sandoz, G., Lérin, J., Pham-Delègue, M.H. and Jouanin, L. (1998). Effects of transgenic oilseed rape expressing proteinase inhibitors on pest and beneficial insects. *Proc. Int. Symp. Brassicas, Acta Hort.* **459**, 235–239.

17 Jouanin, L., Bonadé-Bottino, M., Girard, C., Morrot, G. and Giband, M. (1998). Transgenic plants for insect resistance. *Plant Sci.* **131**, 456–463.

18 Wilkinson, J.E., Twell, D. and Lindsey, K. (1997). Activity of CaMV 35S and nos promoters in pollen: Implications for field release of transgenic plants. *J. Exp. Bot.* **48**, 265–275.

19 Jouanin, L., Bonadé-Bottino, M., Girard, C., Lerin, J. and Pham-Delègue, M.H. (2000). Expression of protease inhibitors in rapeseed. In: *Recombinant Protease Inhibitors* (Michaud, D., Ed.). Landes Bioscience, Texas, pp. 182–194.

20 Schuler, T.H., Poppy, G.M., Kerry, B.R. and Denholm, I. (1998). Insect-resistant plants. *Trends Biotechnol.* **16**, 168–175.

21 Losey, J.E., Rayor, L.S. and Carter, M.E. (1999). Transgenic pollen harms monarch larvae. *Nature* **399**, 214.

22 Eady, C., Twell, D. and Lindsey, K. (1995). Pollen viability and transgene expression following storage in honey. *Transgenic Res.* **4**, 226–231.

23 McBride, K.E., Svab, Z., Schaaf, D.J., Hogan, P.S. and Maliga, P. (1995). Amplification of a chimeric *Bacillus* gene in chloroplasts leads to an extra-ordinary level of insecticidal protein in tobacco. *BioTechnology* **13**, 362–365.

24 Kota, M., Daniell, H., Varma, S., Garczynski, S.F., Gould, F. and Moar, W.J. (1999). Over expression of the *Bacillus thuringienis* (Bt) Cry2Aa2 in chloro-plasts confers resistance to plants against susceptible and Bt-resistant insects. *Proc. Natl. Acad. Sci. USA* **96**, 1840–1845.

25 Black, L.T. and Bagley, E.B. (1978). Determination of oligosaccharides in soy-beans by high pressure liquid chromatography using an internal standard. *J. Am. Oil Chem. Soc.* **55**, 228–232.

26 Mesquida, J., Pham-Delègue, M.H., Marilleau, R., Le Métayer, M. and Renard, M. (1991). La sécrétion nectarifère des fleurs de cybrides mâle-stérile de colza d'hiver (*Brassica napus* L.). *Agronomie* **11**, 217–227.

27 Pierre, J., Mesquida, J., Marilleau, R., Pham-Delègue, M.H. and Renard, M. (1996). Nectar secretion in winter oilseed rape (*Brassica napus*): Quantitative and qualitative variability among 71 genotypes. *Plant Breed.* **115**, 471–476.

28 Pham-Delègue, M.H., Blight, M.M., Kerguelen, V., Le Métayer, M., Marion-Poll, F., Sandoz, J.C. and Wadhams, L.J. (1997). Discrimination of oilseed rape volatiles by the honey bees: Combined chemical and biological approaches. *Entomol. Exp. Appl.* **83**, 87–92.

Index